Reintegrating Fragmented Landscapes

R.J. Hobbs
D.A. Saunders
Editors

Reintegrating Fragmented Landscapes

Towards Sustainable Production and Nature Conservation

With 72 figures

Springer-Verlag
New York Berlin Heidelberg London Paris
Tokyo Hong Kong Barcelona Budapest

Richard J. Hobbs
Denis A. Saunders
CSIRO Australia
Division of Wildlife and Ecology – Perth
LMB4, P.O. Midland
Western Australia 6056
Australia

Library of Congress Cataloging-in-Publication Data
Reintegrating fragmented landscapes: towards sustainable production
and nature conservation/R.J. Hobbs and D.A. Saunders, (editors).
p. cm.
Includes bibliographical references and index.
ISBN-13: 978-1-4613-9216-3 e-ISBN-13: 978-1-4613-9214-9
DOI: 10.1007/978-1-4613-9214-9
1. Agricultural ecology – Australia – Western Australia.
2. Agriculture – Australia – Western Australia. 3. Nature conservation – Australia – Western Australia. 4. Land use, Rural – Environmental aspects – Australia – Western Australia. 5. Landscape ecology – Australia – Western Australia. I. Hobbs, R.J. (Richard J.) II. Saunders, Denis A.
S478.W4R45 1992
333.76'16'09941 – dc20 92-5865

Printed on acid-free paper.

Production managed by Dimitry L. Loseff; manufacturing supervised by Jacqui Ashri.
Typeset by Best-set Typesetter Ltd., Chai Wan, Hong Kong.

9 8 7 6 5 4 3 2 1

ISBN-13: 978-1-4613-9216-3

To Katie, Hamish, Samantha and Romayne:
May they live in hope for,
not fear of, the future.

Foreword

Social historians will look back on the 1980s as a period when a global consciousness of the environment developed. Stimulated by major issues and events such as oil and chemical spills, clearing of rainforests, pollution of waterways, and, towards the end of the decade, concern over the greenhouse effect, concern for the environment has become a major social and political force.

Unfortunately, the state of the environment and its future management are still very divisive issues. Often, at a local level, concern for the environment is the antithesis of development. The debate usually focusses on the possible negative environmental impacts of an activity versus the expected positive economic impacts. It is a very difficult task to integrate development and conservation, yet it is towards this objective that the sustainable development debate is moving.

The issues in the central wheatbelt of Western Australia are typical of the environment versus development debate. It is undoubted that the development of the area, which involved clearing the native vegetation, has had a major impact upon the original ecosystems. Many of the natural habitats are threatened and local extinction of flora and fauna species is a continuing process. Moreover, there are clear signs that land degradation processes such as dryland salinity are depleting the land resource. On the other hand, it is also undoubted that the international success of the agricultural industries of Western Australia has provided the basis for the economic development and the standard of living enjoyed by Western Australians today. While the argument as to which of these two is the most important will undoubtedly continue, it is a futile exercise. To date, it is clear that the environment and agriculture have been competitive and a quite different approach is required if they are to be complementary.

The future must focus on developing a land use approach in the wheatbelt that both maintains the agricultural industry and provides for nature conservation. In terms of this book, the challenge is to reintegrate the fragmented landscapes. In the central wheatbelt, a sustained research effort by biological, physical, and agricultural scientists, together with the involvement and contributions from the community and land users of the region, have clearly focussed the challenges. Much of the current state of knowledge is presented here and serves as a base on which to develop the scenario for the future and chart the appropriate paths for achievement. The successful reintegration of the landscape in the central wheatbelt will be important for both agriculture and nature conservation in Western Australia. However, if successful, the approach will be equally of benefit to the many other areas around the world where a similar challenge is now being recognised.

Graeme Robertson
Deputy Director General,
Western Australian Department of Agriculture,
and Commissioner for Soil Conservation.

Western Australian Department of Agriculture
Baron-Hay Court
South Perth
Western Australia 6151

Contents

IV: Conclusions

Contributors

J. Chatfield c/-P.O. Tammin, Western Australia 6409.

C. Denby Department of Agriculture, Baron-Hay Court, South Perth, Western Australia 6151.

D. Couper c/-P.O. Traying, Western Australia 6488.

P. Farrington CSIRO, Division of Water Resources, Private Bag, P.O. Wembley, Western Australia 6014.

R.J. George Department of Agriculture, North Boyanup Road, Bunbury, Western Australia 6230.

A. Haagensen Department of Agriculture, Marine Terrace, Geraldton, Western Australia 6530.

R.J. Hobbs CSIRO, Division of Wildlife and Ecology, LMB 4, P.O. Midland, Western Australia 6056.

A. Kubicki Department of Agriculture, Baron-Hay Court, South Perth, Western Australia 6151.

E.C. Lefroy Department of Agriculture, Baron-Hay Court, South Perth, Western Australia 6151.

L.A. Lobry de Bruyn Department of Geography, University of Western Australia, Nedlands, Western Australia 6009.

W.M. McArthur 9 Davy Street, Alfred Cove, Western Australia 6154.

D.J. McFarlane Department of Agriculture, Albany Highway, Albany, Western Australia 6330.

A.R. MAIN Department of Zoology, University of Western Australia, Nedlands, Western Australia 6009.

B.Y. MAIN Department of Zoology, University of Western Australia, Nedlands, Western Australia 6009.

R.A. NULSEN Department of Agriculture, Baron-Hay Court, South Perth, Western Australia 6151.

J. SALERIAN State Energy Commission, 363-5 Wellington Street, East Perth, Western Australia 6004.

D.A. SAUNDERS CSIRO, Division of Wildlife and Ecology, LMB 4, P.O. Midland, Western Australia 6056.

M. STEVENS Department of Agriculture, Marine Terrace, Geraldton, Western Australia 6530.

Participants in the workshop "Reintegrating Fragmented Landscapes" held on Rottnest Island, Western Australia, May 1990. From left to right, standing: Judy Eastham, Bob Nulsen, Tony Kubicki, Graeme Arnold, Paul Brown, Don Couper, Jos Chatfield, Peter Farrington, Barbara Main, Ted Lefroy, Ros Prinsely, Bert Main, Chris Margules, John Ive, Odile Poulequin, Bill McArthur, John Salerian. Seated: Richard Hobbs, Lisa Lobry de Bruyn, V Saunders, Denis Saunders, Don McFarlane.

(Photo: R. Hobbs)

Section I

Background

Prior to development for agriculture, the Western Australian wheatbelt landscape consisted of a complex mosaic of vegetation types whose distribution was closely related to landform and soil type. The woodland communities, occupying the lower-lying areas with better soils, were quickly developed for agriculture following settlement. This photograph shows a relatively undisturbed area of *Eucalyptus salmonophloia – E. capillosa* woodland characteristic of the central wheatbelt.

(Photo: R. Hobbs)

Chapter 1

Introduction

R.J. Hobbs and D.A. Saunders

Nature put proudly in our care
a land for all of us to share.
We've raped it and spoiled and plundered it
and stripped its bounty bare.
And when our consciences said ''stay''
we threw our consciences away
for tomorrows wasted deserts
are money in the bank today.

("Goodbye Lucky Country," Eric Bogle)

1.1 Landscape Fragmentation

Conservation and production are at odds with each other in every part of the world, and the situation will only become worse as the human population increases and pressure on natural resources grows. Land transformation by human activity is increasing worldwide (Wolman and Fournier 1987; Turner et al. 1990). In almost all cases, development has been driven by economic imperatives with little or no account taken of ecological constraints. Natural ecosystems are becoming increasingly restricted to relatively small remnant areas in an increasingly fragmented landscape (Saunders et al. 1987; Saunders et al. 1991). Landscapes are now composed of mosaics of differing land uses, many of which are mutually incompatible. Management of these landscapes is also fragmented, with many different land owners and management agencies involved in the management of the various segments of the landscape. Until recently, management of these different segments has proceeded more or less independently of what was happening in other adjacent segments. In many parts of the world, for instance, there is a strict separation between national parks and nature reserves on one hand and adjacent productive land (agricultural or forest) on the other. It has long been recognised that the legal boundaries imposed on the landscape by humans do not necessarily match the biotic boundaries (Newmark 1985). It is also apparent that the different components of the landscape do not function independently of one another, but interact with one another in numerous ways (Forman and Godron 1986; Saunders et al. 1991). Despite the arbitrary allocation of portions of the landscape to one land use or another, the different portions remain strongly interlinked. Events in one segment are very rarely restricted to that one segment, but impact on adjacent segments too. Management of individual landscape components in isolation ignores these interactions and leads to serious environmental problems (Myers 1983; Hales 1989).

It is now clear that this fragmented approach to landscape management will not succeed if we are to manage our planet on a sustainable basis. It is for this reason that we have pursued the possibility that management should be able to recognise and adequately deal with the interrelatedness of landscape segments. We have been spurred on by the recognition that scientists can either act simply as recorders of the current environmental decline or they can actively help in trying to reverse the decline. Our approach to reintegrating fragmented landscapes is a first attempt at an active remedy to the problems of degrading agricultural landscapes. The concept of reintegration consists of a recognition that the different segments and land uses in a landscape are closely interlinked ecologically and that the management needs to be integrated across the whole landscape rather than related to isolated landscape segments. It implies a return to landscapes which are integrated entities, rather than collections of artificially segmented subcomponents. We believe that without such an approach, conflicts will increase and the environment will degrade further.

1.2 Sustainability and Integration

Recently there has been a considerable amount of attention paid to the concept of sustainability, particularly in a global context (e.g., Clark and Munn 1986; World Commission on Environment and Development 1987; Hawke 1989; Pearce 1989; Pearce et al. 1989). It has become the buzzword for managers, environmentalists, and politicians alike.

The goal of global sustainability is a very grand concern and it may be seen by the majority of the world's population as an unattainable goal. However, global sustainability has to come about through the development of programs at a local level. Each region and locality needs to be responsible for its own sustainability (although, of course, the influence of world markets and global environmental issues have to be taken into account). "Think globally, act locally" is the catchword for sustainable development and, when the problem is reduced to a smaller scale, solutions begin to become more apparent. Brokensha and Riley (1989) have suggested that there will be no general solutions to current environmental problems, but that solutions generated at the local level will have the best chance of success. Stokes et al. (1989) illustrated how local approaches to rural conservation have been initiated in the United States. In this book, we approach the question of sustainable production and nature conservation at the landscape level within one particular region in the wheatbelt of Western Australia. Although we limit most of the analysis to this small region, we believe that the approach has much wider relevance.

Definitions of sustainability have generally been rather diffuse, although recent attempts have been made to focus on this problem (Brown et al. 1987; Dovers 1990; Shearman 1990). Shearman (1990) contended that the concept of sustainability simply infers "a continuity through time" and that confusion over its meaning arises from its implications when it is applied to any given context. Here, we regard sustainable land uses as those which permit the maintenance of future options. Land degradation and loss of nature conservation values essentially reduce the number of future options available by, for instance, reducing soil fertility or removing species.

Brown et al. (1987) argued that the term sustainability could be used in an ecological, social, or economic context and the meaning depended on the area of concern. However, it is our contention here that ecological, social, and economic contexts cannot be separated and, following from this, sustainability cannot be achieved in one context without also being achieved in the others.

This means that the development of land use practices which are sustainable in one context will be successful only if they are sustainable in the other contexts. Thus, ecologically sustainable practices must be economically and socially acceptable to be adopted. Many of the current problems of land degradation have arisen because the ecological sustainability of present land use practices has not been considered. However, the current call for ecological sustainability to be the prime criterion for judging land use practices ignores the economic and social imperatives which operate in the real world. Unfortunately, there are few instances where economic, social, and ecological criteria have been balanced successfully. Here we present a first approach to this problem.

Discussions of sustainability usually involve the separation of productive systems from areas set aside for conservation. "Sustainable agriculture" is concerned only with the productive system (Gleissman 1990; Ikerd 1990; Parr et al. 1990; Reganold et al. 1990) while sustainability, in the conservation sense, has tended to focus mainly on the preservation of biotic diversity, particularly in national parks and nature reserves (Wilson and Peter 1988; but see Walker 1989). Since productive and conservation systems frequently lie adjacent to one another and interact strongly, we argue here that it is impossible to achieve sustainability in one system without also achieving it in the other. For instance, a conservation network cannot be isolated from the surrounding production landscape and will not persist if the landscape is degrading rapidly (Dasmann 1983; Myers 1983; Hales 1989). Conversely, the agricultural component will also be considerably less stable without the ameliorating influences of remnant native vegetation (on soil stability, water movement, etc., as explained in subsequent chapters).

1.3 Outline of the Book

Following a preliminary attempt to develop a conceptual framework for the problem of integrating management on a landscape scale (Hobbs and Saunders 1991), we organised a workshop in May 1990 on Rottnest Island, Western Australia, to explore the problem further. The aim of the workshop was to select a group of researchers from the fields of ecology, agriculture, hydrology and economics and a group of farmers and land managers, all of whom had worked within the central wheatbelt of Western Australia, and isolate them at a remote field station for three days. Within that time, and following a loosely defined agenda, the disparate components were drawn together and two separate publications were planned. The first was a practical guide to landscape management and restoration (Lefroy et al. 1991). The second was a scientific publication on reintegrating fragmented landscapes, using the central wheatbelt as an example. As a result, the outline of this book was developed.

In the chapters that follow, we examine the factors which have led to the current state of affairs within the agricultural area of southwestern Australia, particularly the central wheatbelt (centered on Kellerberrin), and attempt to develop an approach by which both agricultural production and nature conservation can be placed on a sounder ecological footing and coexist more comfortably. The wheatbelt of Western Australia (Fig. 1.1) is a very ancient weathered landscape (Chapter 2) which has been developed rapidly and become one of the major farming areas of Australia. It produces many millions of dollars worth of wheat, wool, and other products each year (Chapter 3). The wheatbelt is also part of one of the richest nontropical botanical provinces in the world and interspersed throughout the agricultural land are remnants of native vegetation within which the remains of that rich flora and fauna survive. These are also undergoing continual decline due to species losses, invasions, and lack of regeneration (Chapter 4). At the same time, it has increasing problems of land degradation caused by the salinisation of soil and water, erosion, waterlogging, and changing soil structure (Chapters 5 and 6). Chapters 5 and 6 not only discuss these problems, but provide management options for the individual problems

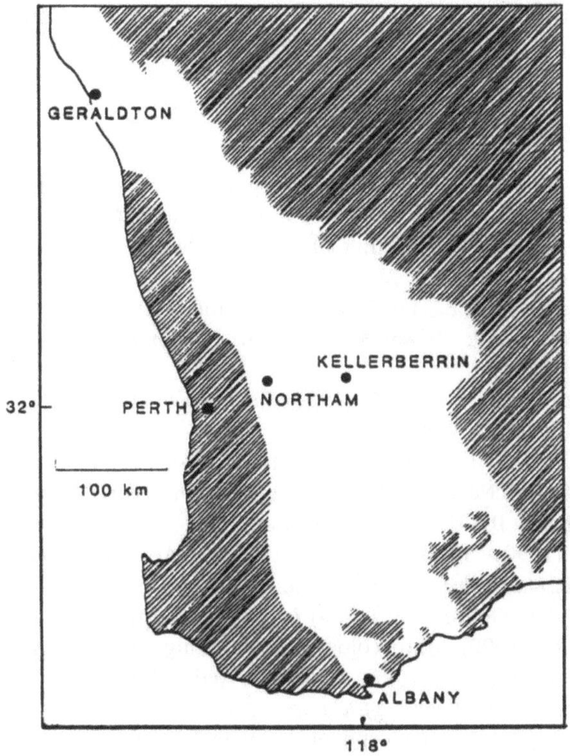

FIGURE 1.1. The wheatbelt of Western Australia (unhatched area).

discussed in each. Chapters 7 to 10 then develop an approach which offers solutions to these problems by considering the landscape as in integrated entity rather than as a collection of separate entities.

The resulting book represents the current knowledge of the effects of rapid clearance for agriculture in the central wheatbelt of Western Australia, the potential solutions to these problems, and an initial attempt to integrate the solutions to individual problems in a rational way. Resource management and conservation is not a simple process but is made up of finding solutions to complex, often poorly understood, problems. Long-term success will depend on the development of integrated approaches such as that developed here.

Acknowledgments. We thank all the contributors to this book and the other workshop participants, Ros Princely, Graeme Arnold, Paul Brown, John Ive, and Chris Margules, for their input at all stages. We also thank V Saunders, who catered for the workshop, and Brian Walker, CSIRO,

Division of Wildife and Ecology, who provided financial support. We are grateful to Eric Bogle for permission to quote from his song "Goodbye Lucky Country", Copyright Larrikin Music Publishing Pty. Ltd., Sydney.

References

Brokensha, D. and Riley, B.W. (1989) Managing natural resources: The local level. In D.B. Botkin, M.F. Caswell, J.E. Estes, and A.A. Orio (eds.) *Changing the Global Environment. Perspectives on Human Involvement.* Academic Press, San Diego. pp. 341–366.

Brown, B., Hanson, M.E., Liverman, D.M., and Merideth, R.W., Jr. (1987) Global sustainability: Towards definition. *Environmental Management* 11: 713–719.

Clark, W.C. and Munn, R.E. (eds.) (1986) *Sustainable Development of the Biosphere.* Cambridge University Press, Cambridge.

Dasmann, R.F. (1983) The relationship between protected areas and indigenous peoples. In J.A. McNeely and K.R. Miller (eds.) *Proceedings of the World Congress on National Parks.* IUCN, Gland, Switzerland.

Dovers, S.R. (1990) Sustainability in context: An Australian perspective. *Environmental Management* 14:297–305.

Forman, R.T.T. and Godron, M. (1986) *Landscape Ecology.* Wiley, New York.

Gliessman, S.R. (1990) Agroecology: Researching the ecological basis for sustainable agriculture. In S.R. Gliessman (ed.) *Agroecology: Researching the Ecological Basis for Sustainable Agriculture.* Springer, New York. pp. 3–10.

Hales, D. (1989) Changing concepts of national parks. In D. Western and M.C. Pearl (eds.) *Conservation for the Twenty-first Century.* Oxford University Press, Oxford. pp. 139–144.

Hawke, R. (1989) *Our Country, Our Future: Statement on the Environment.* Commonwealth of Australia, Canberra.

Hobbs, R.J. and Saunders, D.A. (1991) Reintegrating fragmented landscapes: A preliminary framework for the Western Australian wheatbelt. *Journal of Environmental Management* 33:161–167.

Ikerd, J.E. (1990) Agriculture's search for sustainability and profitability. *Journal of Soil and Water Conservation* 45:18–23.

Lefroy, E.C., Hobbs, R.J., and Atkins, L. (1991) *A Revegetation Guide for the Central Wheatbelt.* Western Australian Department of Agriculture, CSIRO and Greening Australia Perth.

Myers, N. (1983) Eternal values of the parks movement and the Monday morning world. In J.A. McNeely and K.R. Miller (eds.) *Proceedings of the World Congress on National Parks.* IUCN, Gland, Switzerland.

Newmark, W.D. (1985) Legal and biotic boundaries of Western North American national parks: A problem of congruence. *Biological Conservation* 33:197–208.

Parr, J.F., Stewart, B.A., Hornick, S.B., and Singh, R.P. (1990) Improving the sustainability of dryland farming systems. *Advances in Soil Science* 13:1–8.

Pearce, D. (1989) Sustainable futures: Some economic issues. In D.B. Botkin, M.F. Caswell, J.E. Estes, and A.A. Orio (eds.) *Changing the Global Environment. Perspectives on Human Involvement.* Academic Press, San Diego. pp. 311–323.

Pearce, D., Markandya, A., and Barbier, E.B. (1989) *Blueprint for a Green Economy*. Earthscan Publications, London.

Reganold, J.P., Papendick, R.I., and Parr, J.F. (1990) Sustainable agriculture. *Scientific American* 262:112–120.

Saunders, D.A., Hobbs, R.J., and Margules, C.R. (1991) Biological consequences of ecosystem fragmentation: A review. *Conservation Biology* 5:18–32.

Saunders, D.A., Arnold, G.W., Burbidge, A.A., and Hopkins, A.J.M. (eds.)(1987) *Nature Conservation: The Role of Remnants of Native Vegetation*. Surrey Beatty and Sons, Chipping Norton, NSW.

Shearman, R. (1990) The meaning and ethics of sustainability. *Environmental Management* 14:1–8.

Stokes, S.N., Watson, A.E., Keller, G.P., and Keller, J.T. (1989) *Saving America's Countryside. A Guide to Rural Conservation*. Johns Hopkins University Press, Baltimore.

Turner, B.L., II, Clark, W.C., Kates, R.W., Mathews, J.T., Richards, J.R., and Mayer, W. (eds.) (1990) *The Earth Transformed by Human Action*. Cambridge University Press, New York.

Walker, B.H. (1989) Diversity and stability in ecosystem conservation. In D. Western and M.C. Pearl (eds.) *Conservation for the Twenty-first Century*. Oxford University Press, Oxford. pp. 121–130.

Wilson, E.O. and Peter, F.M. (1988) *Biodiversity*. National Academy Press, Washington, D.C.

Wolman, M.G. and Fournier, F.G.A. (eds.) (1987) *Land Transformation in Agriculture*. John Wiley and Sons, New York.

World Commission on Environment and Development (1987) *Our Common Future*. Oxford University Press, Oxford.

Chapter 2

History of Landscape Development

W.M. McArthur

2.1 Introduction

From the times of early European visitors it was recognised that, in Australia generally and southwest Australia in particular, there were remarkable assemblages of plants and animals in a remarkable landscape. Throughout the early years of settlement, from 1827 to the 1890s, many geologists, botanists, and naturalists reported on different aspects of the country and a body of scientific knowledge accumulated. The geology of southwestern Australia was fairly well known by the 1870s and a geological sketch map was published (Brown 1873). With the discovery of gold in the 1880s, the geological mapping received extra impetus and Woodward (1908) published the first map of Western Australia; this was followed by a more detailed version (Maitland 1924).

At this stage, people from other disciplines were looking at the overall distribution of landscape elements and, with the geological background available, statements on geomorphology (Juston 1934), soils (Teakle 1938), and native vegetation (Gardner 1944) followed. Since those benchmark papers there has been much detailed work on the landscape and its biota and, while there have been refinements and modifications, the main themes are unchanged. Thus, in the 1990s, there is considerable knowledge of the pattern of landscape elements and some data and inferences on past events and processes which have produced the present land surface. The purpose of this chapter is to examine the available evidence and construct a likely landscape history of the central wheatbelt of Western Australia.

2.2 Fragmentation of Gondwanaland

Western Australia is made up of a nucleus of Archaean igneous and metamorphic rocks of the Yilgarn Block, flanked to the south by Proterozoic metamorphics and with marine and terrestrial sedimentary rocks on the western and southern margins (Anon. 1975). These rocks, both crystalline and sedimentary, form the substrata for the Great Plateau of Western Australia. The history of this plateau, as a land surface, extends back at least to the Palaeozoic when Australia was part of Gondwanaland. The west coast of Australia formed in the Early Palaeozoic when a trench developed and sediments began to accumulate in what is now the Perth Basin (Johnstone et al. 1973). This trench persisted throughout the Palaeozoic and into the Mesozoic and then, in the Early Cretaceous, India drifted away to the northwest. The separation was complete with the Darling Scarp forming the western margin of Australia. Similarly, to the south of the Precambrian nucleus, a section of Gondwanaland was down-warped and, as Australia drifted north from Antarctica in the Eocene, the marine sediments now known as the Plantagenet Group were deposited. Thus the gross shape of Australia was established and landscape processes which were originally operating on a supercontinent now became accentuated because of new base levels and oceanic climates.

Before, during, and after the fragmentation of Gondwanaland, the Great Plateau of Western Australia had remained above sea level and so was subject to subaerial weathering and erosion for an immense period of time. The end result of this episode, in the Late Tertiary, was a landscape of very low relief with a mantle of weathered rocks often as much as 50 m thick. This mantle, consisting of a sandy or gravelly (lateritic) soil in the surface and underlain by mottled and pallid zones of weathered material, is commonly referred to as the laterite profile (e.g., Stephens 1946). It is noted that Ollier et al. (1988), writing of the adjacent region to the east, considered this model of landscape development too simple. They postulated that there have been many episodes of planation, erosion and deposition, and formation of lateritic soils.

2.3 Miocene

Bowler (1982), in a study of aridity in southern Australia, considered evidence from several different disciplines and constructed a paleoclimatic sequence extending back to the Miocene period 20 million years ago. While little of Bowler's evidence comes from Western Australia, it is possible to relate some local landscape features to inferred climatic variations. According to the model, the Miocene climate was probably 10°C warmer than at present and the rainfall was high and nearly uniform throughout the year. These conditions continued for about 15 million years, and we can assume that there was a profound effect on chemical weathering, erosion, soil formation, and the biota. The *in situ* weathering of basement rocks continued and the strongly acidic substrata in many wheatbelt landscapes (e.g., Bettenay and Hingston 1961) could have resulted from the long-continued severe leaching. Also in this period, it is probable that silcretes formed in the upper parts of the weathering profile. Silcrete, which occurs sporadically in the central wheatbelt, is a brittle, intensely indurated material formed mainly of often rounded quartz grains in a matrix of amorphous silica. Butt (1985) considers that a humid climate is required to mobilize silica (silicic acid), followed by a markedly seasonal climate to permit precipitation.

The occurrence of sedimentary materials, either as isolated outcrops of conglomerate or occasional waterworn pebbles in upper landscape positions throughout the wheatbelt, points to the existence of high energy drainage systems which could also be assigned to this pluvial period. Whatever the age of the deposits, they are, at least in part, younger than the period of silcrete formation because they include rounded fragments of silcrete. Wilde and Backhouse (1976) have examined many sedimentary materials on the Darling Plateau and have postulated that, on the basis of some fossil evidence, these may relate to the Eocene marine transgression associated with the breakup of Australia and Antarctica.

During this long pluvial period the main vegetation across southern Australia was rainforest which was dominated by *Nothofagus* spp., conifers such as *Dacrydium* spp., and *Podocarpidites* sp., and many ferns. Fossil evidence of these species has been found in valley fill deposits in the wheatbelt and the goldfields area immediately to the east; some of these deposits are more than 100 m thick (Balme and Churchill 1959). The main elements of the present flora were also present, possibly in sites of low fertility (Lamont et al. 1984).

2.4 Pliocene

With the passage from the Miocene to Pliocene five to six million years ago, there was a change to intense winter aridity across southern Australia. According to Bowler (1982), this caused the rainforest to contract and finally disappear and the present sclerophytic vegetation became established. Associated with this was a change in the status of some soil constituents with seasonal wetting and drying causing ferruginous gravels to form in the profile. In the Late Pliocene (two to five million years ago), there was a further change to erratic wet-dry oscillations which Bowler (1982) interpreted as transitional towards a winter rainfall regime. By the end of the Tertiary the gross morphology of the Great Plateau was established and, in the wheatbelt, the landscape consisted of a gently undulating surface with about 100 m relief; the main trunk valleys, with functional drainage, were in place (Playford

1954). The soil on this surface was yellow sand, with gravel in the subsoil, overlaying mottled yellow clay.

2.5 Pleistocene

With the onset of the Quaternary period, the detail of the present landscape began to develop. The Pleistocene period was characterised by wide fluctuations in pluviality and temperature which had implications in landscape development. The present pattern of rainfall, with a marked winter incidence, became established in the Early Pleistocene. On evidence from eastern Australia, Bowler (1982) deduced that there was widespread landscape instability and rejuvenation of drainage systems, possibly due to continental uplift, but influenced by the variations in rainfall. Thus the Plateau surface, with its underlying weathered mantle, was differentially eroded, and various weathered and unweathered strata were exposed to become soil parent materials. Bowler (1982) noted that, at some time later than 0.7 million years ago, there was the first production of saline and calcareous materials and dune successions. These materials, which would have been distributed by wind, are responsible for the widespread calcareous soils of the interior and the apparently anomalous situation of being strongly acidic substrata overlain by calcareous and gypsiferous surface soils (e.g., Bettenay and Hingston 1961; Burvill 1988). Landscape instability apparently continued through the Middle and Late Pleistocene. Wyrwoll (1979) has shown that, on the basis of geomorphology and stratigraphy, the past 80,000 years have been characterised by marked changes in climate. He stresses that, in southwestern Australia, there was an intensification of aridity culminating in the dune-building phase of 20,000 to 15,000 years ago. Differential erosion of the Plateau proceeded through the Early and Middle Pleistocene, but the pattern of erosion was by no means regular. Some remnants are several thousands of hectares in extent while others may be minor. Also, some are preserved on local interfluves while others extend downslope to the edge of a valley floor. The soil on these remnants, generally referred to as yellow sandplain, shows extremes in soil formation in terms of texture differentiation, depletion of plant nutrient elements, and segregation of iron and aluminium oxides.

By the Late Pleistocene (about 30,000 years ago), the land had taken on the present characteristics. The landform-soil units, with names proposed by Bettenay and Hingston (1964), are distributed in relation to modifications of the old plateau. The plateau surface (Ulva unit) was fragmented and the Booraan and Collgar units, with red or yellow duplex soils, developed in dissections where mottled and pallid zone materials were exposed. The Danberrin unit, with red or yellow duplex

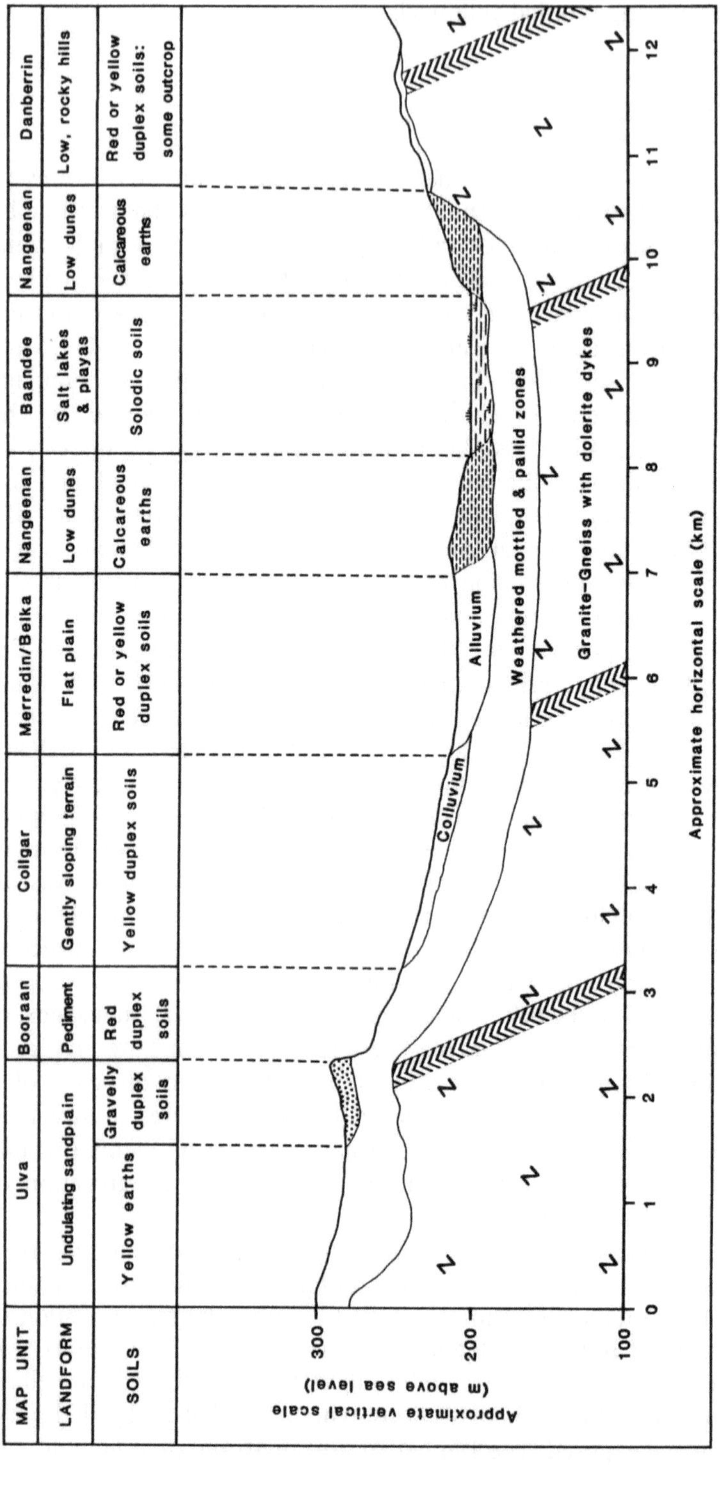

FIGURE 2.1. Landform units along a catenary sequence in the central wheatbelt of Western Australia.

soils associated with rock outcrops, often developed in lower slope positions. The detritus generated was transported into broad valleys to form the Merredin unit with red duplex and the Belka unit with yellow duplex soils. These, in turn, joined the main trunk valley floors of the Baandee unit consisting of salt lakes, playas, and lunettes. The Baandee unit thus was ineffective as a drainage line and became a repository for salts. The Nangeenan unit, which is adjacent and usually to the southeast of the Baandee unit, consists of calcareous clays which originated in the lake beds and was mobilised by wind during times of relative aridity. The units are shown diagrammatically in cross section in Fig. 2.1. Descriptions of the landscape units, in terms of landform, soils, and vegetation, are given in Section 2.8. The major soil of each unit is classified using the United States Soil Taxonomy (Soil Survey Staff 1990) and the Legend of the Soil Map of the World, 1:5,000,000 (FAO-UNESCO 1988).

The lakes of the Baandee unit provide the first opportunity for the absolute dating of events. When water evaporated the coarser materials on the lake floor were blown to the edge to form a crescent-shaped dune, or lunette, always on the southeastern side. Aboriginal peoples lived on the lunettes, because, for part of the year, the lakes had water and supported some animal life. As the lunettes accumulated, evidence of Aboriginal occupation was incorporated and, in some instances, charcoal was found associated with artefacts making radiocarbon dating possible. In some lunettes, several layers of Aboriginal artefacts have been found and these have been interpreted in terms of climatic changes.

The climatic fluctuations during the past 30,000 years have been characterised by Bowler (1976) and Bowler et al. (1976), based on evidence of former lake levels, dune successions, and soil stratigraphy. These data indicate that, after a period of high rainfall, there was a gradual decline, with the climate reaching maximum aridity in the period 17,000 to 15,000 years ago. Bowler considered that this coincided with the last great dune-building phase in southern Australia, which began as early as 25,000 years ago in some areas and persisted until 12,000 years ago in others. Radiocarbon dates show that lunettes formed in the central wheatbelt in the period 20,000 to 15,000 years ago. The calcareous clay materials of the Nangeenan unit accumulated in two phases of activity (Bettenay and Hingston 1964). The older phase, with clear evidence of downward movement of carbonates, salts, and clay, probably dates from the period of aridity 20,000 to 15,000 years ago. The younger phase can probably be assigned to a further brief period of aridity 4000 to 3000 years ago. Also during this latter period, it is likely that some of the lunettes became unstable and the detritus was reworked to form small parabolic dunes. Such dunes are common adjacent to lakes and valley floors throughout the southwest of western Australia (e.g., McArthur et al. 1977) and their southeasterly orientation means they formed in a wind regime different from that at present.

2.6 Holocene

The climatic fluctuations during the Holocene are known in some detail. Churchill (1968) studied pollen records in swamps in the southwest and, using the present distribution of *Eucalyptus* species to characterise climates, constructed a table showing trends during the past 7000 years. His data showed that, following a wet period 6000 to 5000 years ago, rainfall declined and maximum aridity developed in the period 4500 to 2500 years ago; this correlates with the data of Bowler (1976). From that time (500 B.C.) to about 700 A.D., the climate was once again wetter. Another dry period extended from 1200 to 1500 A.D. and, from that time, the climate has become gradually wetter. It has not been possible to equate landscape events in the wheatbelt to these later fluctuations in climate, but it is likely that there was a reworking of sandy materials during the arid phases and a transport of slope deposits in the wet periods.

2.7 Historic

Since European settlement, there have been records of climatic events, especially extremes of rainfall or drought and, in the 1880s, meteorological records were commenced. Many workers (e.g., Gentilli 1971; Pittock 1975; Tucker 1975) have used the available data to characterise the variability of climate. These and other published data have been summarised by Iwasaki (1985) who, using residual mass (RM) graphs, showed the incidence of continuously wet or dry periods since 1840. For southwest Australia, the latter part of the 1800s was relatively dry and reached maximum aridity in 1914. Since then, the trend has been for wetter conditions, with a short dry period about 1940. The rainfall began to decline in 1960 and a relatively dry period continued until 1980.

Up to this point, we have considered the changes of climate in isolation, but it is probably relevant to consider now the additional influence of humans, especially when looking at the reason for the onset of erosion. Firstly, it was suggested by Bowler (1982) that, during the past 100,000 years, when humans probably were present in Australia, the effects of climate may have been complicated or obscured. Then, with the arrival of the Europeans, changes have been rapid and severe. For example, in the Lake Grace area, since records began in 1915, the mobilisation of sand to form dunes and hummocks was first recorded around 1940, coinciding with a statewide dry period. However, it was doubtless exacerbated by the clearing of native vegetation. Indeed, it is likely that no instability would have occurred without the intervention of the Europeans. In this context, it is possible that the most recent erosion gullies were initiated by increased runoff due to the clearing of native vegetation; possibly this began with the long continuous wet period following the 1914 drought, when the area of agricultural land was increasing rapidly and there was a change from horsepower to mechanisation.

2.8 Current Landforms and Vegetation of the Kellerberrin Area of the Central Wheatbelt

Figure 2.1 shows a typical cross section of a catenary sequence of the central wheatbelt landscape, illustrating landforms and soil types. Landscape units are named following the mapping nomenclature used by Bettenay and Hingston (1964).

The Ulva unit, consisting of flat or gently sloping remnants of the lateritic sandplain, commonly forms drainage divides but may sometimes extend to lower slopes. In the pristine state, the unit has about 2 m of yellow sand or sandy loam, often with gravel in the lower horizons, overlying a gravelly mottled sandy clay (Xanthic Acrudox; Xanthic Ferralsol). However, the profile is often modified by erosion and so the depth of sand is variable and may be entirely lacking; at the edges of the unit, the surface often consists of gravelly clay. In some instances, the unit includes a broad shallow valley and, in the lowest topographic position, the sand may be grey and very deep due to the accumulation of erosion products.

The sandplain heath vegetation (or kwongan) is renowned for its endemism and species richness, and these have often been discussed. There are usually three structural layers, the composition of which may vary over short distances. The upper layer of small trees, mallees, and shrubs is 2.5 to 3.5 m high and includes *Leptospermum erubescens*, *Eucalyptus burracoppinensis*, *Acacia assimilis*, *Allocasuarina acutivalvis*, *A. huegeliana*, *Xylomelum angustifolium*, *Hakea coriacea*, and *Santalum acuminatum*. The second layer, at 0.5 to 1 m, is often very dense and includes the shrubs *Beaufortia bracteosa*, *Isopogon scabriusculus*, *Melaleuca conothanmoides*, *M. seriata*, other *Melaleuca* spp., *Persoonia quinquenervis*, *Andersonia lehmanniana*, *Astroloma serratifolium*, *Verticordia* spp., *Acacia* spp., and *Grevillea* spp. The sedges *Ecdeiocolea monostachya*, *Lepidosperma angustatum*, and *Mesomelaena preissii* occur in more open situations. Ground cover consists of *Borya constricta*, *Loxocarya* spp. and *Lyginia* spp., mat plants, and many lichen encrustations. These soils also support *Acacia* thickets or wodjil, dominated by *Acacia neurophylla*, *A. assimilis*, and *A. stereophylla*. In those areas where the sand has been removed and the surface is gravelly, often on the fringes of the sandplain, the vegetation is dense shrubland or tamma, at about 2 m, dominated by *Allocasuarina campestris*.

The Booraan unit is developed on the upper slopes of dissections in the lateritic sandplain landscape; often the boundary between the two is marked by a low scarp. The Booraan unit is typically formed from weathered granitic materials with the soils consisting of grey-brown gritty materials with sandy loam over brown or red-brown clay; lime may occur in the lower horizons (Natric Haploxeralf; Plinthic Lixisol). The profile is generally shallow in the upper landscape position, often with weathered granite outcropping, and becomes deeper down slope. As mapped, the unit may contain small areas of gravelly soils where the old laterite surface has not been completely stripped. The vegetation typically is a woodland with scattered *Eucalyptus salmonophloia* and *E. wandoo* or *E. capillosa*, 15 to 20 m tall, scattered *Callitris canescens* trees, and a dense shrub layer, at about 1.5 m, dominated by *Melaleuca uncinata* but also with other *Melaleuca* spp., *Allocasuarina campestris, Olearea muelleri, Beaufortia bracteosa, Acacia hemiteles*, and *Templetonia sulcata*.

The Collgar unit occurs on very gentle lower slopes of dissections, often down slope from the Booraan unit; the boundary between the two is diffuse. The soil consists of a grey sandy surface, a white subsurface which is often indurated, and a mottled yellow and brown clay at about 30 cm (Natric Haploxeralf; Albic Luvisol). The vegetation is mallee scrub 4 to 6 m tall, with the main species being *Eucalyptus transcontinentalis, E. cylindriflora, E. capillosa, E. erythronema*, and *E. hypochlamydea*. The shrub layer is sparse and includes *Melaleuca uncinata, Acacia hemiteles, Westringia cephalantha*, and *Olearia muelleri*; locally there may be dense thickets of *Melaleuca cardiophylla*.

The Merredin and Belka units are broad, flat, alluvial landscapes which are either fringed by or occur downstream from the Collgar unit. The surface is typically marked by many shallow prior stream channels. The soils of the Merredin unit have a brown loamy surface and a red clay subsoil (Natric Palexeralf; Luvic Calcisol); these are the so-called "heavy" soils of the wheatbelt. The Belka unit has paler and more sandy soils; the surface is light grey-brown sand or sandy loam over mottled yellow and brown clay. The vegetation is woodland with *Eucalyptus salmonophloia, E. salubris*, and *E. capillosa*, at least 20 m tall, and a shrub layer which appears to vary from place to place. In some cases, there is a sparse heath, at 1.5 m, including *Melaleuca* spp., *Acacia hemiteles*, and other *Acacia* spp., with some ground cover of *Loxocarya* and *Borya* spp. In other cases, there is a dense layer of halophytic plants, to 1 m, consisting of *Atriplex paludosa, Maireana carnosa, M. appressa*, and *Enchylaena tomentosa*. This difference probably relates to soil conditions with the halophytic plants associated with highly calcareous soils. In some cases, the Belka unit appears to have less vigorous trees than the Merredin and, in places, mallee vegetation replaces *Eucalyptus salmonophloia*-

E. salubris woodland. Thus, on both soils and vegetation, the transition from Belka to Collgar is often very difficult to determine.

The Baandee unit represents the ancient drainage zones which have become filled with detritus and, with very low gradients, have ceased to function as drainage lines. These zones are now occupied by lakes, swamps, and playas, usually with lunettes around the southeastern margins. The whole landscape is dominated by salts which have accumulated over a very long period. Each year, drainage water containing salts enters the lakes and swamps and, during summer, evaporates, leaving behind the salts which, together with sand from the lake floor, are blown out to form lunettes. The soils are thus very saline and sometimes gypseous or calcareous and the vegetation is mainly halophytic. The most saline areas have samphire communities while less saline parts have *Atriplex* spp., *Maireana* spp., and *Enchylaena* spp. The lunettes may have a woodland of *Callitris canescens, Casuarina obesa,* and *Melaleuca* spp. Ground cover consists of succulent plants such as *Mesembryanthemum* spp., *Enchylaena* spp., *Halosarcia* spp., and *Portulaca* spp.

The Nangeenan unit occurs as a veneer of salt lake materials – clay, gypsum, calcium carbonate, and salts – blown over the adjacent landscapes. Thus, it is most common on the alluvial terrain or the lower gentle slopes. The soils are generally fine-textured and highly calcareous (Natric Paleseralf; Luvic Calcisol). They are prone to develop saline patches when cleared; the surface often shows gilgai microrelief. The vegetation is a woodland with *Eucalyptus salmonophloia, E. salubris,* and *E. longicornis,* at least 20 m tall, and a dense shrub layer, 0.5 m high, of halophytes including *Atriplex paludosa, Maireana carnosa,* and *M. appressa.*

The Danberrin unit consists of irregular low hills and gentle slopes, occurring in all topographic positions but most commonly at lower levels adjacent to drainage lines. This unit represents the maximum degree of erosion with all of the pre-weathered mottled and pallid zone materials removed. The unit is dominated by a shallow soil of brown gritty sandy loam, often 30 to 40 cm thick over granite, (Typic Haploxeralf; Albic Luvisol). Rock outcrops are frequent but are generally not present in sufficient size or frequency to prevent agriculture. There may be mafic dykes which locally produce red, fine-textured soils (Vertic Haploxeralf; Luvic Calcisol). The vegetation is typically a woodland of *Eucalyptus loxophleba* and *Acacia acuminata* with a sparse shrub layer and many introduced grasses.

Rock outcrops or areas dominated by rock outcrops are generally associated with the Danberrin unit. Most soils are similar to those already described, but there may also be shallow, coarse-textured soils formed from weathering granite detritus around the edges of large outcrops. The vegetation is often *Eucalyptus loxophleba*/*Acacia acuminata* woodland, but there may also be dense *Allocasuarina huegeliana* fringing the rock outcrops. The sparse ground cover consists of *Lepidosperma gracile, Stypandra glauca, Waitzia acuminata,* and *Borya nitida*. Rock surfaces have an almost complete cover of lichens with mosses, liverworts, and *Borya sphaerocephala* in shallow depressions. The ferns *Adiantum aethiopicum* and *Cheilanthes austrotenuifolia* grow in clefts in the rocks.

2.9 Conclusion

This consideration of landscape history and paleoclimatology, however fragmentary, provides a framework in which to study the present land surface and its biology. It has been shown that, within the study area, the mapping units form an orderly pattern when seen as differential erosion of an ancient land surface. Similarly, the flora may be seen in the light of the long period of isolation. During this period, there were dramatic fluctuations in climate which would produce stress, especially due to moisture, and so a highly diverse sclerophytic vegetation has evolved. In addition, the soil was subjected to long and severe leaching processes, thus producing nutritional stress. This flora is still present on remnants of the ancient sandplain landscape, with different communities developing on the newly formed erosional landscapes. All of these facets (the various soils, topography, and vegetation) have combined to provide habitats for a wide range of vertebrate and invertebrate fauna.

References

Anon. (1975) *The Geology of Western Australia.* Geological Survey of Western Australia, Memoir 2.

Balme, B.E. and Churchill, D.M. (1959) Tertiary sediments at Coolgardie, Western Australia. *Journal of the Royal Society of Western Australia* 42:37–43.

Bettenay, E. and Hingston, F.J. (1961) Soils of the Merredin area, Western Australia. *CSIRO Australian Soils and Land Use Series* No. 41.

Bettenay, E. and Hingston, F.J. (1964) Development and distribution of soils of the Merredin area, Western Australia. *Australian Journal of Soil Research* 2:173–186.

Bowler, J.M. (1976) Aridity in Australia: Age, origins, and expressions in aeolian landforms and sediments. *Earth Sciences Review* 12:279–310.

Bowler, J.M. (1982) Aridity in the late Tertiary and Quaternary of Australia. In W.R. Barker and P.J.M. Greenslade (eds.) *Evolution of the Flora and Fauna of Arid Australia.* Peacock Publications, Adelaide, South Australia. pp. 35–45.

Bowler, J.M., Hope, G.S., Jennings, J.N., Singh, G., and Walker, D. (1976) Late Quaternary climates of Australia and New Guinea. *Quaternary Research* 6:359–394.

Brown, H.Y.L. (1873) General report on a geological expedition of that portion of the Colony of Western Australia lying southward of the Murchison River and westward of Esperance Bay. Paper No. 1, *Votes and Proceedings, Legislative Council, W.A.*, for 1873, pp. 1–20. Perth: By Authority, 1873.

Burvill, G.H. (1988) The soils of the Salmon Gums district, Western Australia. *Western Australian Department of Agriculture Technical Bulletin* No. 77.

Butt, C.R.M. (1985) Granite weathering and silcrete formation on the Yilgarn Block, Western Australia. *Australian Journal of Earth Science* 32:415–432.

Churchill, D.M. (1968) The distribution and pre-history of *Eucalyptus diversicolor* F. Muell, *E. marginata* Donn. ex sm., and *E. calophylla* R.Br. in relation to rainfall. *Australian Journal of Botany* 16:125–51.

FAO-UNESCO (1988) *Legend of the Soil Map of the World, 1:5,000,000*. FAO, Rome.

Gardner, C.A. (1944) The vegetation of Western Australia with special reference to climate and soils (Presidential Address 1942). *Journal of the Royal Society of Western Australia* 28:11–87.

Gentilli, J. (1971) Climatic fluctuations. In *Climates of Australia and New Zealand*, ed. J. Gentilli, World Survey of Climatology, Vol. 13:189–211 Elsevier Amsterdam.

Iwasaki, K. (1985) Regional classification of climatic change and variability in Australia since 1840. In H. Toya, K. Takeuchi, and H. Ohmori (eds.) *Studies of Environmental Changes due to Human Activities in Semi-arid Regions of Australia*. Department of Geography, Tokyo Metropolitan University. pp. 19–27.

Johnstone, M.H., Lowry, D.C., and Quilty, P.G. (1973) The geology of southwestern Australia – A review. *Journal of the Royal Society of Western Australia* 56:5–15.

Jutson, J.T. (1934) The physiography (geomorphology) of Western Australia. Second edition revised. *Geological Survey of Western Australia Bulletin* 95.

Lamont, B.B., Hopkins, A.J.M., and Hnatiuk, R.J. (1984) The flora – Composition, diversity, and origins. In J.S. Pate and J.S. Beard (eds.) *Kwongan – Plant Life of the Sandplain*. University of Western Australia Press, Nedlands. pp. 27–50.

Maitland, A.G. (1924) The geology of Western Australia. *Geological Survey of Western Australia Bulletin* 89:9–89.

McArthur, W.M., Churchward, H.M., and Hick, P.T. (1977) Landforms and soils of the Murray River catchment area of Western Australia. *CSIRO Division of Land Resources Management Series* No. 3.

Ollier, D.C., Chan, R.A., Craig, M.A., and Gibson, D.L. (1988) Aspects of landscape history and regolith in the Kalgoorlie Region, Western Australia. *Bureau of Mineral Resources Journal of Australian Geology and Geophysics* 10:309–321.

Pittock, A.B. (1975) Climatic change and the patterns of variation in Australian rainfall. *Search* 6:323–328.

Playford, P.E. (1954) Observations on laterite in Western Australia. *Australian Journal of Science* 17:11–14.

Soil Survey Staff (1990) Keys to soil taxonomy (Fourth Edition). *SMSS Technical Monograph* No. 19. Virginia Polytechnic Institute and State University. Blacksburg Virginia.

Stephens, C.G. (1946) Pedogenesis following the dissection of lateritic regions in southern Australia. *CSIRO Bulletin* 206.

Teakle, L.J.H. (1938) A regional classification of the soils of Western Australia (Presidential Address 1937). *Journal of the Royal Society of Western Australia* 24:123–195.

Tucker, G.B. (1975) Climate: Is Australia's changing? *Search* 6:323–328.

Wilde, S.A. and Backhouse, J. (1976) Fossiliferous Tertiary deposits on the Darling Plateau, Western Australia. *Western Australian Geological Survey Annual Report 1976.*

Woodward, H.P. (1908) Geology of Western Australia. *Official Year Book of the Commonwealth of Australia* 1:94–98.

Wyrwoll, K.H. (1979) Late Quaternary climates of Western Australia: Evidence and mechanisms. *Journal of the Royal Society of Western Australia* 62:129–142.

Chapter 3

Social History and Impact on Landscape

Barbara York Main

3.1 Introduction

> *"In a word, land settlement is at present the national policy of Western Australia, and in inviting settlers to cast their lot there she is able to make the invitation well worthy of their acceptance." (Fraser 1906).*

This statement just after the turn of the century confirmed the mood that began with establishment of the colony over seventy years earlier, so that now we should not be surprised that most of the land deemed suitable for agriculture and horticulture in succeeding periods has been largely denuded of its natural vegetation. Rather, since the settlers came, for the most part, from closely cultivated regions, we should be both surprised and grateful that any vestiges of natural landscape remain at all. The following discussion is concerned primarily with the present shires of Tammin, Kellerberrin, and Trayning (for localities in this chapter see Fig. 3.1) and the surrounding areas, in so far as their history has impinged on that of the region under study in this volume.

The study area and environs fall within the tribal boundaries of the traditional Aboriginal Nyaginyagi and Balardany societies (Berndt et al. 1986). Little apparent modification by the Aborigines of the natural landscape was evident to the first European colonisers. It is known that they used water holes generally associated with granite outcrops such as Yorkrakine Rock, Nalyerine, and others. They may have used fire (Hallam 1985) and, of course, they fulfilled a predatory role on the native fauna and used many of the food plants. Their impacts are still being debated. Perhaps the irreversible impact on the landscape by humans was initiated by the pastoralists dating from just after the middle of the nineteenth century, following the explorations of the country eastward from the earlier settled areas at Toodyay (Newcastle), Northam, and York. This chapter discusses the exploration and development of the central wheatbelt, as summarised in Table 3.1.

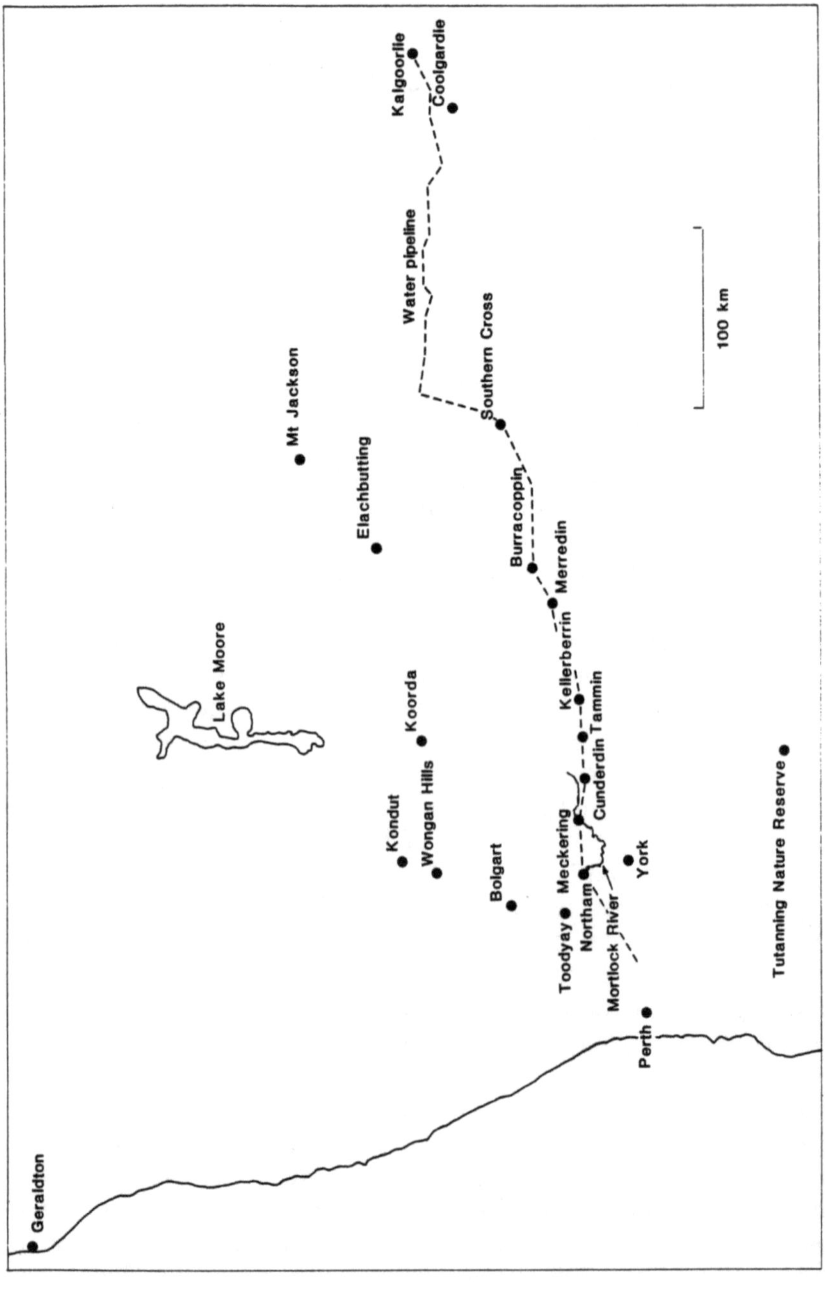
Geraldton
Lake Moore
Mt Jackson
Elachbutting
Kalgoorlie
Coolgardie
Water pipeline
Southern Cross
100 km
Burracoppin
Merredin
Kellerberrin
Tammin
Koorda
Kondut
Wongan Hills
Cunderdin
Meckering
Bolgart
Toodyay
Northam
York
Mortlock River
Perth
Tutanning Nature Reserve

(a)

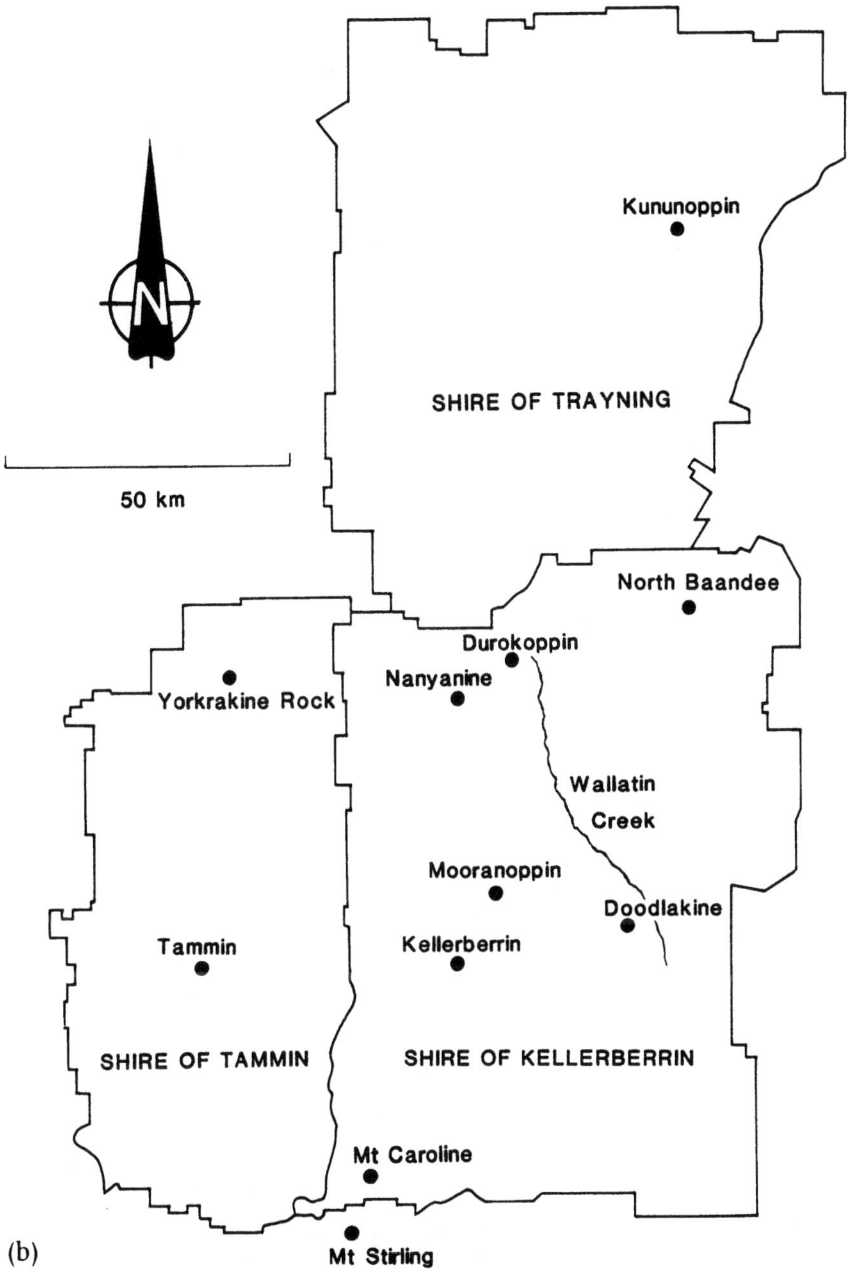

(b)

FIGURE 3.1. (a) Map of central and northern wheatbelt of Western Australia indicating locations referred to in text. (b) Map of Trayning, Tammin, and Kellerberrin shires incorporating the major study area referred to in this volume and indicating locations referred to in text.

Table 3.1. Sociological and environmental consequences of government acts and edicts, interaction with natural phenomena, and the cumulative impact on the landscape. Data derived from various sources, namely: Fraser (1906); Western Australian Year-Books; Bureau of Statistics; and citations as in text. W. McArthur and D. Couper are acknowledged for comments and relevant information (WA refers to Western Australia throughout).

Date	Events and Government Acts Facilitating Settlement	Sociological Impacts	Depressing Events and Circumstances	Date
1830	Dale explored east to Mt. Caroline		No surface water available for stock	
		Hopes of discovery of inland sea and pastures		
1836	Roe explored east to edge of (future) Goldfields			
1845	First exportation of sandalwood	Support for local commerce	Beginning of modifications of vegetation	
1861	Dempster, Clarkson explored east of Northam			
1863	Lefroy discovered good agricultural land	Expansion of sheep flocks eastward onto pastoral leases	Selective grazing of native vegetation	
1862–4	Hunt blazed track from York to Coolgardie			
1864–8	Pastoral leases including Leake pastoral leases at Mooranoppin and Nanyanine			
1880s	Corrugated iron available	Used for domestic roofing and water catchment		
1888	Wire available for fencing	Enclosure of crops, protection of stock from "natural" predators, e.g., dingoes	Restricted movement of larger animals	
1888	Gold discovered at Southern Cross	Influx of population to WA Goldfields	Economic depression world wide and eastern Australia	1890s

1891	Responsible Government for WA	Sandalwood "pulling" increased		
	Royal Commission on Agriculture		Droughts in early 1890s	
1892	Gold discovered at Coolgardie			
1893	Homesteads Act	Free land of 160 acre blocks for settlers		
	"Special occupation leases"	Pastoral lessees acquired water holes and good land for agriculture (Packham's farming at Tammin)		
1894	Railway to Southern Cross	Tammin and Doodlakine	Rabbits at Eucla	1894
	Superphosphate imported and obtained from Abrolhos	Agricultural areas open for selection		
		Selection encouraged by railway and fertilisers improved crop yields		
1895	Agricultural Bank	"Starting" finance for clearing and establishment of farms		
1896	Railway extended to Coolgardie			
1898	Land Act, conditional purchase 100–1000 acres (6 months residency required)	Diversification pastoral to agriculture		
1901–3	Goldfields pipe line, opened at Kalgoorlie 1903	Water available along line for agriculture	Rabbit proof fences constructed due to westward advance of rabbits	1901–7
1906	Liberalisation Agricultural Bank			
1907–8	Terry's classification of land north of Yilgarn railway (Tammin, Doodlakine, etc)	Unemployment	Collapse of gold mining and economy	1908–12
	Recommendation resumption pastoral leases	Interest sparked in land selection; farms of mixed land classification (up to 1000 acres)		
1908		Yorkrakine settlement for unemployed		
1909	Agricultural Lands Purchase Act	Availability of previous pastoral leases for repurchase for agriculture		
1910	Fertiliser manufactured locally			
1911	Northern railway "loop," Dowerin to Merredin	Transport of produce and water	Severe drought	1911

(*Continued*)

Table 3.1. *Continued*

Date	Events and Government Acts Facilitating Settlement	Sociological Impacts	Depressing Events and Circumstances	Date
1912	Land settlement assistance for unemployed, including loans for clearing	Schools established		
prior to 1913	Tammin site of annual military training manoeuvres	Disturbance of natural bushland		
	Flour mill at Kellerberrin	Domestic advantage for farmers		
1914		Crop failure, no seed for next season, shortage of stockfeed	Severe drought	1914
	Railways cart water			
	Telephone	Improved communications	World War I	1914–18
	Industries Assistance Board	Salvaged drought-devastated farmers		
		Shortage of manpower as men left country for war		
1915	Importation grain/hay from South Africa	Seed and stockfeed	Introduction of weeds, e.g., cape weed and double gee	post 1915
1915–21	Good rains	Temporary reprieve		
1917	Land Act Amendment Act	Easing of repayments	Clearing of second and third class land	
1918	Establishment Correspondence Classes	Primary education available for rural children beyond reach of schools		
1919–20	Soldiers' Settlement Scheme	"Free" farm blocks available to returned soldiers		
	Vehicular transport (cars) but wagons still used for farm work			
1920	Rabbit proof fencing	Partial alleviation only of rabbit problem	Rabbits continue to spread	1920
1922	Introduction of small tractors	Useful where water is short		
1924	Radio 6WF	Social amenity-information	Initial salinity on cleared land	
			Sandalwood pulling reducing natural stocks	

1920–28		Population expansion peak about 1925, augmented by immigration from South Australia		
1925–35		Lowering of morale and severe destruction of farm lands and disturbance of bushland (by rabbits and burning etc., as control)	Rabbits at peak from 1928 Salinity increasing (some scientific recognition of cause)	
1929	Sandalwood Act	Limiting "pullers"		
1929–35		Decline in rural population – farms abandoned, some absorbed into surviving properties	Economic depression; Agricultural Bank foreclosures	1929–35
		Reduced rural education opportunities	Closure of some rural schools	
			Emus moved in from northeast due to inland drought	1932
1933	Royal Commission on Agriculture	Abandonment of farms and drift to city; low morale	Unviability of many farms (poor land, extended loans)	1933
	Land Act – Conditional Purchase	Mechanisation of farms – phasing out of horses	Decline in *number* of families	1933 onwards
	Cooperative Bulk Handling, Ltd. (for wheat)	Accelerated handling/transport of grain		
	Introduction of sheep	Farm diversification		
1935	Last dingo sighted (Trayning)	Expensive high fencing no longer required	Low rainfall 1935–38 and 1940	
1935–39	Resurgence in economy	Consolidation of holdings		1935–39
			Dry years, grasshopper plagues, dust storms	1935–38
	More vehicular transport (cars, trucks) and tractors	Decline in number of rural holdings and decline in rural population Abandonment of small rural community centres		
1939	World War II	Gas producers prevalent	World War II Shortage of essential items, e.g., petrol, tyres, machinery, and superphosphate Destruction of timber (salmon gum and wandoo) for charcoal	1939–45

(*Continued*)

TABLE 3.1. *Continued*

Date	Events and Government Acts Facilitating Settlement	Sociological Impacts	Depressing Events and Circumstances	Date
1942		Italian prisoners of war recruited as farm workers	Shortage of farm "manpower"	
		Towards *end of war* "boom" in production of commodities, particularly wool, and increase in farm incomes and mechanisation		
1945	War Service Land Settlement Scheme	Availability of farm land for returned soldiers – increased vocational "opportunities"		1945
1945–47	Availability war surplus heavy machinery	Further clearing	Land degradation	1945–47
1946	Education Department School Bus Service	Closure of most rural schools, children transported to town centres – better educational opportunities		1946
1946	Modified Comprehensive Scheme (water supply)	Kellerberrin/Trayning and surrounding areas connected with northern pipe line from Goldfields pipe line		
1947–50	Soil Conservation Service (Department of Agriculture)	Some absentee "owners"/leasing of farms	Continued clearing of bush/woodland 1950s/1960s	
1947–8	Introduction of pesticides	Control of blowflies, some weeds		
1950	Korean War and world demand for wool	Increase in sheep stocking, sudden rural affluence		1950
1951	Formation of Agricultural Protection Board	Noxious weeds and vermin control		
1952	Myxomatosis control of rabbits	Increased productivity cereals and pastures – boosting of morale and general euphoria		1952
1955	First salt land survey by Department of Agriculture	Reduction in arable land	Salinisation of cultivated land	1955

1957–8	Crawler tractors	Clearing light lands		
1958	Extension of Comprehensive Water Scheme north from Kellerberrin and Doodlakine	Improvement of domestic life and increased stock (Taryning area)		1958
	Television introduced to WA	Improved communication but not extended (directly) to rural areas		
late 1950s/ 1960s	State Electricity grid extended in Wheatbelt	Improved domestic amenities and farm mechanics and maintenance		
1960	Landboom, compound fertilizers make "light" lands available for agriculture, introduction of new pasture plants	Clearing of previously "third class" land, expansion of productive area of individual farms	Increasing salinity	1960
		Increase in individual affluence and farm consolidation		
1962	Second salt land survey	Contraction of earlier arable land		1962
1960s	Television to rural areas	Increased educational opportunities and communication amenities		
	Establishment of junior high schools in smaller rural centres			
			Drought, introduction of wheat quotas	1969
1970 | 1990		Collapse of business in towns between major centres (Northam/ Merredin)	Progressive decline in population	1970 | 1990
			Low wool prices (early 1970s)	
		Expansion of some individual holdings (retraction of others, with associated social distress)	Many poor seasons	
			Increasing salinity, erosion, land degradation	
		Increased affluence amongst surviving farmers		
1974	Third salt land survey	Families with greater educational opportunities (including Tertiary) which encouraged females to leave (population changes partly reconciled by marriage of local farmers to professional women from outside area)	Continuing degradation of farmland, increasing salinity, compaction of soil by heavier machinery	1974
1979	Fourth salt land survey			
	Continuing advice on salt land management and reclamation			

(Continued)

Table 3.1. *Continued*

Date	Events and Government Acts Facilitating Settlement	Sociological Impacts	Depressing Events and Circumstances	Date
1984	Fifth salt land survey Establishment: Land Care Units; soil conservation districts (Department of Agriculture)	Increased awareness of land degradation; greater sense of responsibility to rehabilitate land because of loss of land use and productivity Encouragement of tree planting	Drought	1980 1984
1988–89	Remnant Vegetation Protection Scheme	110 farmers subsidised to fence off 7855 ha ($290,000) to manage for 30 years	Rural population continuing to decline	1988–89
	Save the Bush Programme (Commonwealth) Greening Australia and other tree planting programmes	Consciousness of induced degradation of landscape; some rehabilitation and partial reprieve of landscape	Continuing rise in bank interest rates, above 20% in 1989 – farmers unable to meet debts	
1989	Prime Ministerial statement, "Our Country, Our Future: Statement on the Environment"			
1990		A new "conscience" and land and nature ethic developing	Collapse of wheat and wool prices (1990–1), rise in bank interest rates	1990
		Prognosis		
		Because of increased higher educational opportunities, some women of farming families *may assume* managerial role and exercise greater sensitivity towards landscape of the future	Alternative possibility for larger holdings with "company" and impersonal ownership and administration	

3.2 Exploration Period

The first explorer to penetrate the country east of the Darling Range, which impinges on the region under discussion, was Dale (Dale 1830). Dale made a circuit in October 1830 from Mt. Bakewell (previously named by him in August) to as far east as Mt. Stirling and Mt. Caroline which he named. He recorded as new the gimlet (*Eucalyptus salubris*; termed by him "cable gum") and the quandong (not named as such by Dale but described as a "new" variety of plant with "a red round fruit, about the size of a cherry, and containing a stone within." The country he recorded as alternating between "open downs or wolds" (clearly, the stretches of sandplain) and "open forest land, characterised by its growth of timber, with little brushwood below," (presumably York gum, *Eucalyptus loxophleba*, salmon gum, *Eucalyptus salmonophloia*, and gimlet, with possibly some wandoo, *Eucalyptus wandoo*). He also described the salt lake country and mentioned some freshwater sites. Dale's excursion appears to have evaded most of the landscape hazards; nevertheless, his discoveries did not immediately attract appropriation for pastoral or agricultural pursuits mainly because of the general absence of surface water.

In hindsight, it is easy to understand why the early to midcentury explorers tended to circumscribe or pass northwards of the present central/western wheatbelt. The early settlement focus, east of the Darling Range encompassing Toodyay, Northam, and York, was the nexus for the debouchment of the salt lake chains, that is, the three converging arms of the Mortlock River. Between the branches of unnegotiable salt flats (bogs for drays and wagons) were the tamma thickets or breakaways clothed with poison and prickly bushes, again difficult terrain for horses. Hence, it was not until the latter half of the century that land was taken up for pasturing sheep.

The explorations beyond the Darling Range, eastward of York, were initiated by the need for pastoral and agricultural land to support the growing Swan River colony. Furthermore, early enthusiasm to explore the hinterland was sparked by a belief of an inland sea, surmised from stories of the Aborigines, and hopes of grassy plains. In 1836, a short expedition was undertaken by G.F. Moore, accompanied by George Leake and R.M. Brown, to the north of where Meckering is now (Fraser 1906). Apart from these putative forays, the earliest expedition into what was later to become the wheatbelt was that led by the first Surveyor-General, John Septimus Roe, who was accompanied by G.F. Moore. Roe's trek in 1836 covered a broad circuit east-southeast from York to the edge of the (later) Goldfields, then north and west to Wongan Hills and the Moore River and south to the Swan River (see Fraser 1906; Braid 1975). This trek must have circumscribed the central wheatbelt area of the present study site, but perhaps the nearest Roe got was Lake Cowcowing, Mangowine and a little north of the present town of Koorda.

Neither Roe nor Moore could have been happily impressed by the landscape and vegetation, although Roe gave what must have been amongst the earliest descriptions of many of the trees which characterise the wheatbelt today, e.g., salmon gums. And, like Dale, Roe was intrigued with the gimlets. Moore somewhat morosely noted his impression of the landscape as comprising "vast gloomy forests, dense prickly scrub, and dreary uninhabited plains." Not until people were born into the landscape was there to be an unselfconscious acceptance of the indigenous vegetation as aesthetically pleasing nor for the wide expansive countryside to be comfortably embraced.

Nevertheless, Roe's designations of many physical features and sites became the lasting names. At the same time, he reported on the nature of the soil and vegetation in respect to its suitability for cultivation. The landscape at that time appeared doubly inhospitable because of the absence of surface water. Neither Dale nor Roe were given much credit for being the first to explore around what was later to become the very core of the wheatbelt. In spite of the apparent failure of Roe's search for an inland sea and grassy plains, ten years later in 1846 the three Gregory brothers (A.C., F.T., and H.C.) again explored the hinterland, in a northeasterly direction from Bolgart to Mt. Jackson (named on this trip) and around Lake Moore (Gregory and Gregory 1846). Thus, again, it was country to the north of our region that was the focus.

In 1860–1862 A. Dempster and B.D. Clarkson finally penetrated the shrublands and salt lakes east of Northam, thereby opening up for pastoral establishment this terrain which would have impinged on the northwest of the study area. Although Kodj Kodjin was reached, the absence of water and presence of poison bush made the area unsuitable for pasture. In 1863, H.M. Lefroy extended the known terrain to east of York (apparently north of the route taken by Dale in 1830) and discovered good agricultural land which, however, lacked water, apart from the possibility from wells. Finally, in 1864, B.D. Clarkson, C. Harper, and L. Lukin searched (unsuccessfully) north of Doodlakine for pastoral country while C.C. Hunt in 1863/4 made a long transect from York eastward beyond Coolgardie. The track which he blazed with an axe from York became the main route thereafter right through to the time of the gold rushes (beginning in the late 1880s). Hunt's name is commemorated in one of the water holes south of Tammin (Hunt's Well of today). An alternative route to the north leading eastward from Toodyay to Mangowine was also in use from the early 1860s.

3.3 Grazing Period

Hunt's Track became the main orientation for the eastward migration of the sheep flocks which were established first north of Meckering. In 1864, Morgan and Hardey held several large leases (of which they "overestimated the carrying capacity and also the water supplies" [Leake 1961]) north and south of where Doodlakine and Kellerberrin are now. By 1868, these leaseholds were aquired by the Leake family (e.g., Nanyanine and Mooranoppin north of Kellerberrin) and later other pastoral properties were added, all extending from north of Tammin to Doodlakine. A few years later, in 1871, Forrest also traversed the same region which, from the early 1860s to 1890 and later, was under pastoral leasehold (divided amongst four or five families; see Figs. 3.2, 3.3, and 3.4). It is pertinent to note that both Yorkrakine Rock and Durokoppin Nature Reserves must have been within pastoral boundaries from about the mid-1860s.

The pastoralists employed both white and Aboriginal shepherds to look after the flocks of sheep which were generally folded at night (in yards made of bush timbers) to protect them from dingoes (*Canis familiaris dingo*) and to prevent straying amongst the poisonous vegetation.

Some sort of agriculture, particularly cereal growing, was practised quite early by the pastoralists, long before the land was surveyed, around their main homesteads, as at Mooranoppin by the Leakes and near Doongin Peak north of Tammin on the large lease of John Packham. Certainly the Leakes had cleared about 14 acres (5.6 ha) by the mid-1880s and were growing both wheat and barley (Leake 1961).

So, from the mid-1860s onwards, the whole of this area was grazed, in parts heavily, and some areas (particularly around the water holes) were subjected to not only heavy use by sheep but also to cultivation (by the pastoralists). At the same time, the water holes and rocky outcrops were used as catchments; thus, the very areas which one would now regard as having high conservation value in that they harbour specialised biota were irreversibly modifed at the outset by the colonists. Throughout the same period, the pulling of sandalwood (*Santalum spicatum*) was practised and this supplemented the incomes of both the pastoralists and later the farmers.

Domestic housing at this time was generally of gimlet pole walls lined with mud plaster and with thatch roofs of rushes from the sandplain (Leake 1961), which suggests *Ecdeiocolea monostachya* and possibly other Restionaceae. Corrugated iron became available in the 1880s, meaning it could be used for roofing which also doubled for water catchment and thereby greatly improved the domestic water supplies. Leake mentioned some improvements and enlargement of the homestead in 1895 with a roof of sawn timber and galvanised iron.

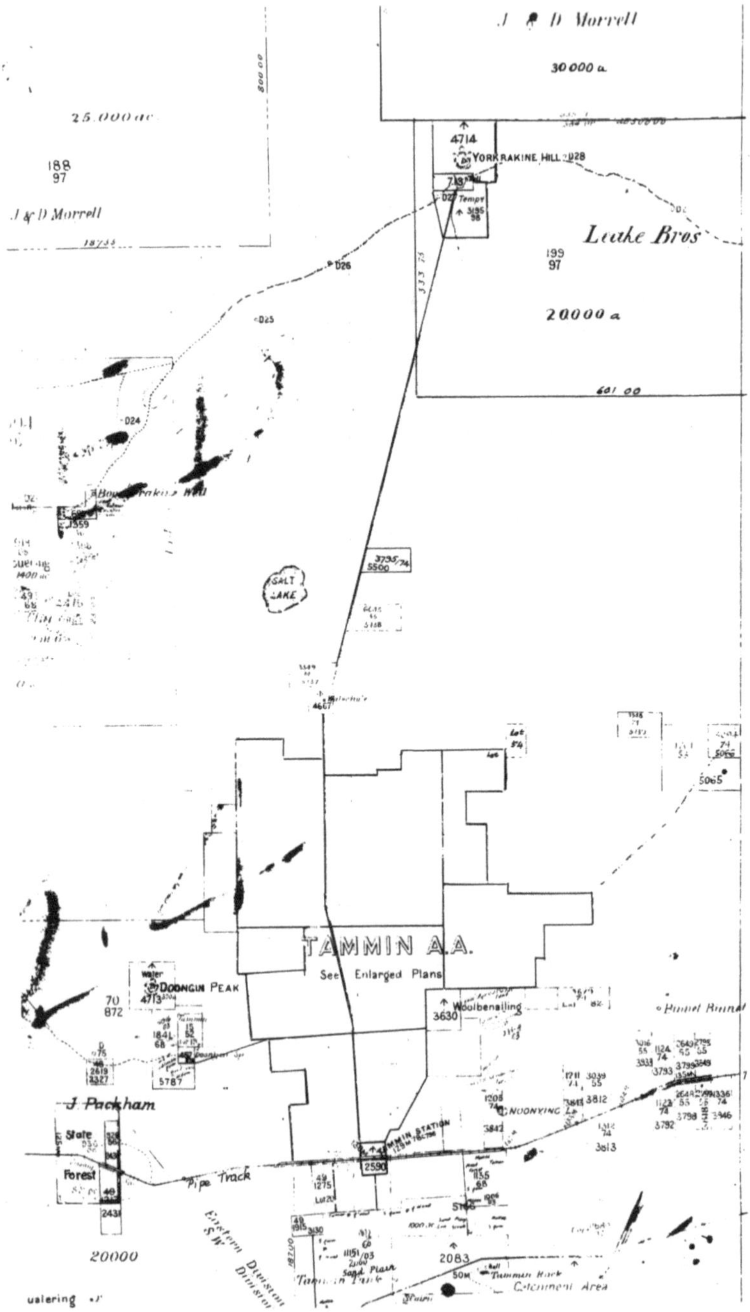

FIGURE 3.2. Excerpt from Department of Lands and Surveys No. 1 Public Plan 26 over 80 chains (cancelled 1904) of part of Tammin and Kellerberrin locations. Note: Pastoral leases still held by Morrell, Leake Bros. (surrounding Yorkrakine Hill) and Packham; and Agricultural Area (Tammin A.A.). Map in Figure 3.3 joins eastern boundary. (Figs. 3.2–3.4 reproduced by permission of the Department of Land Administration, Perth, Western Australia, under copy license 325/92).

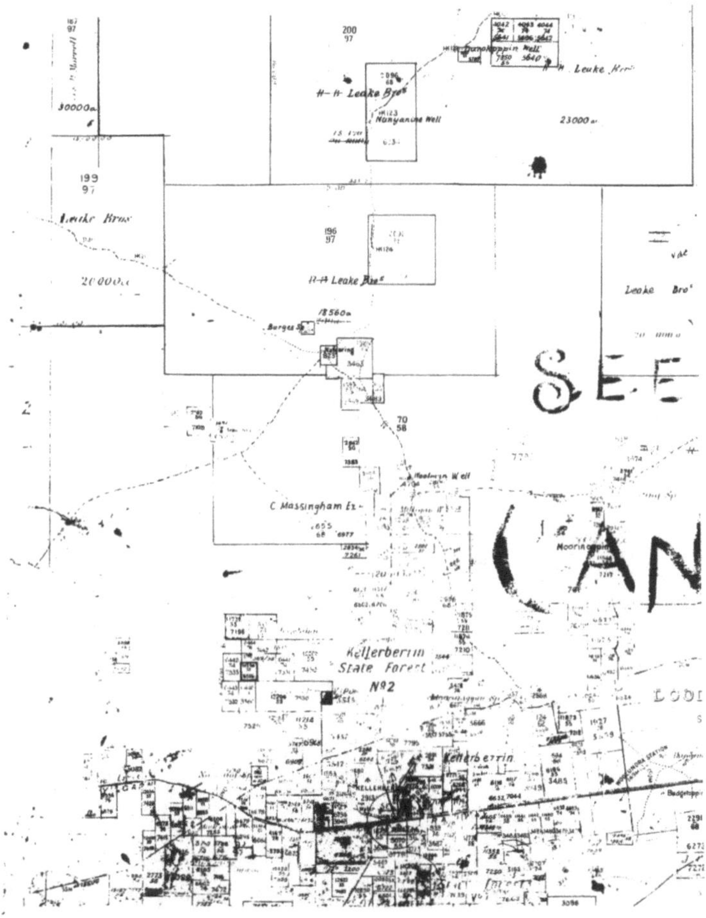

Figure 3.3. Excerpt from Department of Lands and Surveys No. 2 Public Plan 25 over 80 chains (cancelled in 1905) of part of Kellerberrin district. Note: Pastoral leases still held by Leake Bros. (including areas around Durokoppin) and Massingham Executors; old tracks converging on Nalyering Well; Kellerberrin State Forests (north and south of railway); agricultural selections around waterholes within pastoral leases, e.g., Durokoppin, Nanyanine, and Moorinoppin (see text); western section of Doodlakine Agricultural Area; and earlier proposed railway route crossing actual railway line and pipe track. This map adjoins map in Figure 3.2 (note continuity of old track from Yorkrakine Hill). Faded areas on map, particularly southwest section, are areas destroyed by mould and mildew on original. (By permission of Department of Land Administration, Perth.)

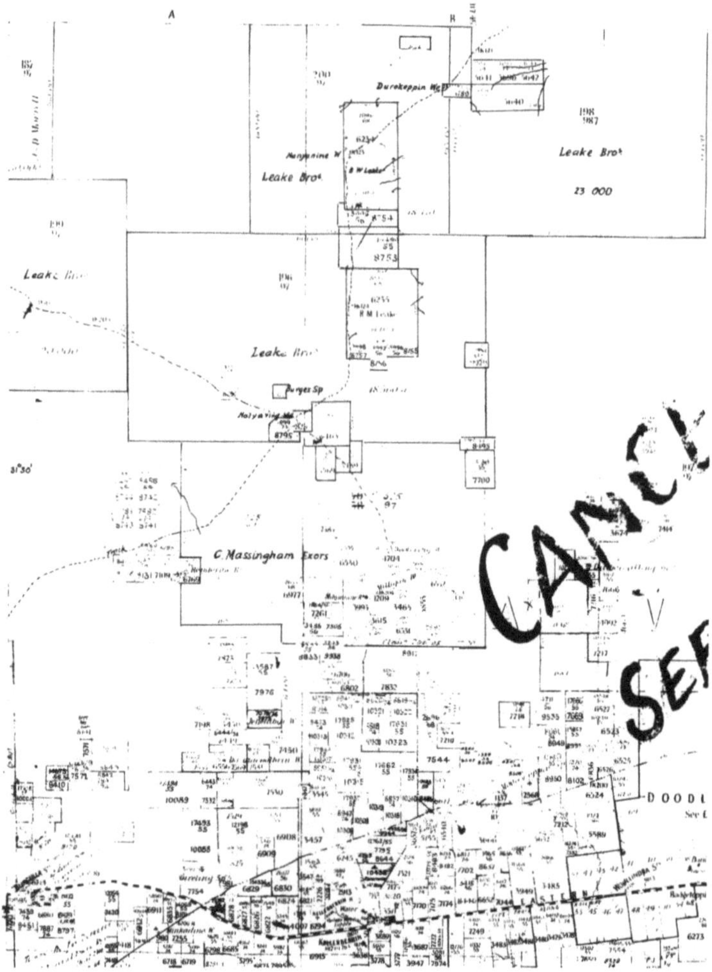

Figure 3.4. Excerpt from Department of Lands and Surveys No. 4 Public Plan 25 over 80 chains (cancelled 1907). Map covers same area as Figure 3.3 but at a later date. Note: Further crown land areas and more of pastoral leases (e.g., Nanyanine surrounds) surveyed for agricultural blocks and relinquishment of State forests similarly for agricultural blocks. (By permission of Department of Land Administration Perth.)

3.4 Settlement

3.4.1 Phase 1: Pioneering (1890–1909)

Selection and settlement followed the discovery of the gold fields (at Southern Cross in 1888, followed by Coolgardie in 1892) which saw an influx of people. This coincided also with the general world-wide economic collapse which, in Australia, was felt most severely in Melbourne, then the economic centre for the country. Many people left the abandoned gold fields in Victoria and even California and migrated to Western Australia. Even before this sudden increase in the population, the colony was not self supporting and primary products were imported.

To encourage settlers to adopt agriculture, the Homestead Act was passed in 1893, which allowed settlers to select farm blocks of 160 acres (64 ha). Conditional purchase land was also made available at this time and residency for a stipulated period was a condition of both types of land settlement. About the same time, Special Occupation Leases of unsurveyed land were also allowed. However, instead of ensuring that more settlers pursued agriculture, this effectively meant that the pastoral lessees permanently acquired the main water holes and best ground within their leases. In 1894 Agricultural Areas north of Tammin (24,000 acres [9712 ha]) and Doodlakine were surveyed for selection (see Figs. 3.2, 3.3, and 3.4).

Surveying for a railway line (later built to Southern Cross in 1894 and extended to Coolgardie and Kalgoorlie in 1896) provided transport for the settlers and their products. The Agricultural Bank was established in 1895, which provided "starting" finance for settlers with loans of up to 50% of the value of improvements, e.g., clearing, ring-barking, and cultivating, up to four hundred pounds (Glynn 1975). In 1898, the Land Act provided for conditional purchase blocks of 100 to 1000 acres (requiring six months residency). Such blocks could however, on application, be converted to crown grants if all the conditions had been met. This began a diversification from pastoral to agricultural activities, which were of more immediate benefit to the sudden population increase associated with the gold discoveries. The telegraph line went through to Southern Cross in 1892, which improved communications along the Goldfields route.

The inauguration of Responsible Government for Western Australia in 1890, which opened the way for development on borrowed money, was fundamental to all the developments of the gold industry, the accelerated population boom, and the concomitant establishment of a farming community. Prior to this, primary products had to be imported and agriculture was so depressed that a Royal Commission on Agriculture was set up, reporting in 1891.

One of the early commodities from cereal growing was hay and chaff, in great demand for the horses used along the route to the gold fields. Agricultural practice of the day meant that the ground was ploughed by single and triple furrowed ploughs but until at least the late 1890s grain was sown by hand. Leake (1961) mentioned an American cultivator with a seed box in which guano was also mixed in use in 1897, but it was not until 1901 that the first drills were generally imported from America. Reapers and binders from Canada were available in Western Australia by about 1894 but generally the crop was cut with a scythe. Although some guano was obtained from the Abrolhos in the 1880s and the first superphosphate was imported in about 1894, it was not until the turn of the century that it became widely available. Local manufacturing companies were established in 1910–11 which facilitated transport and improved crop yields.

Lack of surface water was still a major constraint to farming expansion. Construction of the Goldfields Water Scheme in 1901–3 provided water for farms within reach of the pipe line, thus allowing for greater stocking of sheep on pastoral leases and generally improving domestic life. Forest reserves (see Fig. 3.3) in proximity to the pipe line from then on presumably supplied wood for the pumping stations (in this region at Cunderdin and Merredin) – hence, the continued cutting of woodlands in addition to clearing for agriculture. However, at about this time, rabbits, *Oryctolagus cuniculatus,* (originally introduced to Victoria from Europe), were encroaching westwards (first noted at Eucla in 1894) and must have acted as a deterrent to settlers, as they ravaged crops and pasture. The severity of the rabbits prompted construction by the Department of Agriculture of the three rabbit proof fences between 1901–7 (Glynn 1975). The No. 2 fence ran north/south just east of Cunderdin. A further deterrent was the prevalence of poison bush scattered throughout the potential farming areas. Because of financial difficulties amongst the farming community, the Agricultural Bank Act was liberalised in 1906 to ensure the continuation of farming establishments.

There was no great upsurge in settlement after the initial land releases in the Agricultural Areas of Tammin and Doodlakine (see Figs. 3.2, 3.3, and 3.4) close to the railway line (mid-1890s) until the surveyor M. Terry was commissioned to devise a land classification system (reported on in Terry 1907). Terry classified a large area "North of the Yilgarn Railway, between the two rabbit proof fences" which encompassed the Yorkrakine Rock ("Yorkrakine Hill"), Durokoppin, and Nanyanine Well localities and north and eastward almost to Eluchbutting and east to Burracoppin (see Terry 1907, 1908).

Land was classified as: first class, "forests carrying salmon gum, gimlet wood, and morrel, with smaller areas of jams, scrub jams, and mallee" (Fig. 3.5); second class, "forests of white gum, salmon gum, and mallee,

FIGURE 3.5. Clearing mixed mallee woodland on a farm near Tammin, 1909, with horse drawn scrub-roller. Photograph by G.H. York (Battye Library 21738F.)

light loamy soil, and more or less infested with box poison"; and third class, "plains and stony rough feeding country, carrying a large amount of scrub." The assumed quality of the land which supported salmon gum and gimlet forest had prompted Gliddon (1906) to state "these forests mark the heaviest, richest, and most lasting soils. . . ." Ironically, these are the very lands which later were the first to deteriorate through salinisation.

Terry recommended farm lots of one to two thousand acres (809 ha) (with a mixture of land types including third class which, he stated, carried "valuable feed at certain times of the year"). In all he classified some two million acres (809,700 ha), of which about 692,000 acres (280,000 ha) were recognised as first class forest country and 126,000 acres (51,000 ha) as second class. Salmon gum and gimlet he regarded as the most suitable for crops with morrel, which grew adjacent to salt lakes, of slightly lesser value. Of the 900,000 (364,000 ha) to one million acres (404,000 ha) of third class land, he suggested that "If some quick method of getting rid of the scrub on these plains could be cleared, there is no doubt a great change would soon be effected over the face of the country." Most of these lands were to lie latent until the 1960s when a combination of heavy machinery and the availability of compound fertilisers made them both accessible and productive for cereal growing. There are still some ungrazed "remnants" on private property within the focal region under discussion as well as patches in the nature reserves of the current study.

In spite of the need to provide food for the increased population and the interest in farming due to unemployment resulting from the downturn in the gold industry and timber exports between 1908–12, with repercussions in other areas, e.g., on the wharves, the classified land was not readily available for agriculture as it was mostly within pastoral leases. Although grazing leases were withheld from selection, in effect, 30-year leases of second and third class land of up to 3000 and 5000 acres, respectively, were a form of "conditional purchase," which meant that pastoral lessees became major land owners. "Improvements" included *"clearing, grubbing, draining, ringbarking, tanks, dams, wells, and any other mark upon the land which increases or improves its agricultural or pastoral capability."* [my italics] (Fraser 1906).

Apart from the difficulty of aquisition of agricultural land within the pastoral leases, another deterrent to land occupancy in the Tammin area may have been the occupancy of some 43,000 acres (17,402 ha) by the Commonwealth Military Forces, prescribed for annual "manoeuvring" (Battye 1913). As land grants were only up to a maximum of 1000 acres (404 ha), this excluded a large number of potential farmers. This land was not released until the 1960s.

In 1909, the Agricultural Land Purchase Act provided for the resumption of pastoral leases so that the land could be surveyed for agricultural blocks. However, some of the choicest areas had already been aquired by the pastoralists, e.g., Packhams of Tammin and Leakes north of Kellerberrin. Nevertheless, in 1908 James Mitchell, as Minister for Lands, in adopting a policy of "settling landless people on the peopleless land" offered 50 unemployed men (and their families) from Fremantle 1000 acre (404 ha) blocks at Yorkrakine (Sutton 1952). In spite of contemporary criticism, the scheme was a social success and many descendants of the original settlers are still farming in the district. In 1909, Mitchell, as Minister for both Lands and Agriculture, instigated more widely the policy of surveying farm blocks before selection. Combined with the earlier liberalisation policy of loans from the Agricultural Bank, this, together with the Land Settlement Assistance for Unemployed Act of 1912, gave an added impetus to settlement over the next few years.

Apart from ringbarking, clearing (see Fig. 3.5), and cultivation of the land, domestic and work animals of the early settlement were also to leave a lasting impact on the natural landscape. Although throughout the early settlement period farming meant cereal growing, most early farms had, in addition to horses, a few stock, e.g., cows and goats for domestic use, and some had poultry (Fig. 3.6). The latter were officially recommended not only for domestic use but also as a profitable product for consumption in the Goldfields and Perth since eggs and poultry

FIGURE 3.6. "... it is a mistake not to leave some scrub shelter uncleared near the homestead for the shelter of poultry." (Robertson 1906). Feeding fowls in woodland behind house, Eboracum north east of Tammin, November 1911. Photograph by G.H. York (Battye Library 21750P.)

were, at that time, being imported from South Australia. Robertson (1906) recommended that "Fowls are best kept at liberty on land having grass, grit, clean water, and insect life ... Every settler is advised to keep poultry from the time he first takes up his block, ... Also, it is a mistake not to leave some scrub shelter uncleared near the homestead for the shelter of poultry."

It is perhaps not generally realised what impact domestic poultry have had on the bush remnants associated with farmsteads. Their effect on invertebrate life was noted by Main (1987). Farm fowls (as well as turkeys and other poultry) have undoubdtedly contributed to the destruction of the biological cohesiveness of the remnants associated with farmhouses.

With all the various uses and abuses of the landscape, ranging from open grazing to ringbarking, clearing, the overuse of natural water resources, and the encroachment on shelter fragments of vegetation by domestic animals, we can again marvel that any vestiges of the original landscape remain relatively intact.

3.4.2 Phase 2: Establishment (1910–1920s)

The establishment of new farms proceeded rapidly in the years following the release of previous pastoral leases for agricultural development (Agricultural Lands Purchase Act 1909), the continued assistance through the Agricultural Bank and, in 1912, the Land Settlement Assistance for Unemployed Act (the "experimental" Yorkrakine Settlement through Mitchell's initiative preceded the formal policy). Settlement was spurred by the decline in the gold mining and timber industries with the resultant unemployment and financial loss to the state. Politically, agricultural development in the wheatbelt was envisioned as the state's salvation with, as we now know, repercussions for the landscape.

Shortly after World War I, most of the available land had been appropriated for farms or gazetted for public use, such as water catchment reserves, school sites, recreational areas, or for amenities such as halls and reserves for road materials. Some blocks after the War were taken up under the Soldiers' Settlement Scheme.

However, the forward thrust for agricultural settlement was inhibited by a combination of the War, the enlistment of many men, and devastating droughts in 1911 and 1914. Following the 1914 drought, because of lack of seed wheat and stock feed, hay and grain were imported from South Africa and eastern Australia, with the unfortunate introduction of weeds such as cape weed (*Arctotheca calendula*) and double gee (*Emex australis*) which affected the growth of later crops. Some reports indicate that double gee had previously been introduced during the early years of settlement. Such weeds are still a pest in bushland as well as cultivated areas. Mechanised vehicular transport (e.g. cars, which increased the mobility of people) became more prevalent following the war but horses remained as the principal means of traction and heavy transport (e.g., bagged wheat to the railway, hay on the farm).

Although the ten year period 1910–20 was the main period of expansion and establishment of social amenities (government rural schools and the Correspondence Classes, halls and recreational facilities established by local Progress Associations), the period was also fraught with inhibiting events and phenomena including war, drought, and, finally, increasing numbers of rabbits. Thus, between 1914 and 1919, the area suffered considerably; most farmers were heavily in debt and, the Industries Assistance Act was passed in 1914 as a means of assisting

FIGURE 3.7. "Empty fireplaces and broken windows, termite-ridden floor boards and door jambs, walls with terrible cracks.... There had been voices and singing and quarrelling, joy swelling the rooms, and the comfort of known chairs to contain physical weariness. The house had been full of movement or a hush of temporary sleep." But now "only emptiness and stillness" and "grey, still dust-memory ground to powder... How short a time? That people should live and lovingly build, then leave in the material housing of their emotions only a dreadful melancholy!" (Main 1972). This mud brick house south of Yorkrakine is one of the few of those abandoned in the 1930s Depression years which is still standing. Photograph by B.Y. Main 1990.

farmers in difficulties. Financial assistance was provided to cover miscellaneous costs ranging from mortgage payments to rates and taxes, stores and machinery. In this manner some farms were salvaged but even so a general abandonment of farms threatened. Because of the depressed state of farming a Royal Commission on Agriculture was set up which reported in 1917. This report was critical of government departments, particularly the Department of Agriculture and the pattern of development and land policies. As well as the ensueing government assistance, it was the initiatives of the farmers themselves through the reinforcing of their cooperative associations, such as the Primary Producers' Association, later the Farmers Union, and finally the Westralian Farmers Limited (in 1914), which helped overcome the difficulties posed by drought and debt. By 1918, this organisation was responsible for marketing the state's entire wheat crop (Glynn 1975).

The early twenties marked a burgeoning period in the farming community, partly augmented by a more active advisory role by the Agricultural Department (whose low profile had been criticised by the Royal Commission in 1917) and partly by increased world prosperity and high prices for wheat. Government assistance in various forms was given not only directly to the settlers but also to contract clearers recruited from amongst the unemployed. Establishment of a rural oriented radio programme (6WF) in 1924 must have lifted morale in that its function was to inform the farming community and help remove a sense of isolation. Throughout the twenties domestic life became more comfortable, and many substantial pug brick and stone homes were built to replace some of the makeshift iron or hessian-walled houses. Some of these, now modified, are still in use, while others were abandoned in the 1930s (Fig. 3.7).

However, by 1929, the world economic Depression was already beginning to be felt in this area as elsewhere and, at the same time, rabbits had reached destructive proportions (see Fig. 3.8; Stodart and Parer 1988). In addition, the awareness of soil salinity as an outcome of clearing and its possible deleterious effects on farmlands were formally recorded for the first time (Wood 1924). This was in spite of the fact that this author and the rural citizens of the region just east of the Darling Range around Toodyay/Northam/Goomalling had observed earlier the encroachment of salt on cleared land there.

By the end of the 1920s, the pattern of farm settlement had demarcated the template of agricultural land and bush remnants (already encroached on by rabbits) within a lattice-work of surveyed roads. Today, some sixty years later, attempts are being made to unify the changed landscape into a "new" biological system.

3.4.3 Phase 3: Economic Depression and Recovery (1930's)

. . . he loved the soil,
Running it through his fingers he sensed its dying,
Its blowing away on the winds of time and cut timber,
He saw the salt of its death rising,
He said, "I have a plan," and rode with it into the cities,
A plan for trees, acres of trees blowing by creekbeds,
Forests marching in long green lines to save a country,
Picking up their roots and digging them into the earth,
Holding it fast against the salt and the wind tides.
But the laughter rose in gales from the men in the cities.

(Hewett, 1968)

Following the euphoria of the expansive period, which was facilitated

FIGURE 3.8. Rabbits killed at poisoned watering point, late 1920s. Photograph by G.H. York (York Family Collection.)

by various government acts and the generosity (verging on imprudence) of the Agricultural Bank, which encouraged settlement of sometimes unsuitable land, the subsequent inability to repay loans in the face of falling world markets and economic decline within Australia meant that many farmers were forced to abandon their holdings (Fig. 3.7). The social distress and economic seriousness of the movement of people from agricultural areas to the city (already burdened with unemployed) in 1933 prompted yet another Royal Commission on Agriculture and another Land Act, encouraging the conditional purchase of land. One result in the relevant districts was that surviving farmers enlarged their properties by absorbing abandoned farms. This, in turn, increased the viability of those farms which, at the outset, were a manageable size for a family but were effectively too small to be economically viable. Social consequences of the declining population (from about the year 1935) meant the closure of many small rural schools and reduced educational opportunities for families unable to send their children to boarding schools.

Gradual recovery from the world Depression by the midthirties coincided with the replacement of horses as traction animals by tractors and trucks. At about the same time, bulk handling of wheat was instigated. All this meant that the harvest could be cut and delivered to the depots more rapidly and with less labour. The first diversification in farming began with the introduction of sheep on most farms about 1933. However, these advantages were offset by the uncontrollable

FIGURE 3.9. Rabbits trapped by children (at "Fairfields"), mid-1930s. Rabbits provided meat for families during the Depression and the sale of skins supplemented incomes. Photograph by G.H. York (York Family Collection.)

increase in rabbits and low rainfall seasons between 1935 and 1938 (see Figs. 3.8 and 3.9). Rabbits destroyed young crops and simultaneously disturbed bushland, with lasting effects. Although burrows were fumigated and watering points poisoned with cyanide (Fig. 3.8), destruction of the native landscape was further exacerbated by burning to remove the shelter for rabbits. Fallowing, combined with dry seasons, contributed to severe dust storms and loss of top soil.

Morale was generally low and social life restrained throughout the thirties. The decline in the number of rural holdings and population continued. Abandonment of properties encouraged local grasshopper plagues, which added to the intransigence of the rabbit problem. The population decline and the accompanying disuse of small rural community centres (halls, sporting sites), which was coupled with and due partly to an increase in mechanised vehicular transport, permanently altered the previous social fabric of the region.

Throughout this decade, an insidious change in the landscape was occurring, almost unnoticed except by a few perceptive farmers; signs of salinisation and soil erosion and moribund bushland appeared. The forewarnings of the few observant, apprehensive farmers and their predictions passed unheeded and they continued to be unheeded for almost half a century, in spite of authoritative reports by Teakle (1929) and others.

> *". . . But the Agricultural Department laugh at him. They think he is a dreamer. 'In fifty years,' he says, . . . but they don't care. In fifty years they won't be there."*
> (Hewett 1990).

3.4.4 Phase 4: World War II and Postwar Years (1939–1950)

World War II began just as the farming community was recovering from the Depression. The immediate and continuing effect was a shortage of essential items such as petrol, tyres, and machinery (as factories were diverted to the war effort), which was doubly inhibiting at a time when farms had only just become mechanised. Gas producers were improvised to overcome the shortage of petrol. An unfortunate consequence of this was the further destruction of remaining woodlands as salmon gums (and wandoo) were the favoured trees for charcoal production for the gas producers. Next, the absence of "man power" was critical due to enlistment. Recruitment of Italian prisoners of war helped to alleviate the labour shortage. It is noteworthy that the rapport between some families and the Italians was such that, after the War and following their repatriation, some of the Italians migrated to the Western Australian wheatbelt and eventually took up land for themselves (including some blocks in the current study area). However, in spite of the difficulties imposed during the War, towards the end of the War and into the fifties there was a boom in production due to world demands, particularly for wool. This meant that, for the first time, many families were able to enjoy a comfortable standard of living.

During this period, the Education Department school bus services, which transported children from farms to the towns, were instigated; most remaining small "rural" schools were closed and resources thus concentrated in the schools in the towns with greater educational opportunities being provided for farm children.

The War Service Land Settlement Scheme (1945) was responsible for most of the land not already alienated in the district being developed. Clearing of undeveloped timberland on established farms was also facilitated by the availability of war surplus heavy machinery, e.g., tanks and later 100 hp plus bulldozers. Two bulldozers dragging a ship's anchor chain between them could flatten 400 acres (101 ha) a day (D. Couper, personal communication) compared to about an acre a day which a man with an axe could fell. Subsequent burning was followed by cleaning up operations, again with heavy machinery. Shortly afterwards, heavy tractors were also available. At the same time, the Goldfields Water Scheme was extended through feeder pipe lines to northern parts of the region.

While some detrimental effects of the increased farming activity were apparent in incipient land degradation, it was hoped that this would be arrested through the Soil Conservation Service offered by the Agricultural Department. Although during this period there were some absentee landowners (turned share farmers – virtually forced away through labour shortages and an inability to cope with restricted equipment), generally farmers were better off by 1950 and were able to look forward with an expectation of prosperity, heightened to some extent by the markets (again particularly for wool) created by the Korean War.

3.4.5 Phase 5: Consolidation and Expansion (1950–1960s)

"... our duty is to develop our virgin lands, to complete the development of the partially developed, and to increase the productivity of the developed lands until the maximum of which they are capable is reached." (Sutton 1952).

Such was the tone of exhortations to farmers in the early fifties, which became transmuted into a sense of moral obligation to produce food for the world. Myxomatosis was gaining some control of rabbits while formation of the Agricultural Protection Board further assisted farmers in the control of weeds and other pests, the productivity of cereals generally increased, and world prices for wool were high. Various innovative mechanical modifications (e.g., trailer bins for header harvesters, bulk bins for grain transportation from harvester to depot) introduced in the early fifties reduced labour and speeded grain handling. Effectively, farms could be managed with fewer men. Domestic life was comfortable, the State Electricity Grid extended to the wheatbelt, material amenities were within the economic means of most families, television was extended to rural areas, educational facilities improved, including establishment of junior high schools in the local towns, and thus a general euphoria pervaded rural society.

It was during the early fifties that Sutton (1952) summarised the progress of farming in the wheatbelt (and other agricultural areas) with praise for the fortitude of the pioneers and the tenacity and foresight of the politicians (e.g., James Mitchell) in opening up and consolidating agricultural areas, particularly the central wheatbelt. He spoke of the early settlers who conquered the "forest wilderness" and turned "virgin lands" into "well-ordered farms and gardens." Few, at that time, had a rapport with the natural landscape or had any vision of retaining fragments of it as part of a total landscape pattern.

Sutton (1952) also rather patronisingly mentioned the "splendid part played by the women. The women pioneers responded ably; . . . *they did it cheerfully*." [my italics] (see Figs. 3.10 and 3.11). One wonders from where he got his evidence. Certainly not amongst the bitter memories (of isolation, hardship and deprivation) of the octogenarians of the fifties.

In hindsight, cognisant of the legacy of soil erosion and salinisation with associated remedial costs and the general absence of intact bushland and consequent loss of wildlife, one may now justifiably regard some aspects of the pioneering and consolidation phase as overzealous, unwise, and lacking in vision. Throughout this period, the landscape continued to deteriorate and the salinisation of cultivated land increased, promoting the first of five Salt Land Surveys by the Department of Agriculture. The seriousness of salinisation was still unregarded by many farmers who continued to clear remnant vegetation to compensate for loss of arable land due to salinisation. When compound fertilisers and new breeds of wheat, suitable for marginal areas, became available in the early sixties, the remaining uncleared light lands (patches of sandplain heath, wodjil, and tamma scrub) were rapidly cleared and cultivated, e.g., "Much of the new land has been light sandplain previously regarded as useless country best left to the kangaroos." (Griffith 1964). Thus, virtually the only remnants of ungrazed bushland left by this time were the reserves set aside earlier for public use (schools, halls, water catchments, recreation sites, road materials, and so on) and a few on private property (Main 1967). The social need for these plots had long since gone and shire leases lapsed but towards the end of the sixties a new value for them was perceived – that of nature conservation. Although some sites were acquired earlier by adjacent landholders,

FIGURE 3.10. Stooking hay, Eboracum, northeast of Tammin, 1912. Photograph by G.H. York (Battye Library 21794P.)

FIGURE 3.11. House of bush timber, hessian, and corrugated iron, Eboracum, northeast of Tammin, 1911. Photograph by G.H. York (Battye Library 21753P.)

many from this time onward were gazetted as nature conservation reserves. The new affluence in the established farming community was sustained until 1969 when drought and the introduction of wheat quotas (due to a glut on world markets) combined to counter it.

Table 3.2. Summary of settlement, land alienation, number of farm holdings, and changes in shire populations. Comparable data for each category considered is not available for all shires. Data obtained from various sources, namely: Fraser (1906); Western Australian Year-Books; Bureau of Statistics; Shire records; D. Couper (pers. comm.)

		Tammin	Kellerberrin	Trayning
Date open for selection		1894	probably 1908/9 or earlier*	probably 1908
Agricultural Area 1903. (See Table 3.1 and text)		24,000 acres (9712 ha) Holdings of 100–640 acres (40–259 ha)	–	–
Area of Shire 1990		116,500 ha (106,300 cleared)	178,000 ha (162,800 cleared)	136,800 ha (120,300 cleared)
No. of farm holdings	–1912	–	–	241
	–1960	–	–	192
	–1990	58	101	87
Population of shire including towns	–1925	–	–	2300
	–1967	850	–	–
	–1971	–	–	820
	–1980/82	610	1650	614
	–1990	517	1300	530
Population of towns only	–1966	–	1370	–
	–1976	288	1198	–
	–1990	–	1000	254

* Some lots selected mid-1890s (Leake 1961).

3.4.6 Phase 6: Population Decline – 1970s to Present

The continuing low wool prices in the early seventies, the effects of the 1969 drought, and increasing salinity and soil erosion (which reduced further the arable land) all contributed to a general decline in the viability of some farms and in turn, the collapse of business in the towns. The population of both rural areas and town sites in the three shires Tammin, Kellerberrin, and Trayning all experienced a decline in the population from this time onward; however, Tammin suffered particularly (see Table 3.2). Many farmers sold up and such farms were absorbed into surviving properties. Hence, over this period, in retaining viability, more land became managed by fewer people. This was facilitated by larger machinery. But the productivity of the overall landscape decreased due to land degradation. The third Salt Land Survey was undertaken in 1974 and since then two more surveys, in 1979 and 1984, have been conducted (see Stoneman 1978; Anon. 1988 for summary accounts on salt-affected cereal growing areas). All the surveys have indicated increasing salinity in spite of expert advice proffered by the Department of Agriculture, advice proffered perhaps half a century too late as intimated in several literary works, such as the theatrical work of Dorothy Hewett wherein a character voiced his apprehensions to the Department in the following words:

> *"Dear Sir, I wish to draw your attention to my letter of the 21st April 1929, in which I outlined, once again, my plan for the re-afforestation . . . Dear Sir, on the 30th January 1930, I wrote you again about the need to act quickly if the salt problem in the Avon Valley is to be solved before it is too late . . . Dear Sir, The Avon valley is being slowly inundated with salt that flows down the rivers into the creeks. The native trees along the banks are dying, erosion cuts into the paddocks . . . Dear Sir, I have had no reply to my (many) letters to you regarding the . . . salt problem. May I remind you once again of my plan for providing native trees . . . Dear Sir . . . Dear Sir . . . Dear Sir. . . ."*
>
> (Hewett, D., manuscript of unpublished play "Fields of Heaven," first performed in 1982 in Perth).

Nevertheless, the "successful" farms continued to enjoy an affluence which allowed greater educational opportunities for many families (including girls and women). As a result, the sociological pattern was for women generally to leave the district and find employment elsewhere. This bias was partly offset by farmers frequently marrying "temporary" or transient empoyees in the district – teachers, nurses, and other professionals. The demographic changes begun in the seventies are continuing today, to the point where one can foresee the possibility of fewer and fewer major landholders and the formation of both family and absentee (and impersonal, i.e., with no intrinsic attachment to the landscape) companies managing the land. Alternatively, there is the possibility of women, with their increased educational opportunities, playing a dominant role in farm management. With perhaps a deeper sensitivity towards the landscape, such a switch in managerial bias could demonstrably affect the future shape and pattern of the countryside. In Australia as a whole, women are increasingly contributing in the agricultural work force and, by 1987, about 9000 women were working as farmers or farm managers and about 2500 as farm workers or supervisors (Anon. 1987). Yet, in the Western Australian wheatbelt, although women are employed now in advisory and other areas related to farming, to date the entrenched traditional attitudes towards the inheritance of land have generally precluded women members of families from assuming a positive role in land use on wheatbelt farms and thereby also excluded their political representation. In the rare exceptions where women have exercised a strong influence, it is notable that such farms present examples of a blend of conservation of natural landscape and economic farm use zones within the boundaries of the properties.

In spite of the apparent affluence of the persistent "core" of farmers throughout the 1980s, by 1990 the increasing production costs and the fall in world prices of wheat and particularly wool, continues to cause financial difficulties and debt amongst most farmers (see Fig. 3.12).

Figure 3.12. Social comment on changing farming conditions. (Courtesy of D. Alston and *The West Australian*, 10 October 1990.)

3.5 The New Landscape Ethic and Prognosis

Perhaps partly because of the increased level of education and the stronger profile of advisory agencies (including the Department of Agriculture, CSIRO expertise, and personal farm advisers), a general awareness of the *total* landscape has gradually manifested during this period. In the mideighties Land Care Units and Soil Conservation Districts, established with the help of the Department of Agriculture, heralded a new cooperative attitude amongst farmers to combat the deleterious effects of agricultural use of the land.

Independent, state, and Commonwealth government sponsored programmes, such as the Remnant Vegetation Protection Scheme and Save the Bush, various tree planting schemes (e.g., Greening Australia), and landscaping by the Main Roads Department may all have arisen from utilitarian motives, but there are also indications of a developing aesthetic appreciation of and a cultural identity with the concept of a total, cohesive landscape in which the indigenous natural bushland is a rightful and pleasing component. No longer do enlightened farmers and their families regard bushland as wasted land or simply as a haven for vermin, nor do local shires condone wanton burning or ploughing of road verges. Interest in wildflower societies, participation in bird observation recording programmes, and the application of plant and wildlife motifs in local crafts all demonstrate the development of values additional to the solely economic and utilitarian. All this indicates a change of attitude and empathy from that expressed in the mid-1960s following the height of the land clearing boom when wheat was reported to be grown "on country that a few years ago could sustain nothing but low, ugly and unpalatable scrub." (Quinn 1968; see also Hobbs and Hopkins 1990).

Meanwhile, artists, poets, and other writers continue to project a vision of the countryside as something more than a medium for material gain. Some of this aestheticism undoubtedly "rubs off" onto others or perhaps articulates sentiments already present. In Australia there is no strong statement or development of a "landscape ethic" or a coherent, directed philosophy, but much of our literature and art embodies and expresses values broader than the utilitarian. Nevertheless, the strengths of the conservation movement have been because of what Wright (1973) called "feeling," i.e., because people "felt" emotionally about a place, forest, heath, or coast; other arguments or justifications, including ecological and ethical, were then looked for. Judith Wright's writings about the conservation of nature in Australia actually predated the burgeoning development of ecological ethics, particularly in North America, during the late seventies and eighties (see Nash 1990 for a review of the literature of environmental ethics). Philosophers (e.g., Taylor 1986) finally developed a sound philosophical theory justifying regard for and preservation of nature. This theory, in extending environmental ethics to accept humans as moral agents, expounded the opposing attitudes to nature as either "anthropocentric" (utilitarian, i.e., nature as useful to humans) or "biocentric" (nature with its own intrinsic values and species and the biota existing in their own right). In adopting the latter attitude, the ultimate objective of reintegration would be to retain a viable landscape for *nature* (with agriculture as a "new" added component), i.e., preservation of as broad a diversity of indigenous flora and fauna as is now present, with the potential for a sustainable agriculture grafted to this fundamentally "natural" landscape. In looking forward to an integrated landscape, the philosophical approach espoused by Taylor, encapsulating a "biocentric" rather than "anthropocentric" outlook, may indeed be the only valid directive if we are to retain a bushland/agricultural mosaic rather than create a patchwork of dry grassland/saltbush steppe.

Acknowledgments. I thank W.M. McArthur and D. Couper for helpful discussions, information, and comments on early drafts of the manuscript. I am grateful to the following: Professor R.T. Appleyard for access to the rich resources of his files and library; the Western Australian Historical Society for use of the Society's library; the Battye Library for permission to reproduce photographs in their collection and to R.J. York, copyright holder of such photographs; and the Department of Land Administration for assisting in the search for old maps and use of these for reproduction (as acknowledged in text). I thank R. Wood of Media Services of the University of Western Australia for printing excerpts of relevant maps and S. Hopwood for some photographic printing. I pay tribute to my father, the late G.H. (Harry) York, for his skill with a cumbersome camera, the glass plates of which he must have handled under difficult conditions. I am most grateful to Dorothy Hewett and her publishers for permission to quote from her literary works and to Dean Alston and The West Australian for permission to reproduce the cartoon. I thank the Zoology Department, University of Western Australia, for use of facilities.

References

Anon. (1987) *Women's Work on The Land.* Women's Advisory Council, Perth.

Anon. (1988) *Select Committee on Salinity, First Report.* Parliament of Western Australia, Legislative Council.

Battye, J.S. (1913) (ed.) *The Cyclopedia of Western Australia, Vol. 2.* Cyclopedia Co., Perth.

Berndt, R.M., Douglas, W., Kaldor, S,. and Hallam, S.J. (1986) The first Australians. In N. Jarvis (ed.) *Western Australia: An Atlas of Human Endeavour.* Department of Lands and Surveys, and Education Department of Western Australia, Perth pp. 32–37.

Braid, E. (1975) John Septimus Roe: First explorer of the wheatlands; The search for an inland sea. *Early Days, The Royal West Australian Historical Society* 7:81–97.

Dale, R. (1830) Journal of an expedition to the eastward of the Darling Range in August 1830 *and* Journal of another expedition to the eastward of the Darling Range commenced on the 25th October and concluded on the 7th of November 1830. In J. Cross (ed.) *Journals of Several Expeditions Made in Western Australia During the Years 1829, 1830, 1831, and 1832.* J. Cross, London. pp. 51–72. Facsimile edition, University of Western Australia Press (1980).

Fraser, M.A.C. (1906) *Western Australian Year-Book for 1902–4.* Government Printer, Perth.

Gliddon, E.H. (1906) The Avon District, In M.A.C. Fraser (ed.) *Western Australian Year-Book for 1902–4. Part VII, 1, Land Settlement.* Government Printer, Perth. pp. 568–570.

Glynn, S. (1975) *Government Policy and Agricultural Development. A Study of the Role of Government in the Development of the Western Australian Wheatbelt.* University of Western Australia Press, Nedlands.

Gregory, A.C. and Gregory, F.T. (1846) Diary. In *Journals of Australian Explorers.* Beal: Brisbane (published 1884) Facsimile Editions No. 14 Adelaide Libraries Board South Australia (published 1969).

Griffith, C. (1964) Scientists started new farming era. *The West Australian,* August 21.

Hallam, S.J. (1985) The history of Aboriginal firing. In J. Ford (ed.). *Fire Ecology and Management in Western Australian Ecosystems.* Western Australian Institute of Technology, Perth. pp. 7–20.

Hewett, D. (1968) Legend of the Green Country. In *Windmill Country.* Overland, Melbourne.

Hewett, D. (1990) *Wildcard. An Autobiography, 1923–1958.* McPhee Gribble, South Yarra.

Hobbs, R.J. and Hopkins, J.M. (1990) From frontier to fragments. *Proceedings of Ecological Society of Australia* 16:93–114.

Leake, B.W. (1961) *Reminiscences.* B.W. Leake, Perth.

Main, B.Y. (1967) *Between Wodjil and Tor.* Jacaranda, Brisbane, and Landfall, Perth.

Main, B.Y. (1972) Marginal Country. *Westerly* 2:21–36.

Main, B.Y. (1987) Persistence of invertebrates in small areas: Case studies of trapdoor spiders in Western Australia. In D.A. Saunders, A.W. Arnold, A.A. Burbidge, and A.J.M. Hopkins (eds.) *Nature Conservation: The Role of Remnants of Native Vegetation.* Surrey Beatty and Sons, Chipping Norton, NSW. pp. 29–39.

Nash, R. (1990) *The Rights of Nature. A History of Environmental Ethics.* Primavera Press, Leichhardt, NSW.

Quinn, C.W. (1968) 500,000 bushels of grain. *The West Australian,* January 29, p. 5.

Robertson, F.H. (1906) The Poultry Yard. In M.A.C. Fraser (ed.) *Western Australian Year-Book for 1902–4.* Government Printer, Perth. pp. 606–607.

Stodart, E. and Parer, I. (1988) *Colonisation of Australia by the rabbit, Oryctolagus cuniculatus* (L.). CSIRO Division of Wildlife and Ecology, Canberra.

Stoneman, T.C. (1978) Soil salinity in Western Australia. *Journal of Agriculture – Western Australia* 19:99–100.

Sutton, G.L. (1952) *Comes the Harvest. Half a Century of Agricutural Progress in Western Australia 1900–1949.* The West Australian Historical Society, Perth.

Taylor, P.W. (1986) *Respect for Nature. A Theory of Environmental Ethics.* Princeton University Press, New Jersey.

Teakle, L.J.H. (1929) Soil alkali. III. The problem in Western Australia. *Journal of Agriculture – Western Australia* 6:88–102.

Terry, M. (1907) Appendix D, (and accompanying map, Appendix A *in part*). In *Report by the Surveyor General for the Year Ending 30th June 1907, Department of Lands and Surveys.* Government Printer, Perth. p. 13.

Terry, M. (1908) Appendix A, (and accompanying map, Appendix B *in part*). In *Report by the Surveyor General for the Year Ending 30th June 1908, Department of Lands and Surveys.* Government Printer, Perth. pp. 13–15.

Wood, W.E. (1924) Increase of salt in soil and streams following the destruction of the native vegetation. *Journal of the Royal Society of Western Australia* 10:35–47.

Wright, J.M. (1973) The individual in a new environmental age. *Australian Journal of Social Issues,* 8:1–12. (Reprinted 1975. In *Because I Was Invited.* Oxford University Press, Melbourne. pp. 248–256).

Trayning: Case History of a Shire

D. COUPER

The history of exploration, settlement, and development of the Trayning Shire has followed much the same pattern and sequence as that of the Tammin and Kellerberrin Shires. Some of the pastoral leases occupied in the 19th century were located across present shire boundaries. However, by being further north of the main Goldfields route, settlement of the Trayning area was slightly later and initially slower than that of land adjacent to the Goldfields railway. The first land title was at Kodj Kodjin in 1901 and thereafter the chronology of circumstances and events paralleled that of the other shires:

1901–1908	With termination of grazing leases and surveyance of land for farming, more titles continued to be issued.
1911	The northern railway "loop" extended to Kununoppin. Ninghan Road Board gazetted.
1912	Rate payers now numbered 250. Local schools opened. Local sporting associations initiated, e.g., cricket and football. Towns of Trayning and Kununoppin gazetted and Kununoppin railway dam built.
1913	Progress Association built Trayning Hall.
1914	The first cars arrived in the district. Drought with consequent failure of crops and necessity of carting water. Outbreak of war.
1914–1918	Impact of the 1914 drought and the Great War. Absence of many men at war. General effects on local economy. Until about 1920, because of lack of cash, transactions of goods in the region were often by bartering.
1919–1923	From about 1920 the economy improved, people had some cash, and a community consciousness

	was manifested in construction of halls and further establishment of sporting activities. Many new comfortable homes built. A general impetus to farming and domestic life. The first (small) tractors introduced in 1922. The (third) trotting club in WA registered at Trayning in 1923. A local farmer awarded second prize for the best bushel of wheat in a world-wide competition (Carrabin variety).
1923	Kununoppin-Trayning-Yelbeni Road Board formed from part of earlier Ninghan R.B.
1923–1925	Peak population – up to 2300.
1928	Local water scheme opened, based on interconnected dam catchments on granite outcrops. (Previously farmers were dependant on individual water catchments.) New hall built by Road Board at Trayning. Rabbits rife, causing great distress.
1929	General deleterious impact of Depression.
1930s	Introduction of sheep to most farms, made possible because of availability of water through comprehensive local water scheme.
1931–1935	Exodus of many families from "failed" farms.
1935	Farm pests: although dingoes under control, rabbits at their peak.
1936–1938	Dry years. Tractors with rubber tyres introduced.
1937–1938	Outbreaks of "locusts" (little plague grasshoppers) causing much damage to crops.
1939–1945	Although economy on upturn and a season of good rains, outbreak of World War II further depressed farming life and many men left for the front. General shortages of labour and materials ranging from some food stuffs and clothing to machinery and fuel with rationing of some items. Improvisation including gas producers on tractors and vehicles. Severe drought in 1940. Some farms leased.
1943	The wartime air strip built at Kununoppin (but not used as such) modified for civilian use and used by Flying Doctor Service from 1970s.
1946–1950	Release of unalienated land to ex-servicemen. Ex-army vehicles acquired for farmers and tanks used for clearing bush.

1946–1950	Main roads bituminised; secondary roads in early 1950s.
1948	Insecticides and weedicides introduced (DDT and 24D) with improvement of crops.
1951–1952	Korean War resulted in "wool" boom. Clearing of bush continued. *A new prosperity for farmers.*
1953–1960	"Hydraulics" introduced to farming. Bulk handling of wheat introduced. Very little uncleared land now remained.
1957–1958	Goldfields Water Scheme extended northward from Kellerberrin.
1950s (late)–1960's	State Electricity Grid extended to region.
1961–1970	Availability of new pasture plants and compound fertilisers with increased farm productivity and heavier stocking of sheep.
1961	Shire Council of Kununoppin Trayning Yelbeni established as formation of shires superceded old Road Boards.
1965	Shire of Trayning established.
1969	Drought and consequent reduction in farm "labour" with departure of many employee families (population down to 1000). Severe dust storms.
1971–1980	"Labour" replaced partly with larger machines. Population decline (on farms), accompanied by expansion of surviving farms but decline in local businesses. In spite of imminent population decline (on farms), there was conversely an increase in the town population, new council offices were established, and the library opened in 1972. Sheep numbers declined due to a succession of poor seasons.
1981–1990	Changes in farm practices and type of crops (e.g., cultivation of lupins) and reliance on weed killers. Increase in sheep numbers with profitable returns (until 1990). Towards the end of the decade, farm costs outstrip Australian CPI increases. Nonviability of many farms (number of farm holdings down to 87 in 1990) and general rise in debts with resultant collapse of morale and health. Continuing decline in population (down to 640 in 1983 and 530 (of which almost half was "town" population in 1990). In spite of technological aids (computers), professional agricultural advice,

implementation of land conservation practices, and tree planting programmes to combat soil and environmental problems, the current prognosis for farming in the short term is gloomy with a "crash" in wheat and wool prices.

Section II

Landscape Disintegration

Landscape disintegration. Rising water tables following extensive clearance of native vegetation for agriculture have brought stored salt to the surface, increased areas of secondary salinisation and lead to the decline of native vegetation in wetlands and valley bottoms.

(Photo: R. Hobbs)

Chapter 4

Changes in Biota

R.J. Hobbs, D.A. Saunders, L.A. Lobry de Bruyn, and A.R. Main

4.1 Introduction

Over much of the world, conservation of regional biota depends almost entirely on the retention and management of the remnants of native vegetation left following clearing or extensive modification for agriculture, forestry, or other productive use. These remnants vary in size, shape, degree of isolation, and ownership. The fragmentation process results in many changes in the physical environment and in the biota, as reviewed by Saunders et al. (1991). In some parts of the world, as in Europe, fragmentation has been occurring for long periods (Wilcove et al. 1986; Attenborough 1987), whereas in areas more recently settled by Europeans, such as North America and Australia, the process has occurred mainly in the last 200 years (Hobbs and Hopkins 1990). Land clearance in many parts of the world, including Australia and the tropics, has been very rapid over the past 50 years (Saunders et al. 1985; Malingreau and Tucker 1988; Richards and Tucker 1988), and pressures from human population expansion are likely to see the fragmentation process continue (Western 1989).

It is, therefore, clear that nature conservation will continue to be dependent, or will become more dependent, on small remnant areas. Even where large areas have been set aside as national parks and nature reserves, it is becoming obvious that these are not necessarily large enough to function as complete ecological entities independent of their surroundings (Dasmann 1983; Janzen 1983, 1986; Newmark 1985; Wilcove and May 1986; Harris and Eisenberg 1989). Accordingly, it is essential to understand the effects of ecosystem fragmentation and to develop effective management strategies which increase the likelihood of successful long-term conservation of the world's biota. In this chapter, we examine the changes which have taken place in the ecosystems of the Western Australian wheatbelt since European settlement.

4.2 Ecosystems of the Western Australian Wheatbelt

Prior to clearing for agriculture, the wheatbelt area consisted of a complex mosiac of vegetation types, the distribution of which was largely determined by rainfall, soil, and landform characteristics (Beard 1983, 1984, 1990). The wheatbelt covers an area of about 14 million ha (equivalent to the area of England, Scotland, and Wales combined), stretching from Geraldton in the north to Esperance in the southeast, with rainfalls varying from about 600 mm nearest the coast to 280 mm at the eastern edge (Hobbs and Saunders, Chapter 1, Fig. 1.1; Burvill 1979a; A.R. Main 1987; Wallace and Moore 1987). The areal extent of the wheatbelt, coupled with high levels of species richness and local endemism, produces considerable geographic variation in the vegetation composition and structure (Ericson et al. 1973; Aplin 1979; Lamont et al. 1984). Here, we concentrate mainly on the central wheatbelt region, particularly the Kellerberrin area (see B.Y. Main, Chapter 3).

Within the central wheatbelt area, vegetation distribution is linked closely with landform and soil type, as discussed by McArthur (Chapter 2). The landscape is, in general, one of gentle topography, and a typical catenary sequence can stretch for many kilometers. In this sequence, the vegetation ranges from heathlands and shrublands on the deep sands and laterites of the uplands through mallee and woodland types on the valley slopes to salt lake systems in the valley bottoms (McArthur, Chapter 2, Fig. 2.1). Detailed descriptions of the vegetation are available in Muir (1978), Beard (1980) and Hobbs et al. (1989). The main vegetation types are summarised in McArthur (Chapter 2).

Little is known about the dynamics of most of these vegetation types; in fact, even the ecology of the dominant plant species is poorly known. The vegetation type about which most is known is the kwongan, or sandplain heath (Pate and Beard 1984), but, even in this case, most information comes from areas other than the wheatbelt. A basic problem, therefore, in assessing the effects of fragmentation on wheatbelt ecosystems is the lack of basic ecological data for the systems prior to disturbance. A few historical studies are available which provide some descriptive information (Diels 1906; Gardner 1942; Leake 1962), and information can be drawn from early museum collections (B.Y. Main 1990). Oral histories are also a source of anecdotal information (Sanders 1991). Another approach has been to compare fragmented wheatbelt systems with intact counterparts in the adjacent uncleared Goldfields area (e.g., Lynch and Saunders 1991). We review the available information in subsequent sections.

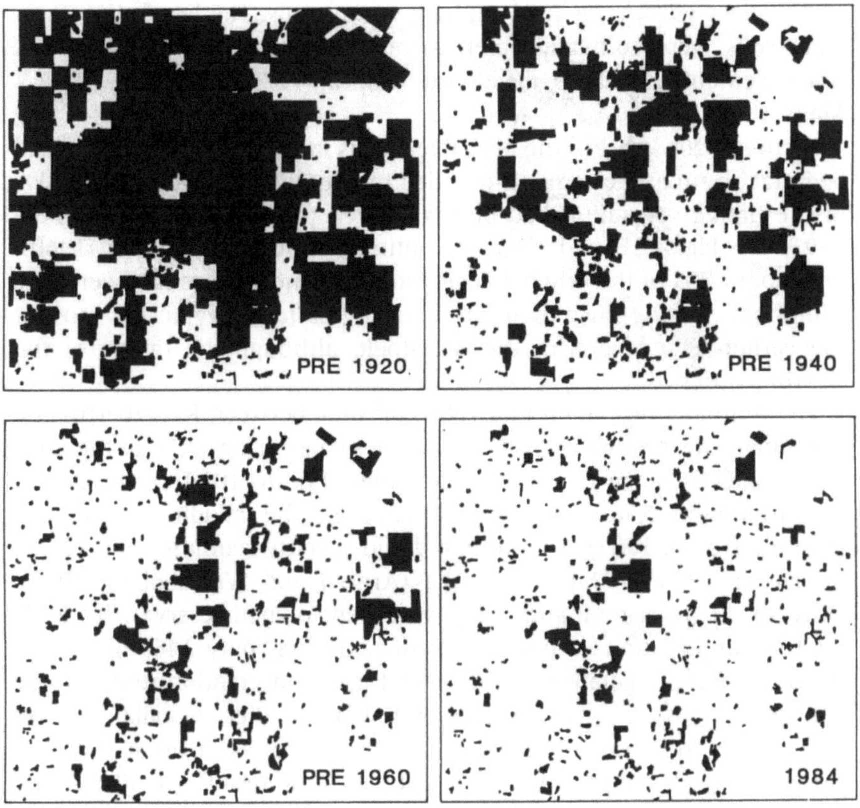

FIGURE 4.1. Clearing history of the Kellerberrin study area. (From Arnold and Weeldenberg 1991.)

4.3 Clearing History

The history of discovery, exploration, settlement, and exploitation given by B.Y. Main (Chapter 3) is central to an understanding of the changes in the biota. The current landscape is the result of changes brought about by the early pastoral phase in the second half of the nineteenth century, followed by the superimposition of an agricultural framework, commencing with the surveys at the turn of the century which laid out subdivisions on the rectangular pattern which is so obvious today (Anon. 1979; Fig. 4.1).

Prior to the release of land for agriculture, surveyors developed a system of land classification based on the native vegetation. Woodland areas of salmon gum (*Eucalyptus salmonophloia*), gimlet (*E. salubris*), and morrell (*E. longicornis*) on heavy soils, regarded as indicators of first class land (Terry 1907, 1908), were taken up first, cleared, and cultivated, followed by the wandoo (*Eucalyptus wandoo* and *E. capillosa*) woodlands and mallees (*Eucalyptus* spp.) on light loamy soils. The less fertile sandplain soils were cultivated later, as an understanding of fertility and trace elements problems developed (Burvill 1979b; Beard 1980; B.Y Main, Chapter 2).

By the 1920s, most of the available blocks in the Kellerberrin area had been taken up and clearing proceeded rapidly. By the 1940s, over 70% of the native vegetation had been removed (Arnold and Weeldenberg 1991; see Fig. 4.1), and clearing continued until, by the early 1960s, only about 7% of the original vegetation remained; there has been little additional clearing since then. This clearing pattern is typical of most of the earlier-settled parts of the wheatbelt, although the timing of the clearing varies.

The remnant vegetation in a 1680 km^2 area between Kellerberrin and Trayning is distributed in over 530 patches comprising 5.7% of the area. These patches range in size from less than 1 ha to 1030 ha; the largest is Durokoppin Nature Reserve, situated in the middle of the area (Saunders et al. 1987). Two hundred and seventy remnants are <10 ha in size and 353 (77%) are <20 ha (Arnold and Weeldenberg 1991). Only six of these remnants are government-owned reserves, set aside for the conservation of flora and fauna. In fact, only 14% of the remaining native vegetation in the district is in conservation reserves; the majority is privately owned. As pointed out earlier, woodlands were regarded as indicators of fertile soil and so were taken up and cleared early in the development of the district. As a result, woodlands are poorly represented on conservation reserves and are best represented on private property or along road verges. Figure 4.2 gives an example of the central wheatbelt landscape as it exists today.

4.4 Effects of Fragmentation on Ecosystem Processes

Fragmentation results in changes in the physical and chemical fluxes across the landscape, including fluxes of radiation, wind, water, and nutrients (Saunders et al. 1991; Hobbs 1992).

Few data have been collected on the radiation fluxes in the Western Australian wheatbelt, but it is clear that significant changes in the radiation balance have resulted from the replacement of the perennial vegetation with pastures and crops which are predominantly annuals. The native vegetation, for the most part, remains green throughout the year since it is composed mostly of evergreen sclerophyllous species, with only a small herbaceous component. The spectral characteristics of the vegetation change during the year depending on the timing of growth and flowering (Hobbs et al. 1989). A few species, such as the shrub, *Allocasuarina campestris*, and a number of ground cover species, turn brown in the summer and regreen with the onset of winter rains

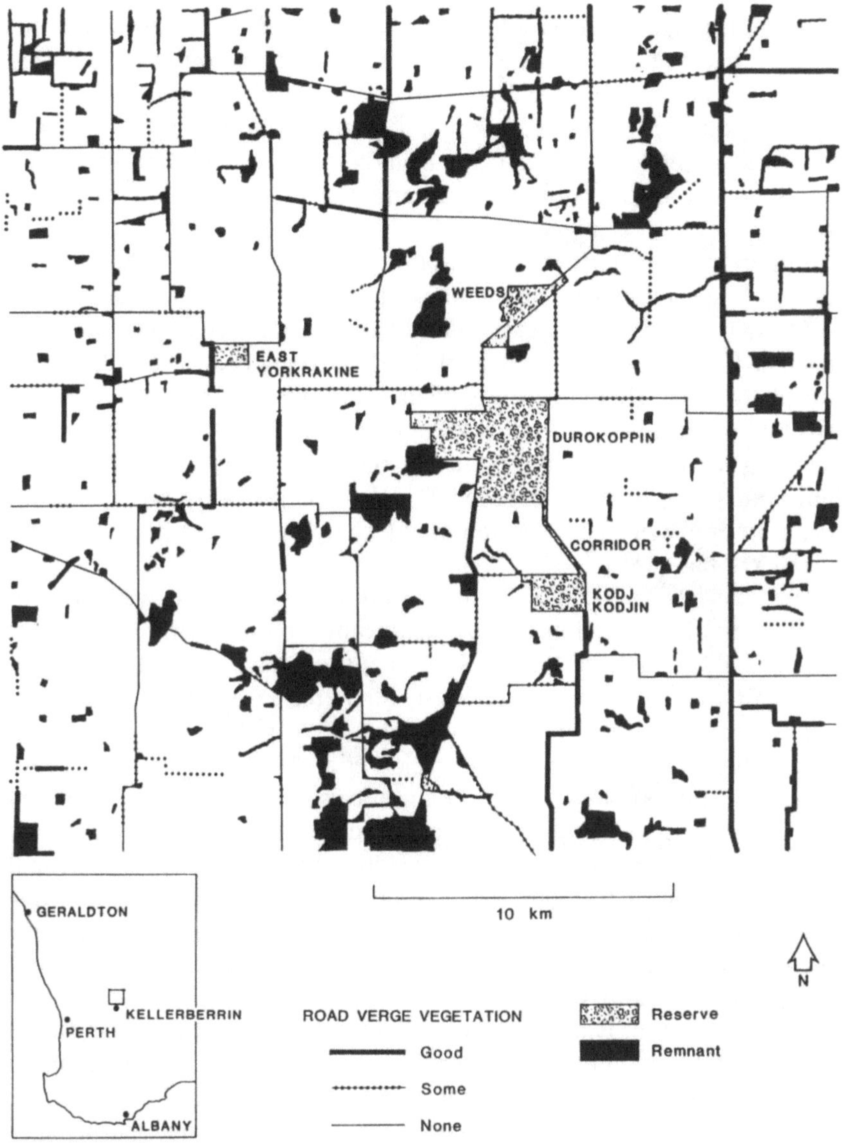

FIGURE 4.2. An example of the present landscape in the central wheatbelt of Western Australia, from within the CSIRO study site north of Kellerberrin, centred on Durokoppin Nature Reserve. Nature reserves, other remnants, and road verge vegetation are indicated. From Saunders and de Rebeira (1991), reproduced with permission of Surrey Beatty and Sons, Chipping Norton, NSW.

(B.Y. Main 1967; Gaff 1981). However, these changes are small compared to the changes which take place in the agricultural lands, where bare soil is gradually replaced by green crop or pasture cover during the winter, reaching maximum cover in spring. With the onset of summer, the crops change colour and are harvested to leave stubbles, while the annual pasture species also dry and change colour, but are mostly removed by sheep grazing. Plant cover on the agricultural land thus varies greatly throughout the year and the radiation balance, particularly the albedo, is affected accordingly (Geiger 1965).

Studies in the Northern Hemisphere and tropics have shown that levels of insolation can be significantly higher at the edges than in the interior of remnant areas, and that this may affect vegetation composition (Wales 1972; Ranney et al. 1981; Lovejoy et al. 1986; Palik and Murphy 1990). There is little evidence for this effect in the Western Australian wheatbelt and one does not observe distinct edge formations. This may be because of the relatively recent isolation of the remnants here, but is more likely to be because light is generally not a limiting factor in the wheatbelt vegetation, in comparison with the northern forests (Hobbs 1989).

The effects of vegetation structure on windspeeds are well known, with an intact vegetation canopy significantly reducing the windspeeds near the ground (Geiger 1965). Several authors have noted the effects of fragmentation in exposing previously sheltered vegetation to increased windspeeds, resulting in tree blowdowns, increased evapotranspiration, and problems for nesting birds (Lovejoy et al. 1986; Reville et al. 1990; Margules, unpublished data). Such effects are visible in woodland remnants in the wheatbelt where numerous trees have been blown down, particularly during summer cyclonic storms (D. Couper, personal communication; R.J. Hobbs, personal observations). Hobbs and Mooney (1992) have indicated that such storms, although regarded as episodic events, are, in fact, frequent enough to have a significant impact on the ecosystems of southwestern Australia.

McFarlane et al. (Chapter 6) discuss in detail the effects of landscape fragmentation on the hydrological balance, particularly in relation to waterlogging and salinity problems in the agricultural land. Rising water tables also have significant impacts on vegetation remnants, particularly those in lower-lying areas, including the salmon gum woodlands (Fig. 4.3) and wetlands and their fringing vegetation (Froend et al. 1987; Bell and Froend 1990; Froend and McComb 1991; Sanders 1991). The soils of many formerly wooded areas are now extremely saline, temporarily or permanently waterlogged, or under water, and the woodland vegetation has either disappeared altogether or remains as standing dead trees. Increased surface flow also results in increased amounts of water moving from agricultural areas into adjacent remnants and bringing with it leached nutrients. This allows nonnative annuals to invade, as

Figure 4.3. Effects of increased salinisation due to rising water tables on woodland formerly dominated by *Eucalyptus salmonophloia*.

discussed in a subsequent section. Nutrients can also be transferred across remnant edges by wind movement of soil particles and stubble, although often soil nutrient levels become elevated only near the remnant edge (Muir 1979; Cale and Hobbs 1991; Hester and Hobbs 1992). Summer cyclonic storms can, however, move considerable amounts of topsoil large distances, hence causing significant redistributions of nutrients (McTainish et al. 1989; Bell 1990). Similarly, animals can move nutrients from agricultural land into remnants by feeding in paddocks and defecating and urinating in remnant vegetation.

4.5 Effects on Vegetation

4.5.1 Reduction in Area

The most obvious effect of clearing for agriculture in Western Australia has been the reduction in the extent of the native vegetation. Over the wheatbelt region as a whole, 93% of the native vegetation has now been cleared. According to Leigh et al. (1984), this has resulted in the known extinction of 24 plant species, although the actual number of species lost is unknown and numbers are still under revision as taxonomic studies and field surveys progress. The wheatbelt region now also has one of the highest numbers of rare and/or endangered plant species in Australia (Hartley and Leigh 1979; Leigh et al. 1984; Hopper et al. 1990). In the region administered as the wheatbelt by the Department of Conservation and Land Management, 348 species are listed as rare or endangered (Hopper et al. 1990). These include species such as *Eucalyptus caesia*, which is restricted to localised populations around granite outcrops (Hopper et al. 1982; Moran and Hopper 1983, 1987), and *Acacia merrickiae*, which is restricted to wandoo woodland in the Kellerberrin area (Hopper et al. 1990). Only 79 of these species have populations within designated nature reserves, 135 are found in road reserves, and 53 are on private land (Hopper et al. 1990). Many species are likely to have had small isolated populations prior to clearance for agriculture; however, in the absence of intensive management, populations in degrading remnant areas or on roadside corridors are unlikely to be viable in the long term.

4.5.2 Changes in Composition and Structure

The remaining vegetation does not provide a true sample of the pre-existing vegetation due to preferential clearing of vegetation types on the better soils. The pattern of clearing for the Kellerberrin area illustrates this, with the early clearing all occurring in the lower parts of the landscape. The sandier soils of the sandplains were avoided initially because of their low fertility but, in the 1940s, it was discovered that these soils could be farmed by correcting trace element deficiencies (Burvill 1979c). However, there now remains a significantly greater proportion of the shrub communities than the woodland communities, both in the Kellerberrin area and in the wheatbelt as a whole (Beard and Sprenger 1984; Arnold and Weeldenberg 1991). This imbalance is reflected in the distribution of vegetation types on designated nature reserves. Most woodland remnants are on private land, and many have been degraded by past or present grazing by stock.

The history of settlement of the central wheatbelt area, with an early pastoral phase followed by mixed crop/pasture systems (see B.Y. Main, Chapter 3), meant that much of the native vegetation would have been grazed by stock at some time or another. It is thus difficult to determine what the original vegetation was like, since there are few areas that can be used as definite baselines. Few vegetation studies were carried out during the early colonisation of the area, and the earliest detailed records of the wheatbelt vegetation came from Diels (1906). His descriptions gave some baseline information against which to compare today's vegetation. It appears that the shrub-dominated vegetation types have survived relatively intact, presumably because they were relatively

FIGURE 4.4. Aerial photograph of the southwest edge of Durokoppin Nature Reserve in the central wheatbelt, with ungrazed vegetation to the right and vegetation subjected to a long history of stock grazing on the left. Note the considerably more open nature of the grazed vegetation. (Photo by Anne Scougall.)

poor for grazing. The woodlands, on the other hand, are mostly degraded, with little shrubby understorey, invasion by nonnative herbaceous species, and no regeneration of the tree species. Similar changes have been reported in woodland remnants in eastern Australia (Buchanan 1989; Landsberg et al. 1990; Hibberd and Soutberg 1991). Where woodland has been fenced for some time, the vegetation structure approximates more closely that described prior to clearance (Fig. 4.4). It is not yet clear whether woodland types will recover from long periods of grazing if the grazing pressure is removed. Stock grazing affects not only the vegetation structure, but also the soil structure through trampling and soil nutrient status through nutrient transfer from the outside. This latter effect has been identified as a possible factor increasing the abundance of defoliating insects and hence rural tree dieback in remnant woodlands in eastern Australia (Landsberg et al. 1990).

A further disruption to woodland vegetation occurred during World War II, when the shortage of oil-based fuels for vehicles led to the widespread use of "gas producers" to power them. These were fueled by charcoal derived from salmon gum, York gum, and gimlet, which were cut and burned in both publicly- and privately-owned woodland. The remains of these activities can still occasionally be detected. Overall, the activities associated with charcoal burning had a further deliterious effect on the woodland canopy, understorey, and soils.

4.5.3 Effects of Changes in Fauna

The grazing regime within remnant vegetation has been changed in ways other than the addition of stock. Clearly, the reduction in populations of the major herbivores, especially kangaroos and smaller macropods (see Section 4.8), in many small remnant areas removes a significant component of the ecosystem and has probably led to changes in nutrient and energy cycling. Whole components of the fauna have virtually disappeared from the area. For instance, accounts of early settlement indicate relatively large numbers of arboreal marsupials. Facey (1981) relates that "The first morning Uncle and the boys caught twenty-two possums." Possums are now virtually absent from the central wheatbelt, but the effects of their removal are little understood. Declining populations of other fauna may also have small but significant effects on the vegetation. For instance, Saunders (1980) noted that Carnaby's cockatoo (*Calyptorhynchus funereus latirostris*) ate seed predators of *Dryandra* affin. *cirsioides*. The disappearance of emus (*Dromaius novaehollandiae*) from most of the agricultural area may have had important implications for the dispersal of large-seeded species (B.Y. Main 1967). Similarly, there may have been significant impacts on pollination regimes through the loss of fauna, but these impacts have been little assessed (Main 1981). The loss of animals which disturbed the soil by burrowing or mound building (e.g., the boodie or burrowing bettong, *Bettongia lesueur*, the billoy or dalgyte, *Macrotis lagotis*, and the mallee fowl, *Leiopoa ocellata*) may also have had unknown effects on the seedling establishment of native species.

While many native animal species have been lost or significantly reduced in numbers, other species have invaded. The rabbit (*Oryctolagus cuniculatus*) has undoubtedly had a huge impact on the vegetation through grazing, seedling removal, and burrowing, as discussed in studies elsewhere in Australia (Lange and Graham 1983; Friedel 1985; Leigh et al. 1987; Wilson 1990). Increases in the number of galahs (*Cacatua rosiecapilla*) also add to the decline of woodland areas through their tendancy to ringbark large eucalypts, especialy salmon gums (Saunders and Hobbs 1989).

Changes in insect abundance may also have significant impacts on the vegetation. For instance, Van Schagen et al. 1992 discussed an outbreak of bag shelter moths (*Ochrogaster lunifer*), whose caterpillars feed predominantly on the tree *Acacia acuminata*. Within a large remnant area, colonies of these caterpillars failed to develop, but, on roadside populations, colonies were abundant and in some cases caused severe defoliation. The reasons for this difference were not clear but may have included altered nutrient status in roadside soils and lack of predators along narrow roadverges. Grasshoppers can also affect seedling regeneration (Whelan and Main 1979), and episodic locust plagues, as were experienced in 1990, can have significant impacts on elements of the native vegetation (notably *Allocasuarina* spp.) and on trees planted for revegetation.

4.5.4 Invasions by Nonnative Plant Species

Invasion by nonnative plant species is a further important consequence of fragmentation. The native vegetation is predominantly perennial, with a relatively small herbaceous component, but this has largely been replaced with an agricultural system dominated by annual crop and pasture species. Invasion into remnant vegetation is predominantly by annual species from the paddocks and other annual weedy species. These are mainly species from the Mediterranean area, such as *Avena fatua*, *Bromus* spp., and *Hypochaeris glabra* which have invaded other regions with Mediterranean-type climates, especially California (Groves 1986; Mooney et al. 1986). Other major invading species, notably *Arctotheca calendula*, originally came from South Africa.

The degree of invasion of the native vegetation varies greatly depending on vegetation type. Shrub communities are relatively resistant to invasion, while woodland communities are generally more susceptible (Hobbs 1989; Hobbs and Atkins 1988). The main reason for this difference is probably the higher level of nutrients present in woodland soils. Hobbs and Atkins (1988) found a marked increase in the growth of weedy species following the addition of nutrients. This effect was accentuated, however, where nutrient addition accompanied soil disturbance. This matches observed patterns in native vegetation, where nonnative species are restricted to areas such as rabbit scrapes or ant nests which combine soil disturbance with elevated nutrient availability (through defecation or debris accumulation). Cale and Hobbs (1991) also found a strong relationship between the abundance of weedy species in corridor vegetation and soil nutrient levels, particularly phosphorus. This study, and results from Hobbs (1989) and Hester and Hobbs (1992), also showed strong gradients of weed invasion at corridor and remnant edges, which could be related to the gradients of soil nutrient levels. Again, however, this was dependent on vegetation type, with a shrub community being invaded only at the extreme edge while a woodland community was invaded, to some degree, throughout. Hobbs and Atkins (1988) suggested that these differences between vegetation types were related to the ability of weedy species to establish and grow, rather than to the availability of propagules. Seed trapping indicated that weed seed is dispersed even into the middle of Durokoppin

Nature Reserve, the largest remnant in the Kellerberrin area (1030 ha), and weedy species will grow anywhere within the area following soil disturbance and/or nutrient addition.

These findings indicate that weed invasion is likely to be an increasing problem within small remnant areas. The juxtapositioning of a nutrient poor system (the native vegetation) with a nutrient rich agricultural system is likely to lead to increased nutrient inputs to the remnant vegetation, and this, in turn, will provide increased opportunities for weed invasion. Weed invasion is minimal where vegetation canopies are intact, as is the case in the shrub and mallee communities, but weed invasion is likely where the canopy has been disrupted. A major question must be whether weed invasion is important. If invasion has little impact on the normal ecosystem and community processes, there would be little cause for concern. However, the indications are that invasion does affect natural processes (Usher 1988; MacDonald et al. 1989). For instance, the regeneration of native species can be inhibited by a cover of weedy species (Panetta and Groves 1990; Hobbs and Atkins 1991), and it is known from revegetation efforts that weed control is essential for successful tree and shrub establishment (Venning 1988; Buchanan 1989; Bird et al. 1990; Loney 1990). Invasion by herbaceous species also significantly affects fire fuel loads and, hence, fire regimes (Wycherly 1984; Panetta and Hopkins 1991).

4.5.5 Changes in Disturbance Regime

Pickett and Thompson (1978) discussed the concept of the "minimum dynamic area" required to ensure that an array of all successional stages was present in any given locality through the operation of the natural disturbance regime. Fragmentation of the natural vegetation into small remnant areas disrupts this disturbance regime, as discussed by Hobbs (1987) and A.R. Main (Chapter 7). Of particular importance is the fire regime, since fire is generally considered to be a major influence on the dynamics of the vegetation in much of southwestern Australia and Australia as a whole (Gill et al. 1981; Bell et al. 1984; Ford 1985). Unfortunately, little is known about fire regimes in the wheatbelt region prior to European settlement, although it appears likely that deliberate burning was carried out by Aboriginal populations (Hallam 1985). This lack of historical perspective again makes it difficult to assess the changes brought about by fragmentation, although it is obvious that many of the small reserves around the Kellerberrin area have remained unburned since the surrounding area was cleared. In some cases, this has resulted in vegetation senescence, with the dominant shrubs dying and nothing being established in their place. A.R. Main (Chapter 7) also argues that the lack of disturbance is leading to a loss of species diversity and increased dominance by a few species. Our studies have shown that there is a suite of heathland species which appear only after a fire and persist as seed between fires and that the germination of some woodland species is stimulated by high temperatures (Hobbs and Atkins, unpublished data). The continued absence of fire is therefore likely to reduce the plant species diversity within small remnant areas, although Muir (1985) argued that it is important to retain some long-unburnt areas as baseline sites. This is illustrated by Heitman's Scrub in Tammin Shire, which has now been unburned for more than 84 years (Gardner 1957; B.Y. Main 1967) and in which the regeneration of some shrubs is occurring amongst the senescent and dead canopy (B.Y. Main, personal communication). Clearly, the timescales involved in growth, senescence, and regeneration may be very long in some of these communities, and there is a necessity for a mixed approach of periodic burning in some areas and retention of unburnt vegetation in others. Certainly, it is necessary to avoid the problem experienced in other parts of southwestern Australia where frequent burning, carried out for fuel reduction purposes, may be detrimental to the vegetation (McGrath 1985; Hobbs and Atkins 1990).

4.6 Effects on Soil Macrofauna

There have been few studies which have documented the changes in soil fauna with the advent of agricultural development in Western Australia (Abbott et al. 1979; Abbott and Parker 1980a,b; Lobry de Bruyn 1990). Here, we confine our discussion to soil macrofauna, which includes ants (Hymenoptera: Formicidae), termites (Isoptera), earthworms (Lumbricidae, Megascolecidae), beetles (Coleoptera), scorpions (Scorpionidae), and spiders (Araneae). Little information is available on the soil microfauna or any potential effects of changes to it.

The decline in soil macrofauna activity and species richness in farmland can be related directly to the clearing of natural vegetation and its replacement with monoculture crops and the introduction of livestock, with the resulting loss of microhabitats and changes in microclimatic conditions and resource availability. The subsequent cultivation of the land generally disturbs and/or destroys much of the soil fauna in the top 15 cm of the soil, while other influences such as trampling by stock also have deleterious effects.

The ant fauna of the Kellerberrin area is very diverse. Van Schagen (1986) found 42, 31, and 36 ant species in small areas of heathland, *Allocasuarina campestris* shrubland, and a revegetated heath area, respectively, while Lobry de Bruyn (1990) found a total of 107 ant species in Durokoppin Nature Reserve and surrounding farmland. In a wandoo woodland on grey sandy loam soils, there were 89 species and, in heath habitats on yellow sand soils, there were 65 species. On equivalent soils

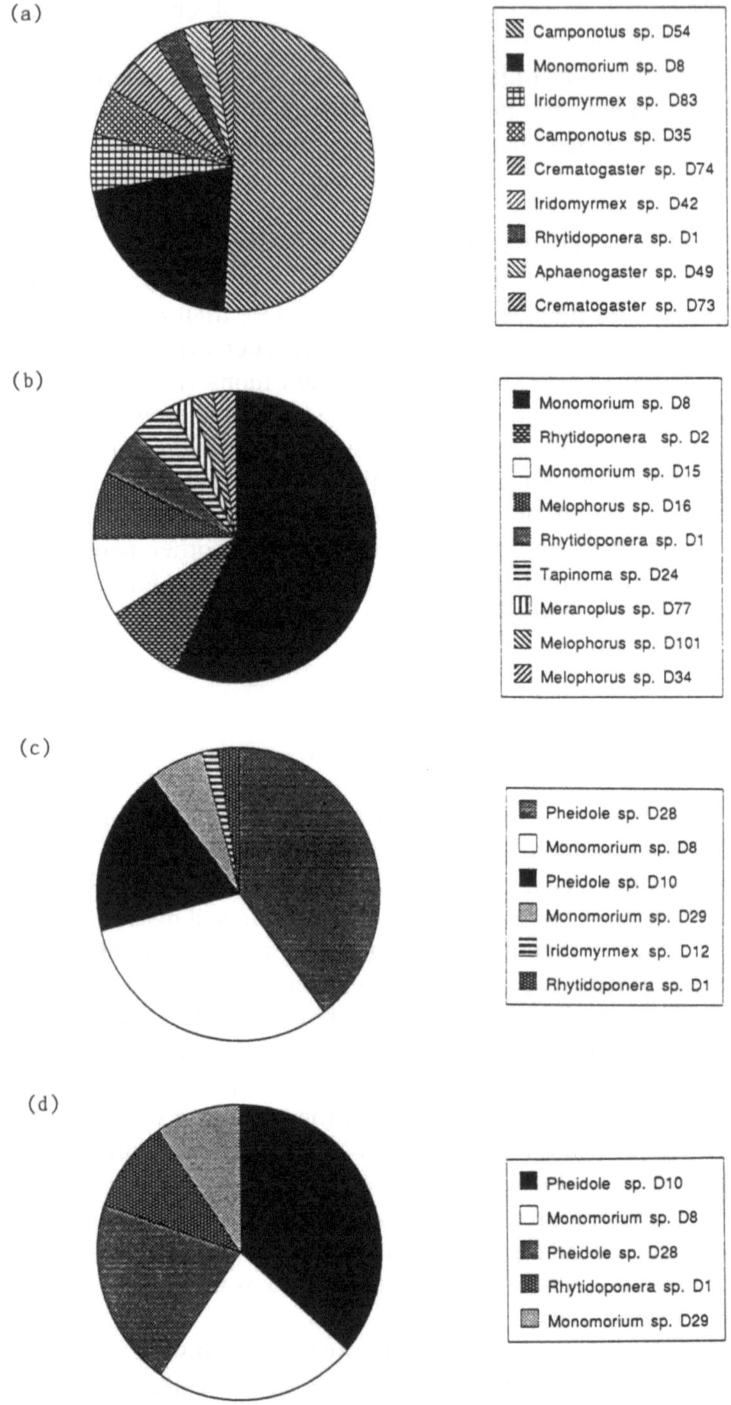

FIGURE 4.5. The most commonly recorded ant species (excluding the ubiquetous *Iridomyrmex* sp D13) on two soil types in natural vegetation and farmland. (a) yellow sand in mixed heath; (b) grey sandy loam in *Eucalyptus capillosa* woodland; (c) yellow sand in farmland; (d) grey sandy loam on farmland. Species are listed and graphed in descending order of abundance, starting at 12 o'clock on the pie chart.

in farmland, there were only 30 species. However, ant activity and nest density was virtually the same in native vegetation and farmland (1.54 and 1.49 nests m^{-2} in the reserve and farmland, respectively). The major species in farmland was found to be a *Pheidole* species, which was found only occasionally in the reserve (Fig. 4.5). Studies by Twigg et al. (1982) and Majer et al. (1987) have also found the ant fauna of agricultural land to be dominated by seed collectors and scavengers such as *Pheidole* spp., *Monomorium* spp., and *Melophorus* spp.

Most of the ant species recorded in the farmland can survive agricultural clearing and cultivation, since the queen and brood are situated well below plough depth; however, other soil fauna are more vulnerable. Ants thus dominate the soil macrofauna in the farmland, despite a reduction in species richness. On grey sandy loam soils in the farmland, 94% of the diggings by soil macrofauna were by ants and 6% were by spiders. In contrast, on equivalent soils in wandoo woodland, 55% of the invertebrate diggings were caused by other invertebrates [scorpions (25%), spiders (6%), and others (24%); Lobry de Bruyn (1990)].

Termites are also adversely affected by agriculture because most of their galleries are located in the top 10 cm of the soil, and the resources and microclimate are changed dramatically. Termites are abundant in native vegetation. The dominant mound builders, *Drepanotermes rubriceps* and *D. tamminensis*, had a density of 52 and 9 mounds ha^{-1} in heath and wandoo woodland, respectively (Abensperg-Traun and de Boer 1990). In contrast, during a three-year study at Kellerberrin, only two *Drepanotermes tamminensis* mounds were found in pasture paddocks, but these were invaded by *Iridomrymex purpureus* ants and subsequently became inactive (Lobry de Bruyn 1990). In areas formerly cleared for agriculture and left to regenerate, the diversity of termite fauna remains low for decades (Abensperg-Traun and de Boer 1990). In addition to the potential impacts of this lower diversity on nutrient cycling, it may also affect other components of the fauna which use termites as a food resource (Abensperg-Traun 1988).

Abbott and Parker (1980b) found four earthworm species in wheatbelt soils (*Aporrectodea trapezoides, A. rosea, Microscolex dubius*, and *Megascolex imparicystis*). The latter species is native while the others have been introduced since European settlement. Earthworms were most abundant in home gardens (100% occurrence), followed by pastures (54%), uncleared areas (24%), and cultivated paddocks (18%). Other small indigenous earthworms have been reported where there is an abundance of litter associated with shrubs such as *Allocasuarina* spp., and have even been found in permanent pastures at Newdegate and Wongan Hills (McCredie and Parker 1988). The native earthworm, *Megascolex imparicystis*, was found at a density of 5 m^{-2} in pasture soil near Lancelin, but this decreased to 1 m^{-2} after two years of cultivation (Abbott et al.

1985). Incongruously, it was not found in adjacent heath vegetation. Other earthworms such as *Aporrectodea caliginosa, A. longa, Lumbricus rubellus,* and *Eukerria saltensis* also persist in agricultural land (Tisdall 1985; Blackwell and Blackwell 1989; Baker and Bukerfield 1990).

However, these studies were mostly carried out in areas other than the central wheatbelt and may have little direct relevance there. Earthworm activity is reliant on adequate soil moisture levels and Barley (1959) estimated that, in a Mediterranean-type climate, earthworms can be active only for 20 to 30 weeks of the year (April to October). The reduction in tillage has had a positive influence on earthworm numbers in South Australia, where numbers of *Aporrectodea caliginosa* under a direct-drilled wheat crop almost doubled compared with a conventionally tilled lupin-wheat rotation (Rovira et al. 1987). Hence, earthworms are able to survive agricultural development where soil moisture and food requirements are met and soil disturbance is minimised. Thus, one might expect exotic earthworms to become part of the fauna of the soil under minimal tillage.

An important aspect of the changes in soil fauna in agricultural land is the introduction of dung beetles in an attempt to speed up the removal of stock dung and hence alleviate the fly problems of rural Australia. For instance, 17 introduced species have become established in pasture soils in New South Wales, in addition to the native dung beetles (Tyndale-Biscoe 1990). These beetles have now established and, while suitable dung is present, will persist as an exotic element in the soil and surface fauna.

While detailed information on most soil macrofauna is lacking, there is a large body of accumulated information on trapdoor spiders within wheatbelt reserves (B.Y. Main 1976, 1987a,b). It is clear from these studies that agricultural practices do not allow the persistence of these spiders, which are now restricted to remnant vegetation. Grazing and trampling by stock also affect these species adversely (B.Y. Main 1976, 1987a,b). B.Y. Main (1987b) pointed out that all but the most common species of these spiders are liable to become extinct and that the current presence of a species within a remnant area does not guarantee its long-term persistence. Some of the spider species are very long lived (e.g., B.Y. Main [1987b] reports individuals of *Anidiops villosus* living for at least 23 years), and current populations are, in some cases, composed entirely of adults; without recruitment, the populations will eventually die out. This work illustrates the importance of long-term detailed studies of populations in remnant areas; without that perspective, inappropriate management decisions are inevitable.

4.7 Effects of Changes in Soil Macrofauna on Soil Properties

Soil macrofauna affects soil properties through the construction of burrows, channels, and nests, the ingestion of soil and organic material (termites and earthworms), and the redistribution of soil and organic material. The impact of macrofauna on soil physical and hydrological properties can influence the severity of such problems as compaction, surface sealing, hardpans, and sheet erosion. The macrofauna also influences soil pH, organic carbon, and available nutrients, which affect soil fertility and structural stability. Lobry de Bruyn and Conacher (1990) have recently reviewed the effects of ants and termites in soil modification. Here, we discuss the results from studies in the Kellerberrin area.

Lobry de Bruyn (1990) concluded that the effects of soil fauna (mostly ants) on surface soil varied according to soil type, habitat type, and time of year in naturally-vegetated sites and accounted for approximately 0.5% of the soil surface area in heath sites and 0.3% in wandoo woodland (*Eucalyptus wandoo* and *E. capillosa*) sites. In farmland, the soil surface area affected by ants varied according to agricultural practices, regardless of soil type, and was approximately 0.2% in cropped paddocks and 1.2% in pasture paddocks. Thus, contrary to expectations, burrowing activity was actually greater under pasture than in native vegetation. The volume of soil in the upper 26 cm layer which had ant nest galleries and chambers was approximately 10 times the soil surface area affected. Where land is cultivated for cropping, there is a significant reduction in the surface expression of ant nests until these are redug in the following months. Ploughing coincides with the beginning of winter and the least active period for soil invertebrates. By the summer months, burrowing activity has reached levels equivalent to those in native vegetation.

Bioturbation (soil mixing) by ants is reduced by a factor of three in farmland compared to native vegetation (0.1 compared with 0.37 t $ha^{-1} yr^{-1}$; Lobry de Bruyn 1990). There is thus a reduction in soil mixing in farmland. A species of *Aphaenogaster* contributes most to bioturbation in naturally vegetated sites but is absent from agricultural land. Ant spoils are generally less compact than the surrounding surface soil, regardless of land use or soil type, and the soil in the spoil can be easily reworked into the topsoil or can possibly enhance seedling germination (Lobry de Bruyn 1990). Ant channels can be important in the movement of water, air, and large soil particles. They are active throughout the year, except for a month after cultivation. They are vertically oriented and continuous for a depth of 60 cm, vary in size from 0.8 to 2 mm, and are maintained by actively removing soil material which has backfilled the entrance. Channels have a negligible influence on water infiltration rate under preponding conditions (i.e., equivalent to the early stages of rainfall). However, under ponded conditions, water infiltration rates are up to six times faster over ant nests than in adjacent soil (Lobry de Bruyn 1990). Majer et al. (1987), also working in the Western Australian wheatbelt, found a similar increase in infiltration rates over ant nests ($8.3 \pm 5.4\,\mathrm{min\,l^{-1}}$) compared with unaffected soil ($27.2 \pm 3.9\,\mathrm{min\,l^{-1}}$). A density of ant nests in farmland of 0.5 to 1.5 nests m^{-2}, as found by Lobry de Bruyn (1990), will significantly aid in conducting water quickly to the subsoil during high intensity storms and, hence, reducing runoff and subsequent sheet erosion (see also McFarlane et al., Chapter 6).

Nests of several ant species contain significantly higher organic carbon than the adjacent soil. In *Rhytidoponera* spp., higher levels of carbon coincide with the location of refuse chambers which contain decomposing plant and animal material. Ant species such as *Pheidole* spp., which dispose of refuse around the nest entrance, have higher carbon levels in the top 2 cm (Lobry de Bruyn 1990). After many years, these nest middens become thick with discarded seeds, dead ants, and the remains of other insects. Periodic ploughing does not allow these middens to accumulate organic material. While no major differences in nutrient levels or pH were recorded between the ant nests and adjacent soil in the Kellerberrin area (Lobry de Bruyn 1990), studies in New South Wales have found higher levels of phosphorus in nests of seed-collecting species from the genera *Chelaner, Pheidole*, and *Rhytidoponera* (Davidson and Morton 1981, Briese 1982).

There is no evidence of large-scale termite activity in the farmland at Kellerberrin, although in naturally vegetated areas, especially wandoo woodland, there is a considerable mass of soil ($3.5\,\mathrm{t\,h^{-1}}$) held in mounds of *Drepanotermes tamminensis*. The soil held in termite mounds has two to three times the amount of clay-sized particles compared with adjacent soil. In natural vegetation, subterranean termites also contribute to bioturbation by constructing protective coverings of soil, referred to as soil sheetings, over food material. These soil sheetings are coarse textured and easily broken up by rain and wind and, hence, become quickly incorporated into the topsoil. Soil held in termite mounds, on the other hand, is not available for incorporation into the topsoil until the mound ceases to be active, which may take 50 to 250 years (Watson et al. 1988). Some erosion of active mounds does occur, however, and echidnas (*Tachyglossus aculeatus*) foraging for termites also redistribute mound soil.

The absence of termites from agricultural land may not have a marked effect on water infiltration, although the effects of subterranean termites on infiltration are unknown and difficult to determine. Termite channels are active mostly during May and October, are sealed when not in use, and are usually horizontally oriented (except for galleries to foraging holes). Nevertheless, data from Noble and Tongway (1988) on semi-arid woodland in New South Wales suggested that feeding sites under dead logs for a variety of termite species had a consistently higher ponded water infiltration rate than soil elsewhere. The major effect of the absence of termites from farmland is likely to be in terms of organic matter and nutrient distribution. Termite-modified soil contains a higher level of organic carbon due to the incorporation of faecal material and saliva as cementing agents in the lining of gallery walls. Soil in mounds of *Drepanotermes tamminensis* has considerably more carbon, available phosphorus, and potassium than the adjacent soil, while the foraging galleries of *Amitermes neogermanus* have twice the amount of carbon than adjacent soil (Lobry de Bruyn 1990). Termite-modified soil tends to be more acidic, possibly as a result of the breakdown of organic matter into organic acids and finally to carbon dioxide which forms a solution of carbonic acid on the surface of minerals (Paton 1978).

The effects of earthworm burrows on soil structure have been examined in temperate parts of Australia (Tisdall 1985; Blackwell and Blackwell 1989). McCredie and Parker (unpublished) provided evidence which suggested that water infiltration rates are faster over earthworm-affected soil than over control soil at Dangin in the Western Australian wheatbelt (about 50 km southwest of Kellerberrin). MacKenzie and Dexter (1987) also found that casts of the earthworm *Aporrectoda rosea* are less compactable than beds of aggregates from bulk soil. There is a scarcity of information from Australia on the effects of earthworm activity on soil chemical properties, although there is a large volume of overseas literature. Earthworms can affect soil chemistry by the deposition of casts (made up of organic and inorganic materials) in the soil and on the soil surface, by the consumption of litter and plant roots, and through interactions with other organisms in the food chain. Pot experiments by Barley and Jenning (1959) in red-brown earth soil established that casts of *Aporrectoda caliginosa* had 83% more available nitrogen than control soil. Lee (1985) concluded that, if earthworms were active for 150 days of the year (Barley 1959), their casts would contribute approximately 22–28 $\mathrm{kg\,ha^{-1}\,yr^{-1}}$ of additional available nitrogen to the soil.

Two important questions arise in connection with soil property modification by soil macrofauna in agricultural and naturally vegetated land in the central wheatbelt. First, at what time of year does the soil macrofauna activity occur and does this coincide with crucial periods such as seed germination and winter or summer storms? In naturally vegetated lands, ant nests and termite mounds are active throughout the year, while termite sheetings occur periodically, probably in relation to rainfall. In farmland, ant nests are active throughout the year, except for a month after ploughing. Earthworms are active over the cooler and moister months (May to October) and predominantly in areas receiving greater than 500 mm yr^{-1} rainfall. The second question relates to the duration of the modified levels of carbon, pH, and cohesion, but at this stage there is little reliable information on this.

The role of ants in soil conservation and increased crop yields is restricted by the existing nest densities, not species richness. Although nest density in farmland is similar to that in naturally vegetated areas, this is insufficient to have a major influence on crop production or a large-scale modification of soil properties. Changes in agricultural practices, including minimal tillage, are required to allow an increase in soil macrofauna activity. Such changes also have direct benefits in terms of soil conservation (Nulsen, Chapter 5). The use of earthworms in agricultural land in the central wheatbelt may also be facilitated by altered farming practices but are probably restricted by the low rainfall experienced in this area.

It is clear from the information presented above that soil macrofauna are an important component of the natural ecosystems of the Western Australian wheatbelt, but that development for agriculture has dramatically altered its composition and activity. Soil macrofauna have an important role to play in the establishment of an integrated approach to agricultural land management (see also Lefroy et al., Chapter 8). The role of the soil macrofauna in soil-forming processes, nutrient cycling, and amelioration of soil problems has been understudied and probably undervalued in Australia (e.g., Parker 1989), and their influence is worthy of greater recognition in theoretical and practical approaches to land management.

4.8 Changes in the Mammal Fauna

The decline in mammal species which occurred in the Western Australian wheatbelt over the past 150 years was also experienced in other parts of Australia, including the arid zone where habitat fragmentation did not occur. Burbidge and McKenzie (1989) found that the highest rates of mammal decline and extinction in Western Australia were recorded from the wheatbelt. Kitchener et al. (1980a) noted that 43 species of native mammals (excluding bats) were believed to have occurred in the Western Australian wheatbelt prior to settlement by the Europeans, yet, at the time of their surveys in the mid-1970s, only 12 of these species were moderately common or abundant. There is continuing debate as to the proximate and ultimate causes of the disappearance of mammals (Burbidge and McKenzie 1989), but it is likely that a combination of factors was responsible (Morton 1990). In the arid zone, stock grazing, rabbits, altered fire patterns, introduced predators, and drought are thought to have interacted to lead to increased probabilities of local disappearance and eventual extinction (Morton 1990). It is likely that the same factors were important in the wheatbelt but with the addition of habitat fragmentation and human activities such as poison baiting and burrow ripping for pest control.

In order to understand the changes in fauna since the European settlement, we need to examine the effects of several phases of the settlement and development as well as the confounding factors resulting from the introduction of exotic species. The distribution and abundance of animals varies naturally over time, and there is evidence that indicates that some species were already in decline when European settlement started. For example, the broad-faced potoroo, *Potorous platyops*, was discovered in 1839 and the type specimen collected from Goomalling in 1842 (Strahan 1983). By 1875, ten specimens had been collected and this species, which had been common and widespread as a fossil or subfossil, may have been in decline at the time of the European settlement. A similar interpretation is likely for the pig-footed bandicoot, *Chaeropus ecaudatus*, which Gould (using Gilbert's notes) reported from east of Northam (Strahan 1983).

4.8.1 Effects of Pastoralism

The pastoral phase followed settlement (B.Y. Main, Chapter 3) with leases being taken up and sheep shepherded across the countryside. Shortridge (1910) summarised the distribution of marsupials in south-western Australia and indicated that declining abundance was apparent then. Some indication of the decline in abundance can be obtained from specimens from the central wheatbelt received by the Western Australian Museum (Table 4.1). However, the following limitations need to be noted. Donations to the Museum were affected by whether the collector believed the specimens were curiosities and whether there were facilities for forwarding the material. Furthermore, from the depression years of the 1930s through the extensive expansion of land settlement following the Second World War, few people had the resources to send material to the Museum, even if they were curious about the specimens collected. These factors mean that the period of accelerated decline of native mammals is poorly documented.

Sheep grazing has had clear effects on vegetation composition and structure (Section 4.5.2) and, no doubt, grazing sheep would have affected thicket-dwelling marsupials like the potoroo or bandicoot. However, thickets containing plants poisonous to sheep would have offered some sort of refuge and might have contributed to their persistence into the early part of this century. Indeed, the late C.G. Jessup (died 1962) recalled that the pig-footed bandicoot was around the Yorkrakine Rock area when, as a railway engine driver early in the century, he stayed over in Tammin during crew changes (B.Y. Main

Table 4.1. Western Australian Museum accessions from the central wheatbelt by year, from Kitchener and Vicker (1981). U = uncommon; E = no longer occurs in area; and ?? = status unknown.

Year	Species	Locality	Status
1913	Echidna, *Tachyglossus aculeatus*	Tammin	U
1914	Southwestern pygmy possum, *Cercartetus concinnus*	Yorkrakine	E
1915	Bilby, *Macrotis lagotis*	Tammin	E
1917	Bilby	Doodlakine	E
1919	Mitchell's hopping-mouse, *Notomys mitchelli*	Bencubbin	E
	Numbat, *Myrmecobius fasciatus*	Burracoppin	E
1924	Bilby	Cunderdin	E
1925	Bilby	Cunderdin	E
1925	Bilby	Clackline	E
1928	Bilby	Bolgart	E
	Brush-tailed possum, *Trichosurus vulpecula*	Baandee	??
1930	Fat-tailed dunnart, *Sminthopsis crassicaudata*	Wyalkatchem	??
1931	Bilby	Bruce Rock	E
	Fat-tailed dunnart	Tammin	??
1932	Numbat	Merredin	E
1935	Bilby	Brookton	E
1936	Mitchell's hopping-mouse	Dalwallinu	E
1942	Boodie, *Bettongia lesueur*	Pingelly	E
1944	Mitchell's hopping-mouse	Dalwallinu	E
1945	Southwestern pygmy possum	Merredin	E
1950	Mitchell's hopping-mouse	Wyalcatchem	E
1955	Southwestern pygmy possum	Bolgart	??
1956	Southwestern pygmy possum	Northam	??
1960	Mitchell's hopping-mouse	Burracoppin	E
	Black-footed rock-wallaby, *Petrogale lateralis*	Kellerberrin North	U
1961	Chuditch, *Dasyurus geoffroii*	Tammin	E
1974	Fat-tailed dunnart	Tammin	??
	Common dunnart, *Sminthopsis murina*	Tammin	??
1975	Ashy-grey mouse, *Pseudomys albocinereus*	Kellerberrin	U
1976	Ashy-grey mouse	Durokoppin	U

1967). This was in spite of the fact that the last museum record for this species in Western Australia was in 1842 (Strahan 1983).

Kitchener et al. (1980a) considered that the species which became extinct did so because they had difficulty in coping with the alteration of the patchiness in the vegetation caused by changes in the burning regime from Aboriginal to European pastoral practices. They thought that European burning of extensive areas on a regular basis would have reduced large areas to similar seral stages, and this would have the effect of increasing the distances between the former patches in the landscape.

In addition, the development of the pastoral industry may have alienated watering points from native animals. Because such areas were essential to domestic livestock, they were preferentially taken up (see

B.Y. Main, Chapter 3). Overgrazing and clearing of vegetation round watering points may have made such areas unuseable by some species of native animals.

4.8.2 Habitat Fragmentation

There is no doubt that the clearing and fragmentation of native vegetation during the development for agriculture throughout the wheatbelt did result in a significant loss of native fauna, simply as a result of the reduction in area of habitat. For example, only about 6% of the original extent of native vegetation remains in the Kellerberrin area, and this extensive loss of habitat has been accompanied by the known loss of a large number of native mammals (Kitchener et al. 1980a) and birds (Saunders 1989). The sequence of clearing is shown in Table 4.2. The clearing of native vegetation and grazing by domestic stock on the remnant vegetation would have compounded the effects of the changes in patchiness suggested by Kitchener et al. (1980a). As evidence in support of these ideas, Kitchener et al. (1980a) cited the examples of the woylie or brush-tailed bettong (*Bettongia penicillata*), the boodie or burrowing bettong (*Bettongia lesseur*), the banded hare wallaby (*Lagostrophus fasciatus*), and the western hare wallaby (*Lagorchestes hirsutus*), which formerly occurred in the wheatbelt and had very specialised habitats and/or feeding requirements. These species are now very rare or extinct on the Australian mainland (Strahan 1983).

The result of clearing was to deplete severely the quality and area of habitat available, and this, in turn, limited population size and restricted the ability of both plant and animal populations to recover following drought or wildfires. Fire frequency increased as fires escaped from burning vegetation, which had been stripped from the land during clearing operations, or as a result of fires started by sparks from railway engines (A.R. Main, personal observations). Thus, following every disturbance, whether drought or fire, habitat quality declined further, and populations of those species which were present prior to the disturbance were reduced in numbers.

Despite the loss of some species or a decline in the abundance of others, the boodie, the bilby (*Macrotis lagotis*), the tammar (*Macropus eugenii*), and the western brush wallaby, (*M. irma*) were still present in the central wheatbelt during the 1920s (Table 4.1), the time of the agricultural expansion documented by B.Y. Main, Chapter 3. Most species declined in range and abundance thereafter, lending credence to the common belief that faunal changes were directly due to clearing and associated fragmentation of the native vegetation.

Table 4.2. Summary at the landscape level of the order of development of agriculture, and its effect on biota.

Sequence down slope			Effect on biota	
Soil catena	Vegetation	Development	Where cleared	In remnants
Granite lithosoils	Shrub and tall shrub with sheoaks, York gum, mallee	Later, as heavy equipment became available and fertiliser deficiencies could be corrected	Most lost, with loss of trees and understorey, biota typical of grassland. Aquatic animals persisted until waters became saline	Regeneration following fire hindered by rabbits
Deep sands	Kwongan	As above		As above
Sands over laterite	Mixed kwongan	As above		As above, fauna such as mallee fowl, boodie, and bilby persisted for some time, but foxes and ripping and fumigating of burrows finally killed them. Tammar and brush wallabies declined as habitat destroyed and foxes increased

Laterite	*Allocasuarina acutivalvis* thicket	As above		
Duplex soils	Mallee, woodland, and various species			
Sandy loam	York gum-jam woodland	Grazed in pastoral phase, then cleared for wheat	Clearings used by bustard, bush stone curlew, and banded plover until fox predation intensified	As above, wandoo, jam, and sheoak regenerating
Loam	Wandoo woodland with poison understorey	Avoided in pastoral phase because of the presence of poison plants		Many small marsupials persisted in poison thickets
Heavy loams	Woodland of salmon gum and gimlet	Cleared early as good agricultural land		Understorey lost by grazing, regeneration prevented by rabbits, possums lost to exotic predators
Saline flats, salt lakes, and dunes	Halophytes, York gum, morrel, sheoak, and gimlet woodland	If cleared, subject to flooding	Former biota lost as salt increased	Vegetation killed by salt, all habitat lost

4.8.3 Influence of Rabbits, Foxes, and Other Introduced Fauna

The simple picture of habitat fragmentation is incomplete, however. It is complicated, firstly, by the arrival of the exotic rabbit (Stodart and Parer 1988) and fox, *Vulpes vulpes*, early this century and, secondly, by the growth of their populations until the rabbits were in plague numbers and supporting large numbers of foxes. This rabbit plague led to habitat destruction, the failure of regeneration in remnant vegetation, and the removal of remnant vegetation by burrowing and grazing. The fox preyed on rabbits and also on native mammals whose cover and shelter had failed to regenerate after destruction by rabbits, fire, or drought. The fox also predated reptiles, frogs, scorpions, and many of the larger insects such as the carabid, cerambycid, and buprested beetles and hepialed moths. It should be noted that the arrival of rabbits and foxes in the region pre-dated most of the extensive landscape fragmentation; any effects of fragmentation *per se* are thus likely to be inextricably confounded with the effects of rabbits and foxes.

Other exotic species also had an effect on the native animals (Table 4.3). As farms were established, many farmers retained areas of native vegetation around the house (B.Y. Main, Chapter 3). Initially, these areas retained the native biota typical of the vegetation associations in which they were confined. However, the settlers were accompanied by mice, *Mus musculus*, rats, *Rattus rattus*, and a suite of domestic animals such as dogs, *Canis familiaris*, cats, *Felis catus*, and poultry which had unhindered access to the natural environment. Cats, rats, and mice soon established themselves as feral populations, with mice and cats in particular moving far from farm buildings. Many dogs became feral and bred with the native dingo, *C. f. dingo*. While these wild dogs were a major threat to sheep, they were also likely to be at higher densities than prior to settlement and so probably affected those native birds and mammals which remained in the bush remnants. Once rabbits reached plague proportions, a program of poisoning, fumigating, and ripping of the burrows ensured that mammals such as bilbies and brush-tailed possums, *Trichosurus vulpecula*, both present in the central wheatbelt in the 1930s, were killed along with the rabbits. Thus, it is likely that the native animals, with their lower rates of increase, were readily exterminated while, for practical purposes, rabbit numbers remained unchanged. Many farmers fenced their paddocks with rabbit netting, which not only excluded rabbits but prevented access to the crops by the smaller wallabies, which soon succumbed to fox and dog predation. The effect on the native fauna of the widespread use of arsenic and strychnine in poison baits and water for rabbits and foxes is undocumented.

Despite the foregoing, there was a suggestion in the early 1950s that, in some areas, native animals were increasing in abundance or at least

Table 4.3. Summary of the effects of introduced animals and their impact on native biota in remnants of native vegetation.

Species	Status	Effect
Mouse, *Mus musculus*	Domestic and feral	Feral mice are omnivores and finally replaced native murids, such as *Pseudomys* and *Notomys*
Rat, *Rattus rattus*	Primarily associated with farm buildings	
Cat, *Felis catus*	Domestic	Cats continued to bring small mammals, birds, reptiles, and insects to farm houses long after settlement
	Feral	Predators of native mammals, birds, reptiles, and insects
Dog, *Canis familiaris*	Domestic	Hunted and disturbed nearby native animals
	Feral	Augmented populations of native predators, killed sheep and native animals
Poultry	Domestic but with access to native vegetation	Scratching disturbed litter, depleted invertebrates, prevented regeneration. Attacks by chuditch led to their persecution by farmers
Fox, *Vulpes vulpes*	Feral	Dug lairs or used burrows of bilbies and rabbits. Predated ground nesting birds, reptiles, insects, and mammals
Rabbit, *Oryctolagus cuniculus*	Feral	Burrows more frequent than bilbies and their grazing prevented regeneration. Ringbarked shrubs and small trees. Poisoning, ripping, and fumigating of burrows killed many native animals

their decline had slowed (Serventy 1954). Some species survived in the agricultural land; for instance, while stones and woody material remained in cultivated areas, the fat-tailed dunnart, *Sminthopsis crassicaudata*, persisted (White 1951). As paddocks were cleared of these shelters, this habitat was lost and the species was restricted to remnants. A new burst of agricultural development associated with postwar land settlement, aided by heavy machinery for the clearing of native vegetation, rapidly reduced the size of the remaining fragments of native vegetation. This period coincided with the introduction of myxomatosis (Calaby et al. 1960; Fenner and Ratcliffe 1965), and the resulting decline in rabbit numbers probably diverted the fox populations towards the remaining native species which, restricted as they now were to small patches of native vegetation, rapidly succumbed to this predation. At this time, eviscerated carcasses of bobtail lizards, *Trachydosaurus rugosus*, and gravid females of the burrowing frog, *Heleioporus albopunctatus*, were common sights (A.R. Main, personal observations). The effects of settlement were thus apparently much more extensive than would be expected from simply clearing the land and pasturing sheep in bushland. In the end, the extinctions probably followed the stochastic variation in very small populations.

4.8.4 Remaining Common Elements of the Mammal Fauna

The echidna is still widely distributed over its former range and its populations are apparently stable. Abensperg-Traun (1991) believes that the echidna has coped well with the introduction of exotic predators and the loss and fragmentation of its habitat because it is not restricted in its choice of habitat, it specialises on ants and termites (an abundant and reliable food source), it is metabolically capable of tolerating droughts and fire-induced food shortages, it is semifossorial, predation on adults is low because of its spines, and it is capable of moving between widely separated remnants. However, even such apparently resilient species as the echidna may not be unaffected by predation by foxes because recruits may be susceptible to predation, and their apparent persistence may be due to the presence of extremely long-lived adults which are resistant to predation.

The two largest macropods (Western grey kangaroo, *Macropus fuliginosus*, and euro, *M. robustus*) are still widely distributed in the central wheatbelt where they move between remnants, using corridors of native vegetation, as well as moving over open agricultural land (Arnold et al. 1991). However, a number of the mammals in the size range between the big macropods and the smallest marsupials have disappeared from the area. Chapman and Kitchener (1978) and Kitchener and Chapman (1980) listed 15 species of native mammals, excluding bats, known to have occurred within 20 km of the Quairading and Yorkrakine Reserves or in Durokoppin and Kodj Kodjin Nature Reserves; only nine of these now occur somewhere in the region. Of these, only the Western grey kangaroo, the euro, the fat-tailed dunnart (G.T. Smith, personal communication), and the echidna are probably secure in the medium term, provided no further clearing or degradation of the native vegetation occurs and rabbit and fox numbers are brought under control. The common dunnart, *Sminthopsis dolichura*, is still found on most ungrazed remnants while the ashy grey mouse, *Pseudomys albocinereus*, black-footed rock wallaby, *Petrogale lateralis*, Western brush wallaby, and brush-tailed possum still occur in very isolated populations. However, their survival is in doubt without active management. In fact, the black-footed rock wallaby only survives on several rocky remnants because the Department of Conservation and Land Management has an active fox control program (Kinnear et al. 1988). Without this control, these populations would probably die out. Friend (1990) also showed that, with adequate fox control, the numbat, *Myrmecobius fasciatus*, can be reintroduced into wheatbelt reserves.

Nevertheless, mammals are still being lost from remnants, as illustrated by the brush-tailed possum, which was recorded on the 1030 ha Durokoppin Nature Reserve by Chapman and Kitchener (1978) in 1976 during 16 days of field work; yet, this species was not recorded on this reserve during several years of field work 10 years later (G.W. Arnold, personal communication). The Western brush wallaby may be the next species to be lost from reserve and the central wheatbelt.

4.9 Changes in the Avifauna

Kitchener et al. (1982) found the avifauna of the wheatbelt to be in better condition than the mammal fauna, with a greater proportion of the original species still present in the region. However, the avifauna is now demonstrating a similar pattern of species decline as the mammals (Saunders 1989; Saunders and Curry 1990). Excluding vagrants, 131 species of birds have been recorded in the Kellerberrin district since the European settlement.

Some species were able to increase during the agricultural clearing phase. The expansion of open space, that is, the converse of the habitat destruction and fragmentation discussed above favoured open country species such as the banded plover, *Vanellus tricolor* (Reid 1951; L. Sedgwick 1952; E.H. Sedgwick 1953), bush-stone curlew, *Burhinus magnirostris* (Reid 1952), and bustard, *Ardeotis australis*, (Reid 1952) as areas of pasture and stubble became available. Indeed, these species were common during the earlier phases of development but became increasingly rare from the 1950s onwards as the fox increasingly affected population numbers. Other species, such as the galah, have benefited from agricultural development and have extended their range. The galah is a species which formerly occurred along arid zone water courses (Saunders et al. 1985) but has expanded rapidly into all agricultural areas in Australia because of the availability of grain and water. In the wheatbelt of Western Australia, it is an extremely common species which competes for nest hollows with Carnaby's cockatoo and has been responsible for the failure of nesting attempts of this vulnerable species of black cockatoo (Saunders 1990). This has been a factor in the decline of this cockatoo and its disappearance from much of its breeding range over the last 20 years (Fig. 4.6).

Twelve species of nonpasserines and 29 species of passerines have decreased in range and/or abundance in the wheatbelt as a whole over the last 80 to 90 years. This represents nearly 30% of the recorded avifauna of the area. Of these species, the emu, mallee fowl, Carnaby's cockatoo, regent parrot (*Polytelis anthopeplus*), red-capped parrot (*Pur-*

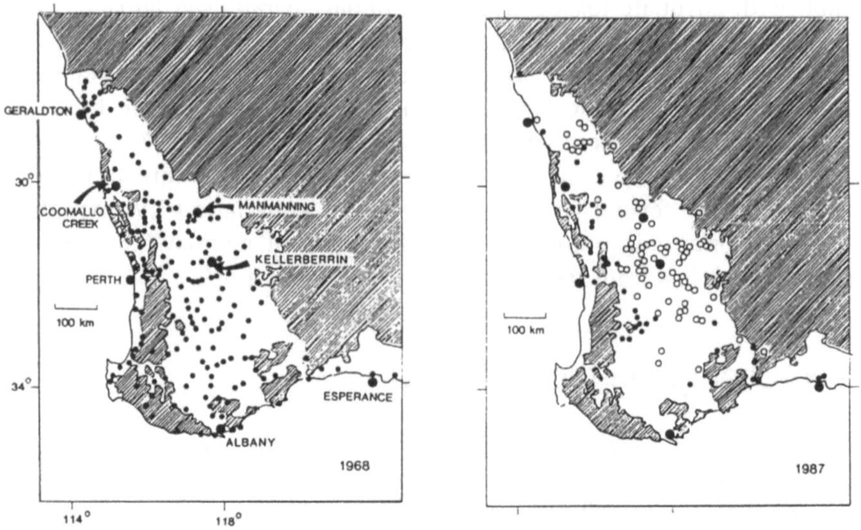

FIGURE 4.6. Change in distribution of Carnaby's Cockatoo, *Calyptorhynchus funereus latirostris*, over the period 1968–1987. Large filled circles are towns, as labelled on left-hand map. Small filled circles are locations of observers recording Carnabys Cockatoo in that year. Open circles are locations of observers not recording Carnaby's Cockatoo in that year. From Saunders (1990), reproduced with permission of Elsevier Applied Science Publishers, Ltd. Barking, Essex.

pureicephalus spurius), Western rosella (*Platycercus icterotis*), ground cuckoo shrike (*Coracina maxima*), hooded robin (*Melanodryas cucullata*), crested shrike tit (*Falcunculus frontalus*), Gilbert's whistler (*Pachycephala inornata*), restless flycatcher (*Myiagra inquieta*), chestnut quail thrush (*Cinclosoma castanotum*), rufous tree creeper (*Climacteris rufa*), yellow-plumed honeyeater (*Lichenostomus ornatus*), and white-naped honeyeater (*Melithreptus lunatus*) have all disappeared from the Kellerberrin district, some within the last few years. It is interesting to note the extent of the decline in some of these species. For example, Leake (1962) noted that the yellow-plumed honeyeater was the most abundant honeyeater in the area and it now no longer occurs in most of the wheatbelt.

Habitat loss and fragmentation have been two of the main reasons for the loss of these species. This is shown by the fact that species that have disappeared from the heavily cleared and fragmented rural landscape of Kellerberrin still occur and are common in the uncleared Goldfields vegetation around Yellowdine, due east of Kellerberrin (see Lynch and Saunders 1991). The purple-crowned lorikeet, Western rosella, Gilbert's whistler, chestnut quail thrush, and rufous tree creeper are commonly recorded at Yellowdine (Lynch and Saunders 1991) but are extinct or occasional vagrants now at Kellerberrin, despite being common there before clearing took place (Leake 1962).

The fact that the remaining native vegetation is not contained in one patch but is distributed in over 500 patches of varying sizes and shapes will lead to the loss of further species because some populations on remnants are too small and too isolated to remain viable (see Saunders 1989). For example, the blue-breasted wren, *Malurus pulcherrimus*, occurs on East Yorkrakine Nature Reserve (81 ha), yet the total population on the reserve is about 10 individuals. This is too small to be viable and there are no populations in a position to recolonise this isolated reserve in the event of loss of the wren population.

As with mammals, introduced herbivores and predators have undoubtedly had a significant impact on the avifauna. Sheep grazing on remnant patches particularly affected species which feed predominantly on the ground. For example, the rufous tree creeper disappears very quickly from remnants subjected to grazing by sheep (G. Chapman, personal communication). Foxes are probably responsible for the disappearance of ground-nesting birds, such as the mallee fowl, from much of the wheatbelt. Cats are known to be significant predators on hole-nesting birds; Saunders (unpublished data) recorded that cats killed 7% of nestling red-tailed black cockatoos, *Calyptorhyncus magnificus*, (total nestlings hatched = 428) over 11 breeding seasons in the Three Springs area of the northern wheatbelt. In the spring of 1978, 17% of nestlings were killed by cats climbing into the nest hollows. Obviously, there is a need for studies on the ecology of feral cats, but there is already enough evidence to suggest that attempts should be made to control them.

4.10 Reptiles and Amphibians

According to Kitchener et al. (1980b), lizards have been less severely affected than mammals and birds by the changes brought about by the development of agriculture. During their surveys on reserves in the mid-1970s, they recorded all but 11 of the 81 species of lizards that had been recorded from the wheatbelt. They added that their observations indicated no widespread loss of lizard species from the wheatbelt, and they believed that there is a good possibility that all species known to have occurred in the region at European settlement are still there. Those they failed to record may represent a marginal wheatbelt distribution rather than extirpations as a result of habitat loss and fragmentation. To quote Kitchener et al. (1980b), "In the Western Australian wheatbelt establishment of nature reserves and their management largely postdate agricultural clearing in the region, so that many nature reserves in the wheatbelt are merely areas unsuitable for farming purposes. However, it appears extremely unlikely, on economic grounds, that if a rational plan for establishment of reserves in the area had been possible, prior to clearing, a single area large enough to carry representatives of all lizard species (let alone all mammals and birds) in the region would have been reserved. It is then of some consolation to note that although the haphazardly spaced wheatbelt reserves are inadequate to preserve entire communities of mammals . . . and apparently also of birds and snakes, we conclude that the reserve system is probably adequate to preserve representative lizard communities."

In addition to habitat loss and fragmentation, the changes in the environment resulting from this extensive and rapid clearing are also having an effect on other groups. For example, the burrowing frog, *Heleioporus albopunctatus*, is disappearing from large areas of its former range, possibly as a result of increases in soil salinity and salinisation of the valley floors where it formerly occurred (Main 1990). There is little information presently available on the changes in other amphibian populations as a result of clearing for agriculture and the resultant environmental changes.

4.11 Conclusions

In this chapter, we have explored the consequences to the native biota of the rapid fragmentation of the natural landscape by agricultural development. Clearly, the loss of vegetation and habitat has had a huge effect on the natural ecosytems of the central wheatbelt. However, it should be pointed out that the biogeographic consequences of fragmentation (isolation in small remnants) have probably been relatively minor compared with the influence of other factors which accompanied (and in some cases preceeded) fragmentation.

The major factors can be summarised as follows:

(a) A possible decline in abundance and range of some fauna species prior to, and leading to extinction shortly after, settlement.
(b) Habitat destruction from local overstocking and prevention of regeneration by sheep grazing in the pastoral phase of settlement, which continued in remnant vegetation following fragmentation.
(c) Denial to native fauna of watering points which were preferentially taken up in the pastoral phase of settlement.
(d) Fragmentation of the landscape which disrupted populations and the distribution of biotic assemblages by clearing, cultivation, and burning. These occurred in three phases: early settlement and the two post-war phases, the first in the 1920s and the second in the 1950s.
(e) The arrival and expansion of rabbit and fox populations at the time of the 1920s agricultural expansion led to the subsequent extinction

of native animals as a result of predation, habitat destruction and ripping, and poisoning and fumigating of rabbit burrows.

(f) Altered hydrologic regimes leading to rising water tables, waterlogging, water erosion, and salinisation with consequent effects on wetlands and woodland vegetation.

(g) Invasion of native vegetation by non native species.

(h) The effect of agricultural practices on the soil fauna, soil structure, and nutrient regimes.

The net result of these events has been the local or global loss of species. The restriction of what remains to remnants of native vegetation of limited size and the restricted populations they can carry necessarily increases the probability of extinctions. This is made more likely by the hostile desert-like environment which surrounds the remnants during summer and contributes to making the remnants' habitat much harsher than it would be on equivalent vegetation assemblages prior to clearing.

For many components of the biota, factors other than those generally considered in island biogeographic theory have been the major cause of population declines and extinctions. While the effects of habitat reduction and isolation are undoubtedly there, and will continue to become evident, the prime importance of these other external factors should not be overlooked. Unless the external factors are recognised and controlled, there is little hope of successfully conserving the biota within the system of remnants still present today. Control of external factors demands that the surrounding landscape has to be managed in a complementary manner; i.e., conservation management has to be integrated into the overall management of the landscape, as discussed further by Hobbs et al., Chapter 10 and Hobbs and Saunders, Chapter 11.

Acknowledgments. We thank Graham Arnold, Chris Margules, Max Abensberg-Traun, Graeme Smith, Jeff Short, and Ken Wallace for constructive comments on the manuscript and C.P. de Rebeira for assistance in diagram preparation.

References

Abensperg-Traun, M.A. (1988) Food preference of the echidna *Tachyglossus aculeatus* (Monotremata: Tachyglossidae) in the wheatbelt of Western Australia. *Australian Mammalogy* 11:117–123.

Abensperg-Traun, M. (1991) Survival strategies of the echidna, *Tachyglossus aculeatus* (Monotremata: Tachyglossidae). *Biological Conservation* 58:317–328.

Abensperg-Traun, M.A. and de Boer, E.S. (1990) Species similarity and habitat differences in biomass of subterranean termites (Isoptera) in the wheatbelt of Western Australia. *Australian Journal of Ecology* 15:219–226.

Abbott, I. and Parker, C.A. (1980a) Agriculture and the abundance of soil animals in the Western Australian wheatbelt. *Soil Biology and Biochemistry* 12:455–459.

Abbott, I. and Parker, C.A. (1980b) The occurrence of earthworms in the wheatbelt of Western Australia in relation to land use and rainfall. *Australian Journal of Soil Research* 18:343–352.

Abbott, I., Parker, C.A., and Sills, I.D. (1979) Changes in the abundance of large soil animals and physical properties of soils following cultivation. *Australian Journal of Soil Research* 17:343–353.

Abbott, I., Ross, J.S., and Parker, C.A. (1985) Ecology of the large indigenous earthworm *Megascolex imparicystis* in relation to agriculture near Lancelin, Western Australia. *Journal of the Royal Society of Western Australia* 68:13–15.

Anon. (1979) *Western Australia. An Atlas of Human Endeavour 1829–1979.* Government Printing Office of Western Australia, Perth.

Aplin, T.E.H. (1979) The flora. In B.J. O'Brien (ed.) *Environment and Science.* University of Western Australian Press, Nedlands. pp. 53–76.

Arnold, G.W. and Weeldenberg, J.R. (1991) The distributions and characteristics of remnant native vegetation in parts of the Kellerberrin-Tammin-Trayning and Wyalkatchem shires of Western Australia. *CSIRO Division of Wildlife and Ecology Technical Memo* 33.

Arnold, G.W., Weeldenberg, J.R., and Steven, D.E. (1991) Distribution and abundance of two species of kangaroo in remnants of native vegetation in the central wheatbelt of Western Australia and the role of native vegetation along road verges and fencelines as linkages. In D.A. Saunders and R.J. Hobbs (eds.) *Nature Conservation 2: The Role of Corridors.* Surrey Beatty and Sons, Chipping Norton, NSW. pp. 273–280.

Attenborough, D. (1987) *The First Eden. The Mediterranean World and Man.* Little, Brown and Company, Boston.

Baker, G. and Bukerfield, J. (1990) Earthworms hold key to pasture productivity. *Soils Brief 6:1–4.* CSIRO, Melbourne.

Barley, K.P. (1959) The influence of earthworms on soil fertility. I. Earthworm populations found in agricultural land near Adelaide. *Australian Journal of Agricultural Research* 10:171–178.

Barley, K.P. and Jennings, A.C. (1959) Earthworms and soil fertility. III. The influence of earthworms on the availability of nitrogen. *Australian Journal of Agricultural Research* 10:364–370.

Beard, J.S. (1980) *The Vegetation of the Kellerberrin Area, Western Australia.* Vegmap Publications, Perth.

Beard, J.S. (1983) Ecological control of the vegetation of southwestern Australia: Moisture versus nutrients. In F.J. Kruger, D.T. Mitchell, and J.U.M. Jarvis (eds.) *Mediterranean-type Ecosystems: The Role of Nutrients.* Springer-Verlag, New York. pp. 66–73.

Beard, J.S. (1984) Biogeography of the kwongan. In J.S. Pate and J.S. Beard (eds.) *Kwongan. Plant Life of the Sandplain.* University of Western Australia Press, Nedlands. pp. 1–26.

Beard, J.S. (1990) *Plant Life of Western Australia* Kangaroo Press, Kenthurst, NSW.

Beard, J.S. and Sprenger, B.S. (1984) Geographical data from the vegetation survey of Western Australia. *Vegetation Survey of Western Australia Occasional Paper* 2. Vegmap Publications, Perth.

Bell, A. (1990) Wind erosion: The winnowing of our soils. *Ecos* 66:4–7.

Bell, D.T. and Froend, R.H. (1990) Mortality and growth of tree species under stress at Lake Toolibin in the Western Australian wheatbelt. *Journal of the Royal Society of Western Australia* 72:63–66.

Bell, D.T., Hopkins, A.J.M., and Pate, J.S. (1984) Fire in the kwongan. In J.S. Pate and J.S. Beard (eds.) *Kwongan. Plant Life of the Sandplain*. University of Western Australia Press, Nedlands. pp. 178–204.

Bird, R., Cumming, K.N., and Kearney, G.A. (1990) Research in direct seeding of trees in the 500–800 mm rainfall zone of western Victoria. In *Sowing the Seeds. Direct Seeding and Natural Regeneration Conference Proceedings*. Greening Australia, Deakin, Australian Capital Territory. pp. 65–74.

Blackwell, P.S. and Blackwell, J. (1989) The introduction of earthworms to ameliorated, irrigated duplex soil in southeastern Australia and the influence on macropores. *Australian Journal of Soil Research* 27:807–814.

Briese, D.T. (1982) The effects of ants on the soil of a semi-arid salt bush habitat. *Insectes Sociaux* 29:375–386.

Buchanan, R.A. (1989) *Bush Regeneration: Recovering Australian Landscapes*. TAFE, NSW, Sydney.

Burbidge, A.A. and McKenzie, N.L. (1989) Patterns in the modern decline of Western Australia's vertebrate fauna: Causes and conservation implications. *Biological Conservation* 50:143–198.

Burvill, G.H. (1979a) The natural environment. In G.H. Burvill (ed.) *Agriculture in Western Australia 1829–1979*. University of Western Australia Press, Nedlands. pp. 91–105.

Burvill, G.H. (1979b) The first sixty years, 1829–1889. In G.H. Burvill (ed.) *Agriculture in Western Australia 1829–1979*. University of Western Australia Press, Nedlands. pp. 4–17.

Burvill, G.H. (1979c) The last fifty years, 1929–1979. In G.H. Burvill (ed.) *Agriculture in Western Australia 1829–1979*. University of Western Australia Press, Nedlands. pp. 47–86.

Calaby, J.H., Gooding, C.D., and Tomlinson, A.R. (1960) Myxomatosis in Western Australia. *CSIRO Wildlife Research* 5:89–101.

Cale, P. and Hobbs, R.J. (1991) Condition of roadside vegetation in relation to nutrient status. In D.A. Saunders and R.J. Hobbs (eds.) *Nature Conservation 2: The Role of Corridors*. Surrey Beatty and Sons, Chipping Norton, NSW. pp. 353–362.

Chapman, A. and Kitchener, D.J. (1978) Mammals of Durokoppin and Kodj Kodjin Nature Reserves. *Records of the Western Australian Museum Supplement* 7:49–54.

Dasmann, R.F. (1983) The relationship between protected areas and indigenous peoples. In J.A. McNeely and K.R. Miller (eds.) *Proceedings of the World Congress on National Parks*. IUCN, Gland, Switzerland.

Davidson, D.W. and Morton, S.R. (1981) Myrmecochory in some plants (F. Chenopodiaceae) of the Australian arid zone. *Oecologia (Berlin)* 50:357–366.

Diels, L. (1906) *Die Pflanzenwelt von West-Australien südlich des Wendekreises*. Engelmann, Leipzig.

Ericson, R., George, A.S., Marchant, N.G., and Morcombe, M.K. (1973) *Flowers and Plants of Western Australia*. Reed, Frenchs Forest, NSW.

Facey, A.B. (1981) *A Fortunate Life*. Penguin, Ringwood, Victoria.

Fenner, F. and Ratcliffe, F.N. (1965) *Myxomatosis*. Cambridge University Press, Cambridge.

Ford, J. (ed.) (1985) *Fire Ecology and Management in Western Australian Ecosystems*. Western Australian Institute of Technology, Perth.

Friedel, M.H. (1985) The population structure and density of central Australian trees and shrubs, and relationships of range condition, rabbit abundance, and soil. *Australian Rangelands Journal* 7:130–139.

Friend, J.A. (1990) The numbat *Myrmecobius fasciatus* (Myrmecobiidae): History of decline and potential for recovery. *Proceedings of the Ecological Society of Australia* 16:369–377.

Froend, R.H. and McComb, A.J. (1991) An account of the decline of Lake Towerrinning, a wheatbelt wetland. *Journal of the Royal Society of Western Australia* 73:123–128.

Froend, R.H., Heddle, E.M., Bell, D.T., and McComb, A.J. (1987) Effects of salinity and waterlogging on the vegetation of Lake Toolibin, Western Australia. *Australian Journal of Ecology* 12:281–298.

Gaff, D.E. (1981) The biology of resurrection plants. In J.S. Pate and A.J. McComb (eds.) *The Biology of Australian Plants*. University of Western Australia Press, Nedlands. pp. 115–146.

Gardner, C.A. (1942) The vegetation of Western Australia with special reference to climate and soils. *Journal of the Royal Society of Western Australia* 28:11–87.

Gardner, C.A. (1957) The fire factor in relation to the vegetation of Western Australia. *Western Australian Naturalist* 5:166–173.

Geiger, R. (1965) *The Climate near the Ground*. Harvard University Press, Cambridge, Mass.

Gill, A.M., Groves, R.H., and Noble, I.R. (eds.) (1981) *Fire and the Australian Biota*. Australian Academy of Science, Canberra.

Groves, R.H. (1986) Invasion of Mediterranean ecosystems by weeds. In B. Dell, A.J.M. Hopkins, and B.B. Lamont (eds.) *Resilience in Mediterranean-type Ecosystems*. Junk, Dordrecht. pp. 129–145.

Hallam, S.J. (1985) The history of aboriginal firing. In J. Ford (ed.) *Fire Ecology and Management in Western Australian Ecosystems*. Western Australian Institute of Technology, Perth. pp. 7–20.

Harris, L.D. and Eisenberg, J. (1989) Enhanced linkages: Necessary steps for success in conservation of faunal diversity. In D. Western and M.C. Pearl (eds.) *Conservation for the Twenty-first Century*. Oxford University Press, New York. pp. 166–181.

Hartley, W. and Leigh, J. (1979) Plants at risk in Australia. *Australian National Parks and Wildlife Service Occasional Paper* 3.

Hester, A. and Hobbs, R.J. (1992) Influence of fire and soil nutrients on native and non-native annuals at remnant vegetation edges in the Western Australian wheatbelt. *Journal of Vegetation Science* 3:101–108.

Hibberd, J.K. and Soutberg, T.L. (1991) Roadside reserve condition 1977–89 in the southern tablelands of New South Wales. In D.A. Saunders and R.J. Hobbs (eds.) *Nature Conservation 2: The Role of Corridors*. Surrey Beatty and Sons, Chipping Norton, NSW. pp. 177–186.

Hobbs, R.J. (1987) Disturbance regimes in remnants of native vegetation. In D.A. Saunders, G.W. Arnold, A.A. Burbidge, and A.J.M. Hopkins (eds.) *Nature Conservation: The Role of Remnants of Native Vegetation*. Surrey Beatty and Sons, Chipping Norton, NSW. pp. 23–40.

Hobbs, R.J. (1989) The nature and effects of disturbance relative to invasions. In J. Drake, F. di Castri, R. Groves, F. Kruger, H. Mooney, M. Rejmánek, and M. Williamson (eds.) *Biological Invasions: A Global Perspective* John Wiley and Sons, Chichester. pp. 389–405.

Hobbs, R.J. (1992) Ecosystem effects of landscape fragmentation in the Western Australian wheatbelt. *Biological Conservation*. in press.

Hobbs, R.J. and Atkins, L. (1988) Effects of disturbance and nutrient addition on native and introduced annuals in plant communities in the Western Australian wheatbelt. *Australian Journal of Ecology* 13:171–179.

Hobbs, R.J. and Atkins, L. (1990) Fire-related dynamics of a *Banksia* woodland in southwestern Western Australia. *Australian Journal of Botany* 38:97–110.

Hobbs, R.J. and Atkins L. (1991) Interactions between annual and perennial vegetation components in a Western Australian wheatbelt reserve. *Journal of Vegetation Science* 2:643–654.

Hobbs, R.J. and Hopkins, A.J.M. (1990) From frontier to fragments: European impact on Australia's vegetation. *Proceedings of the Ecological Society of Australia* 16:94–114.

Hobbs, R.J. and Mooney, H.A. (1992) Effects of episodic rainfall events on Mediterranean-climate ecosystems. In J. Roy, J. Aronson, and F. di Castri (eds.) *Timescales of Biological Responses to Water Constraints*. Springer-Verlag, New York. (In press).

Hobbs, R.J., Wallace, J.F., and Campbell, N.A. (1989) Classification of vegetation in the Western Australian wheatbelt using Landsat MSS data. *Vegetatio* 80:91–105.

Hopper, S.D., Campbell, N.A., and Moran, G.F. (1982) *Eucalyptus caesia*, a rare mallee of granite rocks from southwestern Australia. In R.H. Groves and W.D.L. Ride (eds.) *Species at Risk: Research in Australia*. Australian Academy of Science, Canberra. pp. 46–61.

Hopper, S.D., van Leeuwen, S., Brown, A.P., and Patrick, S.J. (1990) *Western Australia's Endangered Flora and Other Plants under Consideration for Declaration*. Department of Conservation and Land Management, Perth.

Janzen, D.H. (1983) No park is an island: Increase in interference from outside as park size decreases. *Oikos* 41:402–410.

Janzen, D.H. (1986) The eternal external threat. In M.E. Soulé (ed.) *Conservation Biology. The Science of Scarcity and Diversity*. Sinauer, Sunderland, Mass. pp. 286–303.

Kinnear, J.E., Onus, M.L., and Bromilow, R.N. (1988) Fox control and rock-wallaby population dynamics. *Australian Wildlife Research* 15:435–450.

Kitchener, D.J. and Chapman, A. (1980) Mammals of Yorkrakine Rock, East Yorkrakine, and North Bungulla Nature Reserves. *Records of the Western Australian Museum Supplement* 12:49–53.

Kitchener, D.J. and Vicker, E. (1981) *Catalogue of Modern Mammals in the Western Australian Museum 1985–1981*. Western Australian Museum, Perth.

Kitchener, D.J., Chapman, A., Dell, J., Muir, B.G., and Palmer, M. (1980a) Conservation value for mammals of reserves in the Western Australian wheatbelt. *Biological Conservation* 18:179–207.

Kitchener, D.J., Chapman, A., Dell, J., Muir, B.G., and Palmer, M. (1980b) Lizard assemblage and reserve size and structure in the Western Australian wheatbelt – Some implications for conservation. *Biological Conservation* 17:25–62.

Kitchener, D.J., Dell, J., Muir, B.G., and Palmer, M. (1982) Birds in Western Australian wheatbelt reserves – Implications for conservation. *Biological Conservation* 22:127–163.

Lamont, B.B., Hopkins, A.J.M., and Hnatiuk, R.J. (1984) The flora – Composition, diversity and origins. In J.S. Pate and J.S. Beard (eds.) *Kwongan. Plant Life of the Sandplain.* University of Western Australia Press, Nedlands. pp. 27–50.

Landsberg, J., Morse, J., and Khanna, P. (1990) Tree dieback and insect dynamics in remnants of native woodlands on farms. *Proceedings of the Ecological Society of Australia* 16:149–165.

Lange, R.T. and Graham, C.R. (1983) Rabbits and the failure of regeneration in Australian arid zone Acacia. *Australian Journal of Ecology* 8:377–382.

Leake, B.W. (1962) *Eastern Wheatbelt Wildlife.* B.W. Leake, Perth.

Lee, K.E. (1985) *Earthworms: Their Ecology and Relationships with Soils and Land Use.* Academic Press, Sydney.

Leigh, J., Boden, R., and Briggs, J. (1984) *Extinct and Endangered Plants of Australia.* MacMillan, Melbourne.

Leigh, J.H., Wimbush, D.J., Wood, D.H., Holgate, M.D., Slee, A.V., Stanger, M.G., and Forrester, R.I. (1987) Effects of rabbit grazing and fire on a subalpine environment. I. Herbaceous and shrubby vegetation. *Australian Journal of Botany* 35:433–464.

Lobry de Bruyn, L.A. (1990) *The Role of Ants and Termites in Modifying Soil Properties in Naturally Vegetated and Agricultural Environments.* Ph.D. Thesis, University of Western Australia.

Lobry de Bruyn, L.A. and Conacher, A.J. (1990) The role of termites and ants in soil modification: A review. *Australian Journal of Soil Research* 28:55–93.

Loney, B. (1990) Direct seeding in Western Australia – Its status and future. In *Sowing the Seeds. Direct Seeding and Natural Regeneration Conference Proceedings.* Greening Australia, Deakin, Australian Capital Territory. pp. 103–112.

Lovejoy, T.E., Bierregaard, B.O., Jr., Rylands, A.B., Malcolm, J.R., Quintela, C.E., Harper, L.H., Brown, K.S., Jr., Powell, A.H., Schubart, H.O.R., and Hays, M.B. (1986) Edge and other effects of isolation on Amazon forest fragments. In M.E. Soulé (ed.) *Conservation Biology. The Science of Scarcity and Diversity.* Sinauer, Sunderland, Mass. pp. 257–285.

Lynch, J.F. and Saunders, D.A. (1991) Responses of bird species to habitat fragmentation in the wheatbelt of Western Australia. In D.A. Saunders and R.J. Hobbs (eds.) *Nature Conservation 2: The Role of Corridors.* Surrey Beatty and Sons, Chipping Norton, NSW. pp. 143–158.

MacDonald, I.A.W., Loope, L.L., Usher, M.B., and Harmann, O. (1989) Wildlife conservation and the invasion of nature reserves by introduced species. In J. Drake, F. di Castri, R. Groves, F. Kruger, H. Mooney, M. Rejmánek, and M. Williamson (eds.) *Biological Invasions: A Global Perspective.* John Wiley and Sons, Chichester. pp. 215–255.

MacKenzie, B.M. and Dexter, A.R. (1987) Physical properties of casts of the earthworm *Aporrectodea rosea. Biology, Fertility and Soils* 5:152–157.

Main, A.R. (1981) Ecosystem theory and management. *Journal of the Royal Society of Western Australia* 64:1–4.

Main, A.R. (1987) Management of remnants of native vegetation – A review of problems and the development of an approach with reference to the wheatbelt of Western Australia. In D.A. Saunders, G.W. Arnold, A.A. Burbidge, and A.J.M. Hopkins (eds.) *Nature Conservation: The Role of Remnants of Native Vegetation*. Surrey Beatty and Sons, Chipping Norton, NSW. pp. 1–13.

Main, B.Y. (1967) *Between Wodjil and Tor*. Jacaranda Press, Brisbane, and Landfall Press, Perth.

Main, B.Y. (1976) *Spiders*. Collins, Sydney.

Main, B.Y. (1987a) Ecological disturbance and conservation of spiders: Implications for biogeographic relics in southwestern Australia. In J.D. Majer (ed.) *The Role of Invertebrates in Conservation and Biological Survey*. Department of Conservation and Land Management, Perth. pp. 89–97.

Main, B.Y. (1987b) Persistence of invertebrates in small areas: Case studies of trapdoor spiders in Western Australia. In D.A. Saunders, G.W. Arnold, A.A. Burbidge, and A.J.M. Hopkins (eds.) *Nature Conservation: The Role of Remnants of Native Vegetation*. Surrey Beatty and Sons, Chipping Norton, NSW. pp. 29–39.

Main, B.Y. (1990) Restoration of biological scenarios: The role of museum collections. *Proceedings of the Ecological Society of Australia* 16:397–409.

Majer, J.D., Walker, T.C., and Berlandier, F. (1987) The role of ants in degraded soils within Dryandra state forest. *Mulga Research Centre Journal* 9:15–16.

Malingreau, J.P. and Tucker, C.J. (1988) Large-scale deforestation in the southeastern Amazon Basin of Brazil. *Ambio* 17:49–55.

McGrath, M. (1985) Fire planning and management: An overview. In J. Ford (ed.) *Fire Ecology and Management in Western Australian Ecosystems*. Western Australian Institute of Technology, Perth. pp. 219–212.

McTainsh, G.H., Burgess, R., and Pitblado, J.R. (1989) Aridity, drought, and dust storms in Australia (1960–84). *Journal of Arid Environments* 16:11–22.

Mooney, H.A., Hamburg, S.P., and Drake, J.A. (1986) The invasions of plants and animals into California. In H.A. Mooney and J.A. Drake (eds.) *The Ecology of Biological Invasions of North America and Hawaii*. Springer-Verlag, New York. pp. 250–272.

Moran, G.F. and Hopper, S.D. (1983) Genetic diversity and the insular population structure of the rare granite rock species, *Eucalyptus caesia*. *Australian Journal of Botany* 31:161–172.

Moran, G.F. and Hopper, S.D. (1987) Conservation of the genetic resources of rare and widespread eucalypts in remnant vegetation. In D.A. Saunders, G.W. Arnold, A.A. Burbidge, and A.J.M. Hopkins (eds.) *Nature Conservation: The Role of Remnants of Native Vegetation*. Surrey Beatty and Sons, Chipping Norton, NSW. pp. 151–162.

Morton, S.R. (1990) The impact of European settlement on the vertebrate animals of arid Australia: A conceptual model. *Proceedings of the Ecological Society of Australia* 16:201–213.

Muir, B.G. (1978) Vegetation of Durokoppin and Kodj Kodjin Nature Reserves. *Records of the Western Australian Museum Supplement* 7:15–48.

Muir, B.G. (1979) Observations of wind-blown superphosphate in native vegetation. *West Australian Naturalist* 14:128–130.

Muir, B.G. (1985) Fire exclusion: A baseline for change? In J. Ford (ed.) *Fire Ecology and Management in Western Australian Ecosystems*. Western Australian Institute of Technology, Perth. pp. 119–128.

Newmark, W.D. (1985) Legal and biotic boundaries of western North American national parks: A problem of congruence. *Biological Conservation* 33:197–208.

Noble, J.C. and Tongway, D.J. (1988) Termite feeding sites and their influence on soil fertility and herbage production in semi-arid *Acacia aneura* communities. In P.P. Stahler (ed.) *Proceedings of the 5th Australasian Conference on Grassland Invertebrate Ecology*. Melbourne University, Melbourne. pp. 236–242.

Palik, B.J. and Murphy, P.G. (1990) Disturbance versus edge effects in sugar-maple/beech forest fragments. *Forest Ecology and Management* 32:187–202.

Panetta, F.D. and Groves, R.H. (1990) Weed management and revegetation programs. *Proceedings of the Ecological Society of Australia* 16:537–543.

Panetta, F.D. and Hopkins, A.J.M. (1991) Weeds in corridors: Invasion and management. In D.A. Saunders and R.J. Hobbs (eds.) *Nature Conservation 2: The Role of Corridors*. Surrey Beatty and Sons, Chipping Norton, NSW. pp. 341–351.

Parker, C.A. (1989) Soil biota and plants in the rehabilitation of degraded agricultural soils. In J.D. Majer (ed.) *Animals in Primary Succession. The Role of Fauna in Reclaimed Lands*. Cambridge University Press, Cambridge. pp. 423–438.

Pate, J.S. and Beard, J.S. (eds.) (1984) *Kwongan. Plant Life of the Sandplain*. University of Western Australia Press, Nedlands.

Paton, T.R. (1978) *The Formation of Soil Material*. Allen and Unwin, London.

Pickett, S.T.A. and Thompson, J.N. (1978) Patch dynamics and the size of nature reserves. *Biological Conservation* 13:27–37.

Ranney, J.W., Bruner, M.C., and Levenson, J.B. (1981) The importance of edge in the structure and dynamics of forest islands. In R.L. Burgess and D.M. Sharpe (eds.) *Forest Island Dynamics in Man-dominated Landscapes*. Springer-Verlag, New York. pp. 67–95.

Reid, D. (1951) A Bunketch bird list. *Western Australian Naturalist* 3:68–71.

Reid, D. (1952) Further Bunketch bird notes. *Western Australian Naturalist* 3:144.

Reville, B.J., Tranter, J.D., and Yorkston, H.D. (1990) Impact of forest clearing on the endangered seabird *Sula abbotti*. *Biological Conservation* 55:111–142.

Richards, J.F. and Tucker, R.P. (eds.) (1988) *World Deforestation in the Twentieth Century*. Duke University Press, Durham, N.C.

Rovira, A.D., Smettem, K.R.J., and Lee, K.E. (1987) Effect of rotation and conservation tillage on earthworms in a red-brown earth under wheat. *Australian Journal of Agricultural Research* 28:829–834.

Sanders, A. (1991) Oral histories documenting changes in wheatbelt wetlands. *Department of Conservation and Land Management Occasional Paper* 2/91.

Saunders, D.A. (1980) Food and movements of the short-billed form of the white-tailed black cockatoo. *Australian Wildlife Research* 7:257–269.

Saunders, D.A. (1989) Changes in the avifauna of a region, district, and remnant as a result of fragmentation of native vegetation: The wheatbelt of Western Australia. A case study. *Biological Conservation* 50:99–135.

Saunders, D.A. (1990) Problems of survival in an intensively cultivated landscape: The case of Carnaby's cockatoo *Calyptorhynchus funereus latirostris*. *Biological Conservation* 54:277–290.

Saunders, D.A. and Curry, P.J. (1990) The impact of agricultural and pastoral industries on birds in the southern half of Western Australia: Past, present and future. *Proceedings of the Ecological Society of Australia* 16:303–321.

Saunders, D.A. and de Rebiera, C.P. (1991). Values of corridors to avian populations in a fragmented landscape. In D.A. Saunders and R.J. Hobbs (eds.) *Nature Conservation 2: The Role of Corridors*. Surrey Beatty and Sons, Chipping Norton. NSW. pp. 221–240.

Saunders, D.A. and Hobbs, R.J. (1989) Corridors for conservation. *New Scientist* 1649:63–68.

Saunders, D.A., Hobbs, R.J., and Margules, C.R. (1991) Biological consequences of ecosystem fragmentation: A review. *Conservation Biology* 5:18–32.

Saunders, D.A., Rowley, I., and Smith, G.T. (1985) The effects of clearing for agriculture on the distribution of cockatoos in the southwest of Western Australia. In A. Keast, H.F. Recher, H. Ford, and D. Saunders (eds.) *Birds of Eucalypt Forests and Woodlands: Ecology, Conservation, Management*. Surrey Beatty and Sons, Chipping Norton, NSW. pp. 309–321.

Saunders, D.A., Arnold, G.W., Burbidge, A.A., and Hopkins, A.J.M. (eds.) (1987) *Nature Conservation: The Role of Remnants of Native Vegetation*. Surrey Beatty and Sons, Chipping Norton, NSW.

Sedgwick, E.H. (1953) Birdlife at Caron. *Western Australian Naturalist* 4:12–20.

Sedgwick, L. (1952) A Gabbin bird list. *Western Australian Naturalist* 3:152–155.

Serventy, D.L. (1954) The recent increase of the rarer native mammals. *Western Australian Naturalist* 4:128–129.

Shortridge, G.C. (1910) An account of the geographical distribution of the marsupials and monotremes of southwest Australia having special reference to the specimens collected during the Balston Expedition of 1904–07. *Proceedings of the Zoological Society London* 1909, 803–848.

Stodart, E. and Parer, I. (1988) *Colonisation of Australia by the Rabbit, Oryctolagus cuniculatus* (L.). CSIRO Division of Wildlife and Ecology, Canberra.

Strahan, R. (ed.) (1983) *The Complete Book of Australian Mammals*. Angus and Robertson and the Australian Museum, Sydney.

Terry, M. (1907) Appendix D. In *Report by the Surveyor General for the Year ending 30th June 1907, Department of Lands and Surveys*. Government Printer, Perth. p. 13.

Terry, M. (1908) Appendix A. In *Report by the Surveyor General for the Year ending 30th June 1908, Department of Lands and Surveys*. Government Printer, Perth. pp. 13–15.

Tisdall, J. (1985) Earthworm activity in irrigated red-brown earths used for annual crops in Victoria. *Australian Journal of Soil Research* 23:291–299.

Twigg, L., Majer, J.D., and Stynes, D.A. (1982) Influence of seed harvesting ants in annual ryegrass and their possible effects on the epidemiology of ryegrass toxicity. In *Proceedings of the 3rd Australasian Conference on Grassland Invertebrate Ecology*. South Australian Government Printer, Adelaide. pp. 234–243.

Tyndale-Biscoe, M. (1990) *Common Dung Beetles in Pastures of Southeastern Australia*. CSIRO, Melbourne.

Usher, M.B. (1988) Biological invasions of nature reserves: A search for generalisations. *Biological Conservation* 44:119–135.

Van Schagen, J.J. (1986) *Ant-Plant Interactions in a Western Australian Wheatbelt Reserve*. Graduate Diploma Thesis, Curtin University.

Van Schagen, J.J., Hobbs, R.J., and Majer, J.D. (1992) Defoliation of trees in roadside corridors and remnant vegetation in the Western Australian wheatbelt. *Journal of the Royal Society of Western Australia*, 75: in press.

Venning, J. (1988) *Growing Trees for Farms, Parks and Roadsides*. Lothian, Melbourne.

Wales, B.A. (1972) Vegetation analysis of north and south edges in a mature oak-hickory forest. *Ecological Monographs* 42:451–471.

Wallace, K.J. and Moore, S.A. (1987) Management of remnant bushland for nature conservation in agricultural areas of southwestern Australia – Operational and planning perspectives. In D.A. Saunders, G.W. Arnold, A.A. Burbidge, and A.J.M. Hopkins (eds.) *Nature Conservation: The Role of Remnants of Native Vegetation*. Surrey Beatty and Sons, Chipping Norton, NSW. pp. 259–268.

Watson, J.A.L., Barrett, R.A., and Green, J.P. (1988) Growth of the mounds of the Australian harvester termite *Drepanotermes perniger* (Froggatt) (Termitinae). *Sociobiology* 14:217–244.

Western, D. (1989) Population, resources, and environment in the twenty-first century. In D. Western and M.C. Pearl (eds.) *Conservation for the Twenty-first Century*. Oxford University Press, New York. pp. 11–25.

Whelan, R. and Main, A.R. (1979) Insect grazing and post-fire plant succession in southwestern Australian woodland. *Australian Journal of Ecology* 4:387–398.

White, S.R. (1951) Observations of the fat-tailed dunnart in captivity. *Western Australian Naturalist* 3:1–6.

Wilcove, D.S. and May, R.M. (1986) National park boundaries and ecological realities. *Nature* 324:206–207.

Wilcove, D.S., McLellan, C.H., and Dobson, A.P. (1986) Habitat fragmentation in the temperate zone. In M.E. Soulé (ed.) *Conservation Biology. The Science of Scarcity and Diversity*. Sinauer, Sunderland, Mass. pp. 237–256.

Wilson, A.D. (1990) The effect of grazing on Australian ecosystems. *Proceedings of the Ecological Society of Australia* 16:235–244.

Wycherly, P. (1984) People, fire, and weeds: Can the vicious spiral be broken? In S.A. Moore (ed.) *The Management of Small Bush Areas in the Perth Metropolitan Region*. Department of Fisheries and Wildlife, Perth. pp. 11–17.

Chapter 5

Changes in Soil Properties

R.A. NULSEN

5.1 Introduction

Prior to the European settlement of Western Australia in 1829, the soils were subject only to natural weathering processes. They were generally covered with stable, dominantly perennial vegetation and only traversed by soft-footed animals. The only "tillage" was done by burrowing vertebrates and invertebrates. By modern agricultural standards, the soils were nutritionally infertile, being particularly deficient in phosphate, nitrogen, and the plant trace elements copper, zinc, and molybdenum. Nutrient recycling via plants and animals was sufficient to support natural annual biomass production, and the native vegetation used almost all the rain that infiltrated the soil surface. Around Kellerberrin, this pristine environment was disturbed in the 1880s with the granting of pastoral leases and the introduction of hooved animals, i.e., sheep and horses (see B. Y. Main, Chapter 3). Thus began the degradation of the soils of the region brought about by European-style agricultural practices.

Most farmers love their land; they do not set out to destroy the basic resource which gives them their livelihood. However, in deriving a living from the land, farmers in southwestern Australia often imposed what, in hindsight, was not a benign farming system. The result of this system, alien to the old weathered landscape, has been a very rapid change in the basic properties of the soil. To understand why the soil properties have changed, it is necessary to appreciate both the nature of the farming systems imposed on the land and why they were adopted.

The first 20 years of productive activity in the region were limited to sheep grazing the native vegetation. Shepherds tended the flocks and learned by bitter experience to keep the animals away from areas containing poisonous native plants. The presence of York road poison (*Gastrolobium calycinum*), rock poison (*G. callistachys*), box poison (*G. parviflorum*), and Wodjil poison (*G. floribundum*) restricted the areas of natural bush available for grazing. While the productivity of the native vegetation was low and, consequently, stocking rates were low, the presence of the hooved sheep disturbed the natural protective crust on the surface of the fragile soils. Sheep tracks caused some concentration of water flow, resulting in minor rilling and soil transport. So, even at this early stage of development, with an apparently benign land use system, there was a disturbance of the soil surface structure and the beginnings of soil erosion.

Agricultural development began in 1901, with a major expansion in the five years following 1908. Pioneers concentrated on clearing and cultivating the more clayey-textured soils that were naturally vegetated with salmon gum (*Eucalyptus salmonophloia*), gimlet (*E. salubris*), and morrell (*E. longicornis* and *E. melanoxylon*). While these soils naturally supported less grasses than the soils supporting York gum (*E. loxophleba*) and jam (*Acacia acuminata*), they produced good cereal and hay crops. The typical cropping rotation was a cereal crop, grown with the addition of 70 to 100 $kg\,ha^{-1}$ of superphosphate, followed by a volunteer pasture which was grazed, and then bare-fallowed in its second year in preparation for the following cereal crop.

Production from the lighter-textured yellow sands and loamy sands, which naturally supported an *Acacia* community known as wodjil, was disappointing even after the addition of superphosphate. These soils were subsequently found to be deficient in nitrogen, potassium, copper, zinc, and sometimes molybdenum and also to have a low pH at depth.

A three-year rotation of crop, ley pasture, and fallow was sometimes extended to four or five years by increasing the volunteer pasture phase, but it did not maintain soil fertility and crop yields. The volunteer pastures had very little legume component and were unproductive. Invasion of the volunteer pastures by species such as capeweed (*Arctotheca calendula*), wild turnip (*Brassica tornefortii*), and wild radish (*Raphanus raphanistrum*), which were either intentionally or accidentally introduced from overseas, marginally improved pasture productivity but also increased the intensity of cultivation required to control them for the cropping phase. In the 1950s, annual ryegrass (*Lolium rigidum*) was sown to improve pasture productivity. This it did, but it also proved to be a damaging weed in the cereal phase, requiring more intensive cultivation to control it.

The establishment of the ley-farming system in the 1950s resulted from a combination of high wool prices, and hence a reluctance by farmers to leave land fallow for a season; the development of early flowering cultivars of subterranean clover (*Trifolium subterraneum*), which was suited to the light-to medium-textured soils; the release of 'Cyprus' barrel medic (*Medicago trunculata*), which was suited to the heavier and alkaline soils; and the increase in the power and size of farm machinery. Improved machinery meant that cultivation to control weeds and prepare the seedbed could be done quickly and, hence, the fallow period was not necessary. The typical rotation during this period was a cereal crop, followed by two years of grazing on a legume-based pasture. Introduction of legumes into the pasture at least doubled the stock carrying capacity of the farms, and the increase in soil nitrogen lifted crop yields.

Ley farming appeared to be a stable farming system; however, the impact of the system on the soils was very deleterious. Continued multiple cultivations of the heavy soils, combined with the burning of the cereal stubbles to control plant diseases, destroyed the structure of the soil. High stocking rates maintained in dry years forced the sheep to dig for clover burr to survive; the detached soil was very susceptible to wind erosion. The introduction of pasture legumes caused a decline in soil pH, particularly on the lighter-textured soils.

Further major changes to the farming systems of the area began in the 1970s when early flowering, nonshattering pod selections of lupin (*Lupinus angustifolius* and *L. alba*) were released. Lupins were ideally suited to the yellow sands and loamy sands in a two-year rotation with cereals. Despite the significant benefits derived from increased soil nitrogen and organic matter, adverse aspects of this rotation have been increasing acidity and an increase in water repellence due to incorporation of the lupin stubble.

The last decade has seen the introduction of peas (*Pisum sativum*) as a crop for the heavier-textured soils. Management of the fragile pea stubble requires great care to prevent wind erosion. However, the most significant change to the farming systems in the 1980s has been the increased use of herbicides for weed control, thus allowing a reduction in the number of cultivations necessary to sow a crop. Combined with improvements in machinery, which allow stubble to be incorporated rather than burnt, and the greater attention being given to farming by soil type, the wheel is turning towards a farming system which is reconstructing the soil resource.

5.2 Land Degradation

In the past, agriculture was the mainstay of the economy of Western Australia and continues to make a very significant contribution to the economy (Table 5.1). However, the nonmonetary price that has been paid to achieve this gross value of production of $A3.5 billion has been extensive, and there has often been severe degradation of the soils on which this production is dependent (Table 5.2).

The data in Table 5.2 cannot be summed up to derive a figure for the total area of degraded land because some land areas will be affected by more than one form of degradation. For example, an area of yellow

TABLE 5.1. Principal agricultural statistics for Western Australia for the year 1989/90.

Land Use	
Cleared area ('000 ha)	15,578
Cropped area ('000 ha)	5,172
Sown pasture area ('000 ha)	7,730
Livestock	
Sheep ('000 head)	35,879
Cattle ('000 head)	794
Pigs ('000 head)	275
Gross Value of Production	
Total crops ($A millions)	1,617
Livestock slaughter ($A millions)	457
Livestock products ($A millions)	1,391

TABLE 5.2. Area affected by different forms of land degradation in the agricultural areas of Western Australia, 1989.

Form of Degradation	Area ('000 ha)
Land salinisation	433
Waterlogging	
Crops	500
Pastures	1300
Water erosion	
Sheet and rill	700
Gully	50
Wind erosion	2000
Soil structure decline	1750
Subsoil compaction	8500
Water repellence (area susceptible)	5000
Soil acidification	370

(Data from Western Australian Department of Agriculture, except for salinity, from George 1990b.)

sand or loamy sand can be affected by acidity, have a compacted subsoil, be water-repellent in summer, and be prone to wind erosion if the surface vegetative cover is not maintained. Similarly, a heavy-textured, valley soil can have a structural decline and be subject to waterlogging or can be saline, waterlogged, and subject to water erosion.

Land degradation not only affects the land per se. There are many off-site effects such as dust in the cities from wind-eroded farmland; silting of water supply storages from water erosion; salinisation of rivers which serve as potable water resources; and destruction of wildlife habitats due to increased salinity or eutrophication of the water in rivers and lakes.

While the present picture is one of large scale devastation of the landscape, the real concern is the potential for further degradation if the production systems do not become more conservative of the soil resource. Fortunately, farmers are now very aware of land degradation and are adopting farming systems which will stem the tide and eventually reverse the degradation trend. If, for some reason, be it economic, social, or technical, this does not occur, the consequences will be enormous. For example, the Department of Agriculture has estimated that there is a potential for some 2.4 million ha (or more than 15%) of agricultural land to become affected by salinity, and most of this will occur within the next 30 years.

While 0.5 million ha of cropland is affected by waterlogging in an "average" year (Table 5.2), the area doubles in a "wet" year and is increasing as the vertical drainage capacity of the soils decreases due to compaction, water repellence, and structural decline, thus increasing runoff and seepage to the broad valley floors.

Wind and water erosion can affect virtually all agricultural soils if they are mismanaged. The predominance of sandy-textured surface soils in the State makes them extremely susceptible to wind erosion. Soil structural decline, which can reduce productivity by at least 50 percent, has the potential to affect double the current area. The area of subsoil compaction is not expected to increase significantly because most of the susceptible soils have already been cropped for at least five years. Water repellence is increased by leguminous species. Improved legume-based pastures and the expansion of lupin growing has the potential to increase the water repellence on all light textured soils in the State, estimated at five million hectares.

Soil acidity currently affects between 0.25 and 0.5 million ha. As accelerated acidification is a consequence of agricultural production, most agricultural topsoils acidify more rapidly than nonagricultural soils. Western Australian soils are particularly susceptible to acidification because their sandy topsoils are poorly buffered against changes in pH. It is estimated that some 10 million ha (over 65% of agricultural land) could acidify to the extent of causing a significant productivity decrease over the next 100 years.

5.2.1 Perception of Land Degradation

Community perception of the magnitude and severity of land degradation is strongly influenced by the visibility of particular forms of degradation. Driving through or flying over Western Australia's agricultural areas, one is acutely aware of land salinisation. The bare salt-encrusted scalds, often rimmed by the stark skeletons of dead trees, are highly visible. Equally, the urban dweller becomes highly conscious of wind erosion when the city is enveloped in a summer dust storm. The deep scars resulting from severe water erosion are highly visible as they snake down the hillslope. Other forms of degradation are only visible at certain times of the year. Waterlogging, for instance, will only be visible in winter in Western Australia. What is not readily discernible is the cause of the waterlogging. Is it simply due to poor surface drainage or is it due to reduced infiltration resulting from the breakdown of soil structure?

The "invisible" forms of land degradation can often be of far greater consequence than the visible forms. Subsoil compaction, soil structure decline, soil acidification, and water repellence affect vast areas and yet go unnoticed by the casual observer. It is the invisibility of these forms of degradation that makes them so insidious, often unrecognized, and, therefore, not treated.

While the visibility of the forms of degradation influences people's perception of the relative importance of specific types of degradation, the rate of degradation can affect the farmer's awareness of it. Wind erosion can result from a single wind event and the farmer is very aware that there is a problem. Similarly, water erosion can be the result of a single, high intensity rainfall event. Soil salinity, although it is the expression of the accumulation of small changes that have occurred over many years, can quite quickly affect large areas of land. The rapid appearance and high visibility equate with a high level of awareness of the problem. However, unless detailed sampling and analyses are done, land managers are unlikely to be aware that the pH of their soil is decreasing by 0.1 pH units per year, that the structure of their heavy textured soils is declining, or that the failure of the crop to establish on the sandy soil is due to an increase in water repellence.

The data in Table 5.2 show that the most widespread forms of land degradation in Western Australia are not the highly visible forms which readily invoke an emotive response from the general community and call for action. The real "disintegration" of the land resource is invisible and difficult to identify, and its rectification is not spectacular. I would be surprised if the situation is much different elsewhere in the world.

5.3 Changes in Soil Properties – Causes and Impacts

5.3.1 Soil Acidity

The pH of a soil is a measure of the H^+ ion concentration in the soil solution. Acid soils have a pH below 6 with alkaline soils being greater than 7.5. Plant growth is not affected by the pH itself but by the changes in soil chemistry that can occur when the pH changes.

Soil acidification is a natural process resulting from the weathering of rocks and loss of nutrients by leaching. Plants and microorganisms are an important part of the natural process. Sometimes, under agricultural systems, the rate of decline of pH is increased with resulting induced "soil acidity."

5.3.1.1 pH Measurement

Ideally, the soil chemist would like to measure the pH of the soil solution which is in contact with the soil particles and the plant roots. However, the small quantity of this solution in field soils and the difficulties of extracting it make this method routinely impractical. Normally, the soil pH is determined by measuring the pH of a suspension of one part soil to five parts water. A problem arises with this method because the addition of water to a soil changes the H^+ concentration in the soil solution. In an acid soil, the pH rises. If the soil particles have a negative charge, then, on dilution, more divalent ions will be adsorbed and H^+ ions desorbed, resulting in a lowering of pH. Calcareous soils present a further complication in that the pH is governed by reactions between $CaCO_3$, water, and CO_2. The addition of water causes some dissolution of the $CaCO_3$ and the pH rises, only to fall again as respiration by soil microbes increases the CO_2 content of the solution.

To overcome these problems, soil pH is often measured in a 1:5 soil/dilute $CaCl_2$ suspension. The pH measured in the soil/ calcium chloride suspension will be lower than the pH of the same soil measured in a soil/water suspension. Typically, the calcium chloride determined pH is 0.5 to 0.8 pH units lower than that determined in water, but it can range from nil to 1.8 pH units (Wilson 1984). Thus, in discussing soil acidity, it is very important to know which method was used to determine pH. In this chapter, all pH references are to that determined in a 1:5 soil/water suspension unless stated otherwise.

5.3.1.2 Buffering Capacity of Soils

Buffering is a chemical term referring to the capacity of a system to resist a change in pH when an acid or a base is added. Soils are generally viewed as buffered systems, but they vary greatly in their buffering capacity. If a soil has a pH that is more or less ideal for plant growth, then it is desirable that it be highly resistant to pH changes. A well-buffered soil is undesirable if the pH is too high or too low, since it will require large inputs of soil amendment to effect a change in pH.

5.3.1.3 Causes of Soil Acidity

Soil acidity results when organic and inorganic soluble acids accumulate in the soil faster than they can be neutralized or removed from the system. Dissociation of these acids provides most of the free H^+ in the soil solution. The exchangeable ions H^+ and Al^{3+} also contribute to the free H^+ pool; the exchangeable H^+ makes the contribution through direct ionization and the Al^{3+} through ionization and hydrolysis. The hydrolysis of Al^{3+} ties up the OH^- from water and releases an equivalent amount of H^+. The simplified reaction is:

$$Al^{3+} + H_2O \rightleftharpoons AlOH^{2+} + H^+ \quad (1)$$

The $AlOH^{2+}$ may undergo further hydrolysis to form $Al(OH)_2^+$ and an additional H^+, but $AlOH^{2+}$ has a low solubility and thus its contribution to soil acidity is negligible.

Any practice which increases the free H^+ ion concentration in the soil solution will acidify the soil. A common misconception is that agricultural soils are acidified by the application of superphosphate fertiliser. This perception probably stems from the knowledge that superphosphate is manufactured by treating rock phosphate or guano with sulphuric acid. The addition of phosphate fertiliser can hasten the development of other causes of acidity, such as increasing leguminous pasture growth, leading to more nitrogen leaching as nitrate, which is acidifying. The major agricultural practices which cause soil acidity are the build up of soil organic matter, the addition of nitrogen, either as fertiliser or by growing legumes, the removal of produce, especially legume hay, and the addition of elemental sulphur.

The organic matter contents of the soils in the Kellerberrin area are low. Generally, the organic carbon content of the yellow earths of the Ulva surface (McArthur, Chapter 2) have decreased since clearing. Jarvis and Robson (1983) measured the organic C content of a virgin yellow sand to be 0.79 percent in the top 3 cm while the level in the same soil that had been farmed for 17 years was 0.41 percent. Even in the heavy soils of the Merredin surface, the organic C content after some 50 years of farming was only around 1.5 percent (Hamblin 1982). So, in this region, the build-up of organic C is not a major factor in soil acidification.

The main agricultural produce removed under past and current farming systems has been cereal and lupin grains, which have very little acidifying effect. Legume hay removal is only by animals and a large proportion of the nitrogen consumed is recycled via the urine.

As elemental sulphur is not used, the main cause of increasing acidity is the addition of nitrogen both through applied fertiliser and the growing of legumes.

5.3.1.4 Nitrogen Cycling and Soil Acidification

In a closed ecosystem, where nitrogen is cycling through the soil, plants, and animals, the net impact on acidity is nil. This is because, although the acidifying nitrification reaction (Eq. 2) occurs, the reverse reaction of nitrogen reduction (Eq. 3), which metabolises nitrate, occurs within the plant and soil microorganisms (Helyar 1976).

$$\text{Nitrification } NH_4^+ + 2O_2 \rightarrow NO_3^- + H_2O + 2H^+ \quad (2)$$

$$\text{Denitrification } 2NO_3^- + 2H^+ \xrightarrow[-5H_2O]{+4H_2} N_2O \xrightarrow[-H_2O]{+H_2} N_2 \quad (3)$$

There will only be a net change in acidity if the type and quantity of nitrogen gains in the system are different from the type and quantity of nitrogen losses.

The addition of nitrogenous fertiliser can have a strongly acidifying influence on the soil, depending on the type of fertiliser. When ammonium sulphate is added, the NH_4^+ is rapidly nitrified in neutral or acid soils. The reaction is:

$$(NH_4)_2SO_4 + 4O_2 \rightleftharpoons 2NO_3^- + 4H^+ + 2H_2O + SO_4^{2-} \quad (4)$$

If ammonium nitrate is used, the reaction is:

$$NH_4NO_3 + 2O_2 \rightleftharpoons 2NO_3^- + 2H^+ + H_2O \quad (5)$$

In both reactions, the same amount of NO_3^- is available for leaching or plant uptake, but the $(NH_4)_2SO_4$ produces $4H^+$ compared with $2H^+$ from NH_4NO_3. Thus, $(NH)_4)_2SO_4$ has twice the acidifying potential as NH_4NO_3. Whether or not the potential is realised depends on the buffering capacity of the soil.

The addition of urea, $CO(NH_2)_2$, as a nitrogen fertiliser initially raises the pH because the urea is hydrolyzed to NH_4OH and H_2CO_3. The NH_4OH is more highly ionized than the H_2CO_3 and the pH rises. This trend is reversed when the NH_4^+ undergoes nitrification.

Symbiotic fixation of atmospheric N by the *Rhizobium* sp. of bacteria, which establish in nodules on the roots of leguminous plants, can provide those plants with their nitrogen requirements. When the organic matter of these plants is broken down, the N is released from the organic material in a form available to other plants by the process of mineralisation. Mineralisation is a complex process, but it, like the symbiotic fixation process, results in a net change in the form and quantity of N in the soil and hence depresses pH.

Jarvis and Robson (1983) showed that, on poorly buffered, acidic, sandy soils east of Kellerberrin, when subterranean clover was supplied with $(NH_4)_2SO_4$ as the N source, the soil pH depression was 0.9 pH units, 0.5 pH units when N was supplied only by symbiotic fixation, and

TABLE 5.3. Effects of acidity on availability of plant nutrients.

Nutrient	Effects of Acidity
Aluminium	Plant availability increases below pH 5.0 and can be toxic. Damages roots which may limit ability to take up water and nutrients. Reduces ability of roots to take up phosphate.
Manganese	Plant availability increases below pH 5.5. Toxicities can occur but not known in Western Australia.
Molybdenum	Fertiliser molybdenum quickly converted to unavailable forms in acid soils. Deficiencies can occur in soils below pH 6.0.
Phosphorus	Plant availability can be reduced by effects of aluminium.
Calcium	Definciencies can develop because acid soils may contain only small amounts which are poorly available to plants.
Magnesium	As for calcium. Deficiency occurs in some very acid soils.
Zinc and iron	Highly available to plants. No toxicities occur.
Postassium and copper	Little or no effect.
Organic matter nitrogen, phosphorus, and sulphur	Slows the rate of organic matter breakdown because soil microbes are less active in acid soils.

(Adapted from Porter and Yeates 1984.)

negligible when supplied as $NaNO_3$. Thus, growing clover had about half the acidifying effect as the most damaging inorganic fertiliser.

5.3.1.5 Consequences of Acidification

The impact of soil acidity on agriculture is reduced plant production, due not to the low pH but to the changed availability of plant nutrients in acid soils. The marked increase of available aluminium below pH 5.0 damages root membranes and reduces their ability to take up water and plant nutrients, especially phosphate. The plant uptake of Al can, in itself, be toxic. Plants vary markedly in their tolerance to aluminium. Annual medics (*Medicago* spp.) and barley (*Hordeum vulgare*) are extremely sensitive. Wheat (*Triticum aestivum*) and subterranean clover are moderately sensitive, while the narrow leaf lupin (*Lupinus angustifolius*), oat (*Avena sativa*), and triticale (*Triticum* × *Secale*) are tolerant.

The effect of reduced soil pH on the availability of plant nutrients, is summarised in Table 5.3. In a farming system highly dependent on legume-based pastures, one of the major impacts of acidity is on the rhizobia. At low pH, their survival and ability to form nodules and fix nitrogen are reduced. *Rhizobium* spp. vary in their tolerance to acidity. *Rhizobium meliloti*, which nodulates medics, will not tolerate a pH below about 6.0. In contrast, *Rhizobium lupinii*, which nodulates lupins, will tolerate a soil pH down to 4.3 (Howieson and Ewing 1984).

TABLE 5.4. Yield response of wheat ($t\,ha^{-1}$) to applied lime and molybdenum.

	Lime Applied ($t\,ha^{-1}$)		
	0	2	4
No molybdenum	1.05	1.20	1.40
Molybdenum	1.38	1.37	1.39

TABLE 5.5. Effect of wheat yield on liming acid sandy soil profiles to 1.6 m.

	Grain Yield ($t\,ha^{-1}$)	
	Site 1	Site 2
Drums without lime	0.89	0.64
Drums with lime	1.50	1.76
Adjacent, undisturbed plots	1.15	0.48

(From Porter and Wilson 1984.)

5.3.1.6 Responses to Treatment

The effect of reduced soil pH on agricultural production can only be assessed by examining the yield responses to treatments which address the problem. Lime application, to neutralize the acidity, is the most common treatment. However, liming can have adverse effects such as:

- inducing magnesium deficiency because the calcium from lime decreases magnesium uptake by plants;
- inducing manganese deficiency;
- inducing copper deficiency in ruminant animals grazing the pastures because the pastures have a high molybdenum content due to its increased availability in the soil.

In many acid sandy soils of Western Australia's wheatbelt, the most important effect of applying lime is to increase the availability of molybdenum. As it is cheaper to apply molybdenum than lime, this is often a preferred treatment for increasing cereal grain yields. Table 5.4, derived from data given by Porter and Wilson (1984), illustrates the relative impact of lime and molybdenum.

Inherently acid sandy soils in the Kellerberrin region have low pH at depth, typically 4.5 at 30 to 60 cm and 4.0 at 100 cm. The mechanics of liming these soils has not yet been economically resolved, but the response to lime is dramatic, as shown in Table 5.5.

Liming responses can be very variable, taking time to develop and being more manifest in some seasons than in others. Glencross and Clarke (1984) applied lime, molybdenum, and lime plus molybdenum to

Table 5.6. Effect of lime and molybdenum applied in 1981 on pasture growth, expressed relative to the control (= 100).

	Spring 1981	Winter 1982	Spring 1982	Winter 1983	Spring 1983
Lime	82	116	142	147	115
Molybdenum	91	104	114	110	99
Lime + molybdenum	87	114	142	163	111
Nil	100	100	100	100	100

a subterranean clover-based pasture growing on a brown, loamy sand with pH 5.2 in the top 10 cm. Treatments were applied in 1981, but the response did not show until 1982 and in one instance disappeared in the spring of 1983 (Table 5.6).

Other methods of overcoming the productivity decline due to low pH include breeding acid-tolerant cultivars and isolating acid-tolerant rhizobia. While these strategies may enable productivity to be maintained, they do not redress the fact that agriculture has changed an important soil property. To ensure conservation of the soil, it is preferable that the decline in pH be reversed.

5.3.2 Soil Structure

Soil structure refers to the way in which the primary soil particles (sand, silt, and clay) are assembled into aggregates. Aggregation depends on the cohesive nature of the finer particles, particularly the clay fraction, the organisation of the particles, and the abundance of organic and inorganic materials which serve to bind the particles together. The soil aggregates, or "peds," are generally described and classified by their size and shape and the nature of their faces and edges. There are four main structural types: laminar, in which the natural cracking is horizontal (sometimes referred to as platy); prismatic, in which the vertical cracking is stronger than the horizontal and the ped faces are generally smooth and flat; blocky, in which the vertical and horizontal cracking are equally well developed; and spheroidal, in which the ped surfaces are rounded and do not fit snugly with adjacent peds.

Peds themselves can have a well-developed structure, and the large unit can be broken down into smaller clods or crumbs under gentle mechanical pressure. In turn, the crumbs may be broken down to microcrumbs. Crumbs built out of microcrumbs are said to have well developed microstructure. In fine-textured soil, microstructure is related to tilth, which is the physical state of the soil as it relates to plant growth; it is not a natural state of the soil, but one which is developed by various tillage practices. Tilth should not be confused with structure.

While structure is described in terms of the aggregation of soil particles, the important aspect of structure, in terms of gaseous and liquid movement through the soil, is the size and distribution of pores. Structural pores are created in undisturbed soils by plant roots and by soil fauna. Plant roots grow by forcing the root tip into small pores in the soil ped. As the root diameter increases, the pore is enlarged, causing the primary soil particles to move relative to one another. Drying of the soil when the root extracts water may cause the soil to crack and thus create further voids for new roots to penetrate. The greater the density of roots, the more interlinked will be the system of pores, cracks, and voids in the soil, and thus the liquids and gases can move more freely through the soil mass.

Soil fauna can contribute significantly to the formation of structural pores as they burrow through the soil (see also Hobbs et al., Chapter 4). In many temperate climate soils, earthworms are the main soil burrowing species (Payne 1988). In the virgin soils of the Kellerberrin area, Abbott et al. (1979) reported that ants and beetles were the main large soil fauna. Their winter population of 845 per square metre, which was almost twice the summer population, shows that, even in an apparently harsh environment, soil fauna can be abundant.

Structural pores are of little value if their presence is only transient. Pores can only be stable against wetting if some agent holds the soil particles together. Clay particles, bonded together in granules by electrostatic forces, will help keep sand particles apart on wetting, provided the granules retain their integrity. They will retain their integrity (i.e., remain flocculated) only if the concentration of salts in the soil solution is high enough. The concentration required depends on the type of ions on the clay exchange sites. A much higher concentration is needed when sodium ions dominate the exchange sites than when calcium ions are dominant.

Soil organic matter also plays a vital role in stabilising soil structure. Various components of the organic matter act as binding agents which essentially "glue" the mineral particles together. Tisdall and Oades (1982) classified the binding agents into transient, temporary, and persistent on the basis of their longevity. Transient binding agents are rapidly decomposed by microorganisms. Consisting largely of polysaccharides produced by microbial activity and roots, they are produced quickly and decompose within weeks or months. Temporary binding agents are roots and fungal hyphae, particularly the hyphae of the vesicular-arbuscular mycorrhizae. These binding agents build up in the soil over weeks and months as the roots and associated hyphal systems develop and can persist for years. Persistent binding agents are derived from the resistant fragments of roots, hyphae, and bacterial cells and consist of aromatic humic material associated with amorphous iron, aluminium, and aluminosilicates. The chemistry of persistent binding agents is not well understood, but it is thought that they derive part of their persistence by forming complexes with clays and polyvalent metals. Tisdall and Oades (1982) suggested that the transient and persistent binding agents dominate in stabilizing microaggregates, while the roots and hyphae of the temporary binding agents are largely responsible for the stability of macroaggregates.

5.3.2.1 Measurement of Soil Structure

There is no single measurement of soil structure; a number of physical and chemical measurements are taken to give some quantitative characterisation of structure. Traditionally, the characterisation of structure has been in terms of degree of aggregation, aggregate stability, and bulk density. Quirk (1978) considered that, while such measurements are significant in interpreting the behaviour of a soil, the real characteristic of structure is the pore size distribution. Measurement of pore size distribution is not yet a routine task, and structure is largely characterised by bulk density, water stable aggregation, some measure of the porosity, such as infiltration rate, sorptivity or air permeability, a measure of soil strength, such as modulus of rupture or penetrometer resistance, and a measure of the organic matter content.

As most researchers are interested in structural change, rather than in quantifying structure *per se*, the measurement of secondary characteristics appears adequate. In the following sections, it will be seen that there can be relatively large changes in all these measurements when the soil is subject to different cultural and agronomic practices.

5.3.2.2 Deterioration of Soil Structure – Causes

Any action which disturbs the soil, causes the clays to deflocculate (disperse), or reduces the organic matter content has the potential to destroy the porous structure of the soil and reduce its stability under wetting and drying regimes. Cultivation is often singled out as the prime agent of soil structural decline. However, one must be careful in making this allegation because the culprit is not "cultivation" but the type of cultivation and the farming system practised which were obviously not suited to the particular soil type.

In the Western Australian wheatbelt, the impact of agricultural systems on soil structure was illustrated by Stoneman (1962) who compared the structural characteristics of virgin and cultivated soils. He showed that, after only nine years of cultivation, there were significant reductions in water stable aggregation, increases in bulk density, and reductions in noncapillary porosity (Table 5.7).

Organic carbon showed a significant decrease at three locations and a significant increase at one location. Low organic matter contents in the virgin state is a characteristic of these soils and contributes to the

Table 5.7. Effects of cultivation on some physical characteristics of Western Australian soils.

Soil Type (location)	Condition	Organic Carbon (%)	Water Stable Aggregates (% > 1 mm)	Bulk Density ($kg\,m^{-3}$)	Noncapillary Porosity ($m^3\,m^{-3}$)
Loam	Virgin	1.33*	19.64	1240*	0.203*
(Merredin)	Cleared 40 yrs	0.85	18.50	1680	0.093
Sandy loam	Virgin	0.85*	56.48*	1250*	0.236*
(Merredin)	Cleared 9 yrs	1.01	16.47	1390	0.214
Silt loam	Virgin	2.79*	40.40*	1050*	0.185*
(Corrigin)	Cleared 37 yrs	1.19	11.29	1380	0.082
Sandy loam	Virgin	1.34*	37.66*	1570*	0.098*
(Quairading)	Cleared 44 yrs	0.99	15.54	1780	0.040

*Denotes significant differences at $P \leq 0.05$.
(After Stoneman 1962.)

fragility of their structure. Even small losses of organic matter will predispose the soils to structural deterioration.

The data in Table 5.7 only indicate that agricultural development has caused structural decline; they do not indicate which component of the agricultural system was largely responsible.

In a study of structural decline, taken to be represented by the level of water stable aggregation, Stoneman (1973) showed that the cropping phase in a cereal crop-subterranean pasture rotation caused the structural decline (Fig. 5.1). The pasture phase was able to reverse the structural degradation caused by the crop. Stoneman considered that the cultivation associated with the cropping was responsible for the structural decline.

Stoneman (1973) also showed the impact that stock can have on soil structure if they tread wet soil (Fig. 5.2). However, even with the degrading effect of stock, there was an increase in structural stability in the pasture phase, albeit not as rapid as without the stock.

5.3.2.3 Consequences of Structural Decline

Wet, boggy soils with large clods and water ponding on the surface are indicative of structural degradation. The soils are difficult to work when wet, as machinery becomes bogged, and, when dry, the hard-setting surface is difficult to break open. Increased fuel consumption, machinery wear, reduced capacity for timely operations, delayed or decreased crop emergence, and decreased yield can all be significant costs due to soil structural decline.

The detrimental effects of structural degradation on plant growth are due to a lack of aeration and increased waterlogging, due to the collapse

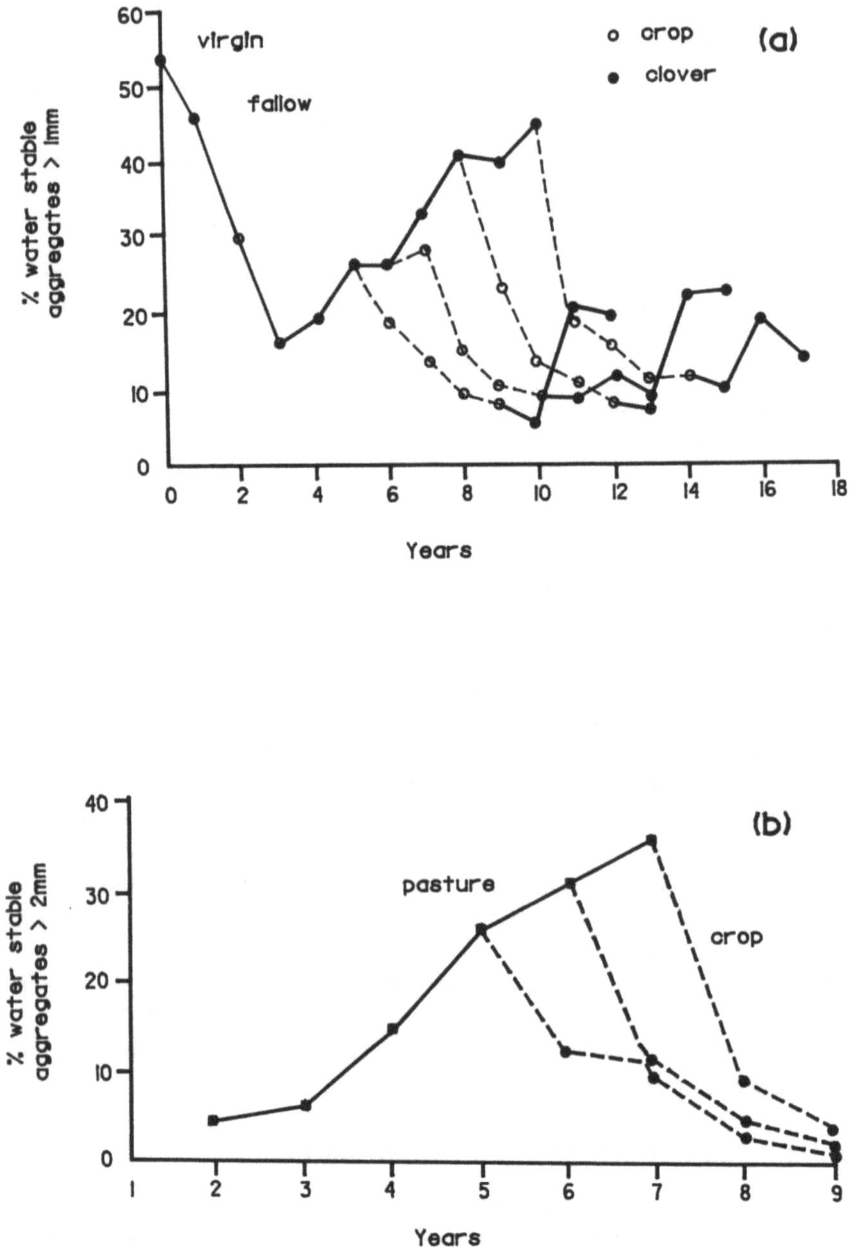

FIGURE 5.1. Changes in water stable aggregation of surface soil under crop and pasture. (a) Differing rotations since clearing on a loamy sand. Year 0 was uncleared, virgin soil; year 1 was fallow before a crop in year 2. (b) Pasture followed by crop on a loam. (After Stoneman 1973.)

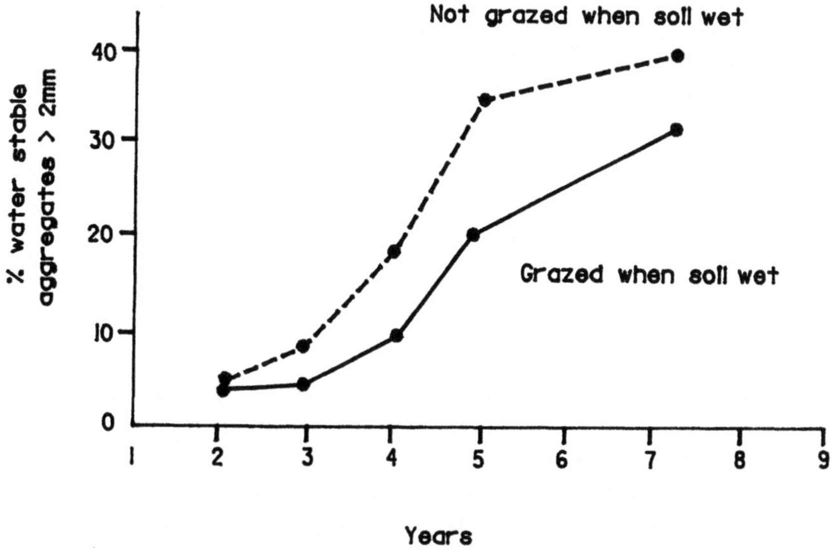

FIGURE 5.2. Effect of sheep treading wet soil on water stable aggregation. (After Stoneman 1973.)

of the soil pores, and to the increased mechanical resistance confronting the emerging roots and shoots. The impact of these effects on plant growth are best illustrated by the response of plants when structure is restored.

5.3.2.4 Responses to Treatment

Structural degradation can be reversed by changing the cations on the exchange sites of the clay, by increasing the organic matter content of the soil, and by changing the tillage practices. Not all soils respond to any particular treatment, and it is difficult to ascribe a plant response to structural change *per se*.

The cation exchange sites on the clays may be changed by using synthetic soil conditioners or by applying gypsum.

Synthetic soil conditioners, which create a stable soil structure without adding nutrients or changing the soil's water holding capacity, have been tested and shown to give wheat yield increases up to 25 percent on the red-brown earths of the Merredin surface (Stoneman 1962). While they are effective, these agents are prohibitively expensive and thus not applicable to dryland agriculture.

Gypsum is used extensively throughout the world to ameliorate soil structural problems. Initial testing in Western Australia gave disappointing results. Stoneman (1973) only got a wheat yield response to gypsum application in three out of 23 trials. He also observed a very

Table 5.8. Average wheat grain yields ($t\,ha^{-1}$) for four tillage treatments with and without gypsum application of $4.7\,t\,ha^{-1}$ in 1983.

Tillage Treatment	1981–86 No Gypsum	1983–86	
		No Gypsum	Gypsum
Triple disc drill	1.27	1.35	1.51
Scarifier/triple disc drill	1.21	1.29	1.47
Direct drill combine	1.38	1.47	1.53
District practice	0.79	0.92	1.36

limited response in pasture production to gypsum application. Despite the lack of plant response, there were responses in the soil physical condition measured in terms of water stable aggregation, dispersion, modulus of rupture, and penetrability. Stoneman concluded that "... the use of gypsum cannot be recommended as an economic procedure for improving yields on most agricultural soils showing physical deterioration."

Although phenoxy herbicides, which selectively control broadleaf weeds in cereal crops, were developed in 1947, it was not until the mid-1960s, with the advent of a greater range of herbicides, that chemical ploughing became a practical reality. By judiciously using herbicides to control weeds, the option to reduce cultivation and to incorporate organic matter from crop stubbles was dependent only on having the appropriate machinery.

Without chemicals for weed control, practical options for reduced cultivation would not be available, and, hence, farmers would be able to do little to restore structure to degraded soils. Critics of the use of chemicals in agriculture should be cautious in their criticism and ensure that they fully understand and appreciate the consequences of banning the use of chemicals.

Structurally degraded soils respond well to combinations of reduced tillage and gypsum application. Jarvis (1988), in summarizing six years of trial data on the red-brown earths at Merredin, showed that the tillage treatments which caused the least soil disturbance and gave the greatest increase in water stable aggregation also gave the greatest yield response. Table 5.8 shows that there was also an interaction between tillage and gypsum. The data in Table 5.8 are taken from Jarvis (1988) and give the average cereal yields for the six years and for the four years after gypsum was applied in 1983. The triple disc drill was the least disturbing tillage treatment and district practice, consisting of two scarifyings and combine seeding, the most disturbing. Jarvis noted that gypsum consistently improved the yield compared to the district practice treatment, but that it only increased yields relative to other treatments in some years.

TABLE 5.9. Soil response to five years of tillage treatment.

Tillage Treatments	Modulus of Rupture	Water Stable Aggregates (% > 2 mm)	Hydraulic Conductivity ($\times 10^{-7}$ ms^{-1})
Triple disc drill	220 c*	8.19 a	1.27 a
Scarifier/triple disc drill	354 ab	2.71 c	–
Direct drill combine	246 bc	4.58 b	0.84 b
District practice	463 a	2.93 c	0.09 a

*Means with a common lowercase letter are not significantly different at $P < 0.05$ (Duncan's multiple range test).
(After Jarvis et al. 1986.)

Tillage treatment directly affects soil structure, and reduced cultivation can result in significant improvements in structure. Table 5.9 shows the results of four tillage treatments after five years on a red-brown earth at Merredin. The least disturbing treatment had the lowest soil strength, measured as modulus of rupture, the greatest aggregate stability, and the highest saturated hydraulic conductivity, which reflects the porosity.

5.3.3 Soil Compaction

Soil compaction occurs when some compressive force brings the solid soil particles closer together. This decreases the bulk volume of the soil and, hence, increases the bulk density. By increasing the bulk density, there is a reduction in the porosity and a change in the pore size distribution. Soil compaction is really one aspect of change in soil structure. While soil structure, as discussed in the previous section, can change as a result of tillage, the distinction between structural change and compaction is usually made on the basis that structural decline occurs as a result of shearing forces whereas compaction results from compressive forces.

Compaction affects agricultural soils throughout the world. The area affected is increasing as the weight of agricultural machinery increases and with the increasing mechanisation in developing countries (Soane 1981). In Australia, where the percentage of tractors with power greater than 75 kW increased from 2 percent in 1970 to 19 percent in 1988 (Makin 1989), compaction limits production in areas ranging from the cotton-growing areas in Queensland (McGarry 1990) and the cereal and subterranean clover areas in northern Victoria (Reeves et al. 1984) to the earthy sands, massive yellow earths, and duplex soils of the Western Australian wheatbelt (Henderson et al. 1988).

In the Kellerberrin area, compaction is mainly a problem associated with the earthy sands and yellow earths of the Ulva surface and, to a lesser extent, the yellow duplex soils of the Collgar surface (McArthur,

Chapter 2). These sandplain soils develop a compacted layer, or traffic pan, at 20 to 30 cm depth (Hamblin and Tennant 1979). Compaction also occurs in the heavier textured soils, but there is not a deep ripping response in these soils for current agricultural crops and pastures.

5.3.3.1 Measurement of Soil Compaction

In sandy soils with no visible macrostructure, it is difficult to identify traffic pans without taking detailed measurements. It is obvious that the compacted layer will have an increase in bulk density and a reduction in the mean pore size. Bulk density measurements are relatively easy to take, but the sample depth must relate to the anticipated thickness of the compacted layer.

Pore volume can be calculated from the dry bulk density, but this gives no indication of the pore size distribution which, as stated by Quirk (1978), is the determinant of the liquid and gas transmissive properties of the soil. As pore size distribution is not easy to determine, a measure of the hydraulic conductivity of the soil can be used to reflect the altered porous state. Measurements of hydraulic conductivity using the disc permeameter, described by Clothier and White (1981), can be made on successive layers of the soil to determine the transmission properties of each layer.

The penetrometer is a simple field technique for detecting compacted layers. Penetrometer resistance is affected by the soil particle shape and size distribution, pore size distribution, soil water content, and nature of the interparticle bonding (Gill 1968). It is therefore necessary to develop a calibration for each soil type, relating penetrometer resistance to the parameter of interest, be it crop growth, trafficability, or workability. Henderson et al. (1988), in a study of 13 Western Australian sandplain soils, concluded that penetrometer resistance could be used in conjunction with other environmental inputs, such as rainfall, to develop probability models relating a single penetrometer measurement to crop growth.

5.3.3.2 Causes of Soil Compaction

Soil compaction is caused by any traffic on the soil surface and by the passage of some tillage implements through the soil. Soane et al. (1982) commented that, even in the days of horse ploughing, compacted layers were observed immediately below the depth of ploughing. The depth of the compacted layer and the degree of compaction depend on the contact pressure of the vehicle, the width of the tyres, and the number of passes.

After a single pass of a small 1.6 t tractor over recently cultivated sandy loam soil, Soane (1970) found that the soil surface was depressed 75 mm and the dry bulk density at 150 mm increased from 950 kg m^{-3}

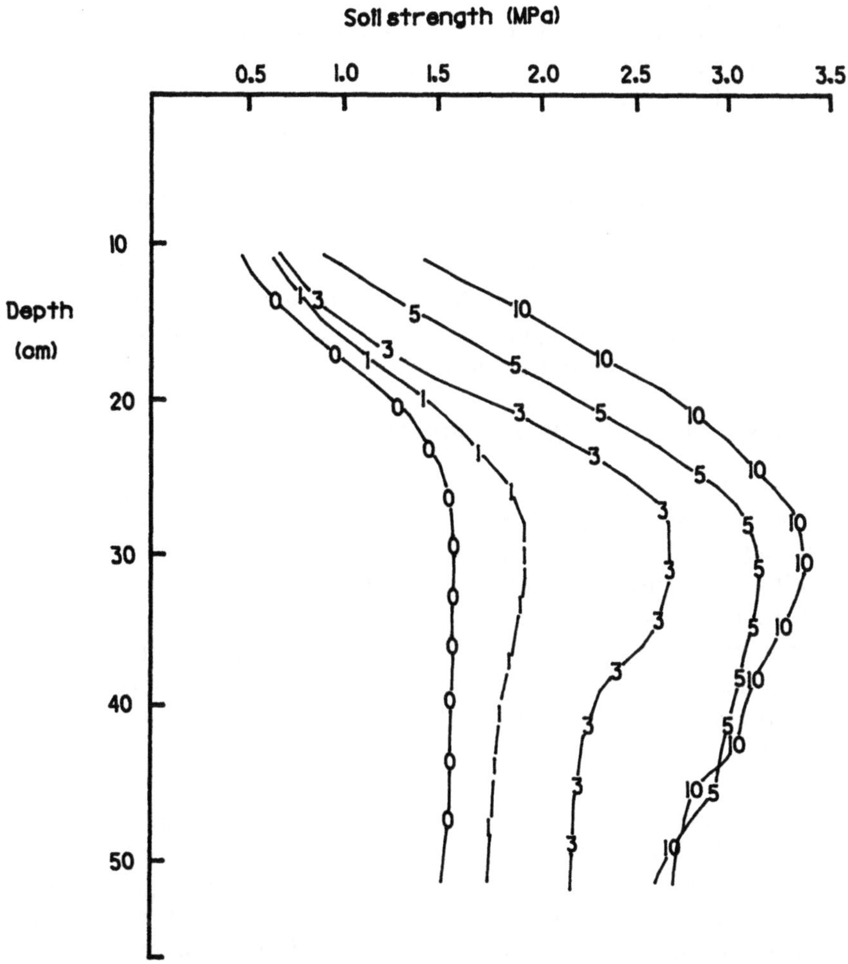

FIGURE 5.3. Effect of the number of tractor wheel passes (0,1,3,5, and 10) on soil strength with depth. (After Jarvis and Porritt 1985.)

to 1400 kg m^{-3}. Jarvis and Porritt (1985) found bulk density to be unreliable in demonstrating the existence of a compaction pan. They found that soil strength, measured with a penetrometer, gave clear evidence of the pan. Using an 8 t tractor on a loamy sand at field capacity, there was evidence of pan development after one pass. Successive passes further compacted the soil and increased soil strength (Fig. 5.3).

Traffic pans can persist in the soil for a long time in the absence of any further traffic. Dickerson (1976) found that compaction under skidding tracks in forest soils completely disappeared after 12 years. In contrast, Greacen and Sands (1980) found that compaction was still evident after 50 years.

Natural amelioration of compacted layers can occur due to physical and biological activity in the soil. In soils with a certain content of smectite clays, swelling and shrinking in response to changes in soil water content will slowly rearrange the soil particles. Frost action is very effective at reducing soil bulk density (Gill 1971), but this is certainly not an active mechanism in the soils of the Kellerberrin area.

While I could not find any data on the biological amelioration of compacted layers in agricultural systems, the evidence from forest soils (Dickerson, 1976 Greacen and Sands 1980), suggests that the process will be slow. Hence, it appears that the only practical way to reduce compaction is by mechanical means.

5.3.3.3 Consequences of Compaction

Compacted layers increase the mechanical impedance to plant root penetration. Rosenberg (1964) cited a number of studies which showed that plant roots of various species could not penetrate soils with bulk density exceeding a critical value, which for most plant roots is $1900\,kg\,m^{-3}$. Corn (*Zea mays*) root penetration into clay decreased linearly with increased bulk density and with decreased depth of needle penetrometer penetration. Since the relationship was independent of free pore space and not affected by moisture levels, it was concluded that the root penetration response was due solely to the mechanical impedence.

In sandy loam soils, Jarvis and Porritt (1985) showed a linear decrease in wheat yield with increasing soil strength at 30 cm. In their trials, they obtained a grain yield of $1200\,kg\,ha^{-1}$ when the soil strength at 30 cm was 1.5 MPa but only $800\,kg\,ha^{-1}$ at a strength of 3.5 MPa. Thus, the impact of compaction on yield can be large, and the question is whether or not compaction can be economically ameliorated.

5.3.3.4 Responses to Treatment

Two aspects of treating compacted soil are destruction of the compacted layer and prevention of development of the compacted layer. It is mechanically possible, by deep tillage, to break up a compacted layer and tillage techniques and methods of working fields can be used to reduce the development of compaction.

Deep ripping to 30 cm on yellow sandplain soils in Western Australia gave an average wheat yield increase, over 21 trials, of $500\,kg\,ha^{-1}$, or 30 percent. The greatest yield increase was $1050\,kg\,ha^{-1}$ (142 percent) and only one trial gave a yield decrease ($-100\,kg\,ha^{-1}$ or -4 percent Jarvis et al. 1985). While the response can be large, it is not necessarily permanent. Results presented in Table 5.10, from Jarvis et al. (1986), show that when the traffic pan at 20 to 30 cm in a loamy sand was ripped to 35 cm, the average wheat yield increase of 34 percent in

Table 5.10. Wheat yield (kg ha^{-1}) response to tillage treatment and deep ripping on a loamy sand.

	TDD*		Tillage Method DCC		DP	
	No-rip	Rip	No-rip	Rip	No-rip	Rip
1981	1516	2171	1881	2543	2041	2648
1982	1766	1974	1998	2154	2075	2089

*TDD = direct drill with triple disc drill; DDC = direct drill with combine with scarifier points; DP = district practice of scarifying, working back, and combine seeding.
(After Jarvis et al. 1986.)

the first year decreased to 6.5 percent in the second year, when no subsequent ripping was done. The residual effect of ripping was greatest when the tillage treatment used caused the least soil disturbance.

The loamy sands at Wongan Hills, where the trial results of Table 5.10 were obtained, formed hard, massive structures on drying. In soils with a developed traffic pan, crops sown with minimal disturbance (TDD and DDC of Table 5.10) encountered increased impedance to root growth both at the surface and at the traffic pan, resulting in reduced early crop vigour. Preanthesis water use was lower in the direct-drilled crops but may be greater during grain filling, providing there is an ample store of soil water. However, the total volume of soil exploited by the roots was often reduced in direct-drilled crops. The result was reflected in reduced yield (Hamblin et al. 1982). With the traffic pan disrupted, roots penetrated to 60 cm after six weeks compared to only 30 cm in the nonripped treatment.

The response to deep ripping depends on a number of factors such as soil type, crop species, time of ripping in relation to soil moisture, and depth and speed of ripping. The interaction between different forms of land degradation is examplified by the fact that Jarvis (1985) did not obtain a yield response to ripping on yellow sandplain with a subsoil pH of 4.3 but did when the pH was 5.2. Obviously, in this situation, it is necessary to overcome both forms of degradation simultaneously.

Since compaction results from traffic on the soil, it should follow that reduced traffic will limit development of a compacted layer. Traffic reduction can be achieved either by reducing the number of passes required to carry out the cropping operations or by confining the traffic to defined paths.

Reducing the number of tillage operations from four to one will drastically reduce the area of a field traversed by tractor and implement wheels. In seedbed preparation and sowing of a barley crop in Scotland, Soane et al. (1982) reported that 91 percent of the field was covered by tractor wheels. As shown in Figure 5.3, a single pass causes compaction at 30 cm. Four passes can form a hardpan which reduces yield by 38 percent (Jarvis et al. 1985). It has been suggested by P. Blackwell (personal communication) that a single tine behind each tractor wheel may immediately break up the compaction. This idea has not been thoroughly tested but seems worthy of further investigation.

The area of a field subject to the compactive effect of tractor wheels can be reduced by restricting the tractor to defined paths. The concept of controlled traffic is not new. Soane et al. (1982) cited that Halkett proposed permanent railway lines to support steam-driven tractors in 1858. Controlled traffic has long been used in many horticultural enterprises, but only in the last 20 years has it been given serious consideration for large scale agricultural crops. While there is considerable evidence that a controlled traffic system results in increased yields (e.g., 30 percent for silage maize, 25 percent for soya beans, 48 percent for cotton; Soane et al. [1982]), the general acceptance and adoption of the system will be dependent on the standardisation of wheel track width by tractor and implement manufacturers.

Compaction effects can also be reduced by spreading the tractor load over a greater area. Wider, low pressure tyres, tracks instead of wheels, dual wheels, and having the front wheels not aligned with the rear wheels are all alternatives to spread the load. While these methods can be effective, they are not necessarily practical where a farm has a range of soil types, of which only some, or one, are subject to severe compaction.

5.3.4 Water Repellence

Water repellence, or nonwetting, is a soil condition which reduces the infiltration of water into the soil and prevents soil wetting at the seed depth. It can be a natural condition of the soil and can be induced, or exacerbated, by certain agricultural practices. In a nonwater-repellent soil, a drop of water placed on the surface can be absorbed within 0.1 s but, in a repellent soil, may stay on the surface for up to three hours.

Water repellence was first noted in Western Australia in 1937 (Teakle and Southern 1937) but had been reported much earlier in other parts of the world. In a survey of 700,000 ha of sandplain on the south coast of Western Australia, Summers (1987) found that 70 percent of all sites sampled were water-repellent. While water repellence is normally associated with sandy soils, it has long been known in Western Australia that the heavy soils of "mallet" hills had severe water repellence (McGhie 1980a). The soils of these hills, originally vegetated with brown mallet (*Eucalyptus astringens*), are naturally water-repellent and, when the land below is cleared, the high runoff from the hill can cause severe soil erosion.

In the 1960s, following rectification of trace element deficiencies in sandy soils, problems of water repellence became increasingly apparent. Bond (1968) attributed poor pasture establishment on sandy soils in

South Australia to their water repellence. Roberts and Carbon (1972) identified some of the causal factors of repellence in Western Australian sands, but it was not until the 1980s that there was a serious study of ameliorative methods (e.g., McGhie and Tipping 1983).

5.3.4.1 Measurement of Water Repellence

Water repellence is a relative property related to the differences in the hydrophobic and hydrophilic characteristics of the soil. The water repellency of a flat surface can be characterised by measuring the contact angle between the liquid and the solid. However, the geometric measurement can be difficult with soil systems because the water drop may not stay on the surface and surface roughness makes the angle difficult to measure. Bond (1968) measured contact angles by photographing a water droplet on the soil surface and then measuring the angle on an enlarged print.

The contact angle (Θ) can be inferred from the capillary radius determined by measuring the height, or rate, of capillary rise. It is necessary to know the rise when the contact angle is zero, and this is obtained by firing the soil sample in a muffle furnace. (There is considerable debate on the appropriate firing temperature and time.) Contact angle is then calculated by Equation (6) where h is the rise in the untreated sample and ho the rise in the fired sample.

$$\cos\Theta = h/ho \tag{6}$$

King (1981) stresses that these laboratory based techniques are not suitable for rapid field assessment of water repellence. In a comparison of field based tests, he concluded that three tests were suitable for rapid assessment. These were:

1. The aqueous ethanol droplet test, wherein 40 µl drops of ethanol solutions at concentration intervals of 0.2 *M* in the range of 0 to 5 *M* were placed on the surface and their entry time recorded. The repellence value was taken as the molarity of the ethanol solution which penetrated the soil surface in 10 s.
2. The water droplet test, which measures the time for a 40 µl water drop to enter the soil surface. King did not consider this test useful if the entry time was greater than four minutes.
3. The infiltration rate test, which measured the time for 5 ml of water to enter the soil through a 25 mm diameter ring infiltrometer embedded 5 mm into the soil. For very repellent soils, King measured the volume remaining above the soil surface after 30 minutes.

The ethanol droplet test is convenient and quick and has been adopted for broad scale surveys. Wetherby (1984) used the method for a large survey in South Australia and developed the rating scale given in Table

Table 5.11. Rating scale of severity of water repellence measured by ethanol droplet test.

Molarity of Ethanol Which Penetrated in 10 Seconds	Rating of Repellence
0.0–0.5	Very low
0.5–1.0	Low
1.0–2.0	Moderate
2.0–3.0	Severe
3.0–4.0	Very severe

(After Wetherby 1984.)

5.11. Wetherby's scale was used in a large scale survey of repellence on the south coastal areas of Western Australia by Summers (1987).

5.3.4.2 Causes of Repellence

It is well established that water repellence in soils is due to a hydrophobic organic coating on the soil particles. Roberts and Carbon (1971, 1972) measured the skins on sand particles derived from the humic acid fraction of the soil organic matter to be about 1 μm thick. They observed that the skins, dehydrated and hydrophobic at the end of summer, softened and swelled when immersed in water for several hours and, if then only partly dried, became hydrophilic. If then dehydrated, the skins again became hydrophobic. While hydrophobic organic coatings are responsible for water repellence, the factors which cause water repellence are generally fungi, fire, and plant residues.

The evidence incriminating fungi as a main causal factor is largely circumstantial. Fungi are universally present in soil and fungal hyphae can be either hydrophobic or hydrophilic. Bond and Harris (1964) found it very difficult to identify specific fungi responsible for water repellence. The variation in water repellence through the season and the increase in water repellence with the age of the soil under a perennial legume pasture were not related to changes in fungal activity (Bond 1964). So fungi may be only partially responsible for water repellence.

Fire as an agent causing repellence was postulated by De Bano and Krames (1966) after studying runoff from watersheds in southern California. The duration and intensity of the fire determines whether the vaporised hydrophobic substances condense on the surface layer or not. Under intense, long-duration fires, the surface can become wettable, but subsurface layers become repellent as the vaporized material moves down through the soil profile and condenses along a temperature gradient.

Plant residues are implicated in all occurrences of water-repellent soils. The type of plant residue appears to be far more important than the quantity. Severe water repellence can be induced in sands with only 0.1 percent organic carbon (Bond 1968). However, in unstable coastal or desert dunes, the sands are wettable, suggesting that there must be some organic matter to induce repellence.

Litter from some plant species is more effective at inducing repellence than litter from others. Water repellence develops under long-term pastures and is reduced when cropped to cereals (McGhie and Posner 1981). The water repellence under lucerne (*Medicago sativa*) increased with age of the stand (Bond 1964). Litter from most dicotyledons increases water repellence whereas organic matter from monocotyledons increases wettability. One reported exception is perennial veldt grass (*Ehrhata calycina*) which, although a monocotyledon, has been found to induce severe water repellence (McGhie 1980b). McGhie showed that the species response was partly due to differences in the cation exchange capacities and the dominant cationic species on the exchange sites of the organic matter. The organic matter formed from wheat, for example, had a low cation exchange capacity dominated by Na^+ and K^+ in contrast to the high exchange capacity of the organic matter formed from subterranean clover, which was dominated by Ca^{2+}. Further, there were differences in the crystalline structure of the cell wall material between hydrophobic and hydrophilic species, with the latter being less crystalline.

While the chemistry of water repellence is complex, the salient feature, in terms of land management in the study area, is that leguminous pasture and crop species induce repellence while cereal crops increase wettability.

5.3.4.3 Consequences of Water Repellence

The major consequences of water-repellent soils in Western Australia are poor germination, late establishment, reduced growth of crops and pastures, and increased soil erosion due to enhanced runoff. Increased wind erosion can be a secondary effect of the reduced ground cover on repellent soils.

The impact of hydrophobic soils on crop and pasture production is simply related to the water relations of the plant. Soil moisture content of repellent soils can be as low as 1 percent (Bond 1968), which is too low for germination. However, because of the variability in repellence over very short distances, adjacent patches of soil will sustain germination. The result is a very patchy crop or pasture. In areas susceptible to repellence, patchiness and slow germination can be used almost diagnostically to delineate problem areas.

On water-repellent sands in South Australia, Bond (1972) found that some barley seeds had failed to germinate three months after seeding. The germination in his test rows was controlled by uneven water penetration. At the end of the season the barley yield from the most repellent soil was a mere 150 kg ha^{-1} compared with 1440 kg ha^{-1} from the wettable soil.

A characteristic of water-repellent soils is that the infiltration rate of water into the soil increases with time. This contrasts with a wettable soil where infiltration rate decreases with time. The impact of this is that hydrophobic soils have a high runoff rate from the start of a rainfall event. This can produce severe soil erosion as found by McGhie (1980a) who, in a survey of mallet (*Eucalyptus astringens*) hills, reported 100 m of gully associated with each 1.6 ha of hill.

5.3.4.4 Responses to Treatment

Many different methods of treating water-repellent soils have been proposed and some tested. However, there is a dearth of published data indicating crop or pasture response to specific treatments. There are difficulties in obtaining such data because of seasonal variability in repellence and because some treatments have secondary effects which confound the results.

Four basic treatments have been tried. These are:

- wetting agents;
- cultivation techniques;
- crop rotations; and
- fine particle amendment.

Most wetting agents are made up of bipolar molecules with a hydrophilic group and a lipophilic group. The hydrophilic group bonds to water and the lipophilic group bonds to the lipid fraction of the organic coating. This chemical reaction then allows water not involved in the bonding to penetrate the soil. Commercially available wetting agents differ in their chemical composition and effectiveness. Table 5.12 illustrates the variability of performance between wetting agents on a sandy soil in Western Australia. Plots were treated with the wetting agent and then allowed to dry before a rainfall simulator was used to apply 2.5 l of water per plot over 10 minutes. Water that had not infiltrated was collected as "excess," and the depth of water penetration at the end of the test measured (McGhie and Tipping 1983). It can be seen that Product C at application rates of 10 and 20 l ha^{-1} is totally ineffective while Product A has a large effect at 10 l ha^{-1}.

There are reports of 160 to 200 percent increases in winter pasture production after sands on the south coast of Western Australia were treated with wetting agents (Summers 1987). The wetting agents remained effective for at least two years.

The cost of wetting agents has prohibited their broad-scale use. However, wetting agents are effective in promoting germination and

TABLE 5.12. Performance of wetting agents on Gavin sand.

Wetting Agent	Rate l ha^{-1}	"Excess" ml	Depth of Penetration mm
Water only	—	1342	3
Product A*	10	138	15
	20	178	15
	50	2	20
	100	0	20
Product B	10	722	15
	20	220	15
	50	0	25
	100	0	25
Product C	10	1358	5
	20	1218	10
	50	720	8
	100	485	10

* Commercial names not given.
(After McGhie and Tipping 1983.)

establishment when applied in a narrow strip over the seed row. Establishment is further improved in sandy soils if press wheels then slightly compact the treated row to improve seed-soil, and hence seed-water, contact. Crabtree and McGhie (1990) got improved establishment of wheat, barley, and lupins using wetting agent application rates below $4\,l\,ha^{-1}$ in conjunction with press wheels (Fig. 5.4).

Deep ploughing and rotary hoeing have been proposed as means of improving wetting by mixing the water-repellent soil with the wettable subsoil. The mechanical action also abrades the organic coating, thus improving wettability.

Accurate placement of the seed at the base of the furrow where water accumulates has also been advocated, but McGhie and Tipping (1983) found no benefit in seed placement or cultivation techniques.

The concept behind using crop rotations is to increase the amount of the wettable, cereal-derived organic matter in the soil. While this concept has merit, other factors such as disease control and plant nutrition dominate rotational decisions and make this option impractical.

Incorporation of fine particles (clay, fly ash) into the surface layer improves the water holding capacity of the soil, and the nature of the surface charges on clays makes them less susceptible to becoming hydrophobic than sands. Incorporation of between 2 and 5 percent by weight of fine particles into the top 5 cm of a sand dramatically improved the emergence and establishment of subterranean clover in glasshouse trials (Roberts 1966). There was no benefit when the amendment was incorporated only in the upper 2 cm. Assuming the bulk density of a sand to be $1200\,kg\,m^{-3}$, then it would require $30\,t\,ha^{-1}$ of amendment to achieve 5 percent in the top 5 cm; clearly, only prac-

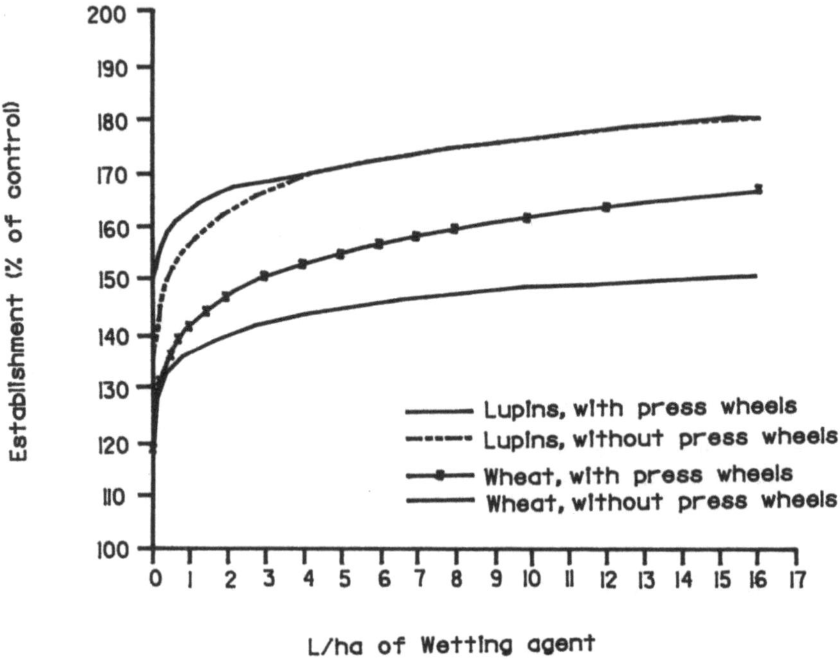

FIGURE 5.4. Establishment of lupins and wheat using wetting agents and press wheels. Control was no wetting agent and no press wheels. (After Crabtree and McGhie 1990.)

tical in broad scale agriculture if there is a close-by source of material, such as a subsoil clay which could be cultivated into the surface soil.

5.3.5 Soil Salinity

The hydrogeology of soil salinity development and current methods of treatment are detailed by MacFarlane et al., Chapter 6.

Saline soils were well developed in the lower parts of the landscape before agricultural development. The soils of the Baandee (salt lake) unit (McArthur, Chapter 2) are highly salinised, being associated with the very low gradient, ancient drainage lines. Adjacent to the Baandee unit are the soils of the Nangeenan (low dune) unit which were formed as a veneer of material blown from the Baandee unit. These soils are fine textured, highly calcareous, and, in parts, high in salt. The highly saline soils are dominated by morrel (*E. longicornis*). A distinguishing characteristic of these saline soils is that the salinity is not associated with a shallow, saline water table. Stoneman (1990) measured the chloride content of a typical morrel soil at 0.37 percent (15 to 45 cm), 0.27 percent (45 to 120 cm), and 0.30 percent (120 to 180 cm). These levels are very high given that Northcote and Skene (1972) classify a soil as saline when the chloride content is greater than 0.1 percent.

After clearing these naturally saline soils, and in the absence of a shallow, saline groundwater, there is downward leaching of salts. Teakle and Burvill (1938) showed that the chloride level in the upper B horizon of a duplex soil decreased from 0.25 percent to 0.05 percent five years after clearing. Such changes are beneficial for agricultural production.

Where the hydrogeological conditions are such that a groundwater table exists, or is formed, and rises (for reasons detailed by McFarlane et al., Chapter 6), the effects on agriculture, or any natural ecosystem remnant, are quite disastrous. When the groundwater level reaches a critical depth of less than about 1.8 m (Nulsen 1981), net accumulation of salt at or near the soil surface is such that the land can no longer sustain normal agricultural pursuits. The impact of salt accumulation on plant growth is due to an osmotic effect; the plants have to expend more energy to extract water from the soil, and a toxic effect occurs due to high ionic levels in the soil water.

Methods are available to:

- obtain production from salt-affected land using halophytic species;
- reclaim saline land using trees or engineering methods such as drainage and aquifer pumping; and
- reduce salt encroachment by redressing the water balance in the landscape, using a combination of appropriate crops and pastures, trees, and earthworks.

5.3.6 Wind Erosion

Wind erosion is a process which mechanically sorts and separates movable soil particles of different weights and sizes. It therefore results in a considerable change in both the physical and chemical properties of the surface soil.

Wind erosion is often only perceived as a problem when there is a major dust storm or when sand accumulates along fence lines and vegetated strips. Any dust blowing indicates wind erosion, and the costs and long-term consequences of even apparently minor events can be significant. Any process which reduces vegetative cover or the natural coherence of the soil surface increases the susceptibility of the land to wind erosion. Conversely, practices which improve cover and surface structure impede wind erosion.

5.3.6.1 Measurement of Wind Erosion

Objective measurement of the adverse impacts of wind erosion is difficult. Measurements of soil surface elevation before and after an event will show the depth of soil removal but will not reflect the differential removal of different-sized soil particles. Entrapment gauges, which sample the downwind air stream and extract its particulate content, do measure the range of airborne particles. However, as they sample the air stream, it is difficult to relate the results to a particular land area.

The aerial extent of severe wind erosion can be determined from timely aerial photographs or satellite imagery. Again, there are difficulties in determining the relative severity of the erosion observed. Economically and edaphically significant events which happen to be less visible will not normally be detectable.

None of the methods which measure either soil removal or the area affected truly reflect the negative impact of wind erosion on plant productivity by sandblasting, fertility loss, or the off-site effects such as sand drift on fences and roads.

5.3.6.2 Causes of Wind Erosion

The interaction of wind flow with surface soil particles to cause erosion is well understood (Chepil and Woodruff 1963). The movement of soil particles is initiated when the wind pressure against detached soil particles is sufficient to overcome the force of gravity. Thus, the prerequisites for wind erosion are a relatively bare soil surface with detached soil particles and a minimum wind velocity.

Wind velocity increases exponentially with height above the soil surface. A wind velocity of 0.5 kmph at 1 cm height will initiate wind erosion on a bare, smooth soil surface containing detached soil particles. This is equivalent to a velocity of 29 kmph measured 2 m above the surface containing detached soil particles (Chepil and Woodruff 1963). The initially displaced particles move over the surface in a series of jumps. On impact at the end of each jump, they dislodge other particles in a process known as saltation. Particles involved in saltation are fine sand sized (0.1 to 0.5 mm). Saltating particles can dislodge and abrade both smaller and larger sized particles. The smaller size fraction, mostly silt and clay aggregates less than 0.05 mm, are light enough to be carried to great heights and considerable distances by the air stream. These form the dust of dust storms, and the process is known as suspension. Larger particles, generally coarse sand, roll along the surface when impacted by saltating particles in the process of surface creep. Thus, the differential mobility of the soil particles causes them to be sorted by the wind and results in different depositions: surface creep results in lag sands, saltation creates sand dunes, and suspension produces deposits of dust or loess.

5.3.6.3 Factors Affecting Wind Erosion

The condition of the soil surface dictates whether or not it will erode in a wind event. Any practice which bares the soil surface, reduces its structural stability, or detaches soil particles will increase the susceptibility of the soil to wind erosion. These factors are manageable. Factors over which we have no control, but which strongly predispose a soil to wind erosion, are soil texture and climate.

Coarse-textured soils are much more susceptible to erosion than fine-textured or uniformly graded soils (Chepil and Woodruff 1963). Coarse-textured soils have low water-holding capacity, drain and dry quickly, and are less cohesive than fine-textured soils. The predominance of sandy-textured surface duplex soils of the Collgar, Booraan (midslope), and Ulva (sandplain) surfaces makes them candidates for wind erosion if mismanaged.

5.3.6.4 Consequences of Wind Erosion

Three major effects of wind erosion are:

- soil and nutrient removal which result in immediate and perhaps longer term reduction in productivity;
- direct damage to plants as a result of sandblasting, which reduces the leaf area; and
- damage to infrastructure such as roads and fences as a result of sand accumulation.

The nutrient profile, even in relatively sandy soils, shows a concentration close to the surface. With the exception of very coarse-textured sands, applied P remains mainly in the surface layers, with only small amounts moving slowly down the profile. Similarly, the immobile micronutrients Mo, Mn, Cu, Zn, and Co remain close to the surface. Only very mobile elements, such as K, achieve a uniform distribution in the profile soon after application (Hamilton et al. 1987). The distribution of N in sandy soils is not well understood, although Marsh (unpublished data) has shown substantial concentrations near the surface of a sandy soil growing pasture.

There are few data on the productivity impact of removing surface soil. Marsh and Carter (1983) removed 8 mm of soil with a road broom to simulate erosion and found that the yield of a subsequent cereal crop was reduced by 25 percent. Observations made after three years of clover harvesting, which resulted in a substantial loss of the "dust" fraction from the surface soil, suggested that productivity was reduced by 50 percent.

Newly emerged plants can be severely damaged by saltating sand grains. The saltating sand can completely remove all emerged leaf material and, in species where the apical bud is above the surface, the crop will require replanting. Any defoliation reduces the plants' photosynthetic area and hence reduces its growth rate. Plants differ in their susceptibility to sandblasting, with broadleaf species such as canola (*Brassica napus*) being very susceptible while hardy grasses such as cereal rye (*Secale cereale*) are relatively tolerant.

For every millimetre of soil removed from a hectare of land, $10\,m^3$, or between 12 and 14 tonnes, of soil is deposited somewhere downwind. Often the sand fraction accumulates along fences, drifts over roads, and blocks culverts under roads. Removing the sand or replacing the fences is a considerable cost to local authorities and individual farmers.

5.3.6.5 Control of Wind Erosion

Retention of a protective vegetative cover on the soil surface, maintenance of a cohesive surface soil structure, and reduction of wind speed are the only practical methods of wind erosion control in extensive agricultural areas. On small, intensively managed areas such as livestock holding yards, artificial soil stabilizers can be effective.

It is essential in the Kellerberrin environment to have some form of vegetative cover over the hot, dry summer months to prevent wind erosion. Erect, anchored wheat stubble is about twice as effective as prostate stubble in preventing erosion (Hamilton et al. 1987). Effective control is achieved with only $400\,kg\,ha^{-1}$ of anchored stubble. This is equivalent to about $1000\,kg\,ha^{-1}$ total stubble (anchored plus leaf and chaff material).

Lupin stubble is coarser and sparser than wheat stubble and requires about twice the weight per unit area to achieve a similar degree of control (Carter 1990). Short stubble, providing it is anchored, is just as effective as tall stubble in controlling wind erosion.

Grazing animals quickly reduce the amount and effectiveness of vegetative cover as they trample the vegetation and detach soil particles. Marsh (unpublished data) estimated that one sheep could loosen 330 kg soil ha^{-1} $week^{-1}$ on sandy loams and up to $660\,kg\,ha^{-1}\,week^{-1}$ on loamy sands and sands. Management of stocking rates to conserve surface cover is therefore an important control strategy.

Cultivation loosens the soil and generally increases the susceptibility to wind erosion. However, tillage also roughens the surface and, if the rough surface is cohesive, susceptibility to erosion can be reduced. Cultivation at high speed, when the soil is dry, increases the pulverisation of the soil and thus its susceptibility to erosion. Minimum or zero cultivation, using herbicides to control weeds, is very effective at reducing wind erosion.

Strips of tall vegetation can reduce the downstream wind velocity for distances up to 30 times the height of the vegetation (van Eimern et al. 1964). Various structures of shelterbelts are used throughout the world, including strips of standing stubble used in strip cropping systems, rows of tall crops such as grain sorghum to protect intervening cereal crops, and trees. Trees represent a more or less permanent windbreak, and their effectiveness depends on the orientation of the strips, the density of the strips, and the height of the mature trees. A tree windbreak, which allows about 60 percent of the wind to move through it, is far more effective as a windbreak than a solid barrier of the same height.

While tree shelterbelts reduce wind velocity downwind, they also significantly change the downwind microclimate by influencing air and soil temperature, relative humidity, evaporation, soil moisture, and dew formation. These changes may have beneficial effects on crop and pasture production *per se*. Such benefits have been measured in the Mediterranean climates of Europe but not in Western Australia.

While shelterbelts are effective in controlling wind erosion, they must not be seen as an excuse for poor land management between the belts. Shelterbelts increase the security of the farming system from devastation by wind erosion and reduce the risks of damage when unforeseen circumstances make the soil surface susceptible to erosion.

5.4 Integrating Solutions

While a particular form of soil degradation will respond to a specific treatment, the treatment may have either adverse or beneficial effects on other forms of soil degradation. For example, a lupin-cereal rotation has been advocated by Nulsen and Baxter (1982) as a means of reducing recharge to the groundwaters under the sandplain of the Ulva surface. However, as has been discussed, lupins may increase soil acidity, induce water repellence, and make the land more susceptible to wind erosion. In addition, the increased cultivation associated with the continuous cropping may cause subsoil compaction. In the context of landscape management, these possible adverse effects must be considered in relation to the possible benefits. From the farmers' point of view, the environmental consequences must be weighed against economic viability.

There should be both economic and environmental benefits from "good" farming practice. It is often considered that water is the main factor limiting production in the Western Australian wheatbelt. Yet very few crops reach their yield potential in relation to the water available.

In a similar environment in South Australia, French and Shultz (1984a) determined the yield potential of wheat to be $20\,kg\,ha^{-1}\,mm^{-1}$ of water used. The average measured yield was $6.9\,kg\,ha^{-1}\,mm^{-1}$ and the highest $12.7\,kg\,ha^{-1}\,mm^{-1}$. French and Shultz (1984b) considered that the main factors reducing the yield were limited fertility, diseases and pests, delayed sowing, weeds, and waterlogging.

Water use can be increased by ameliorating some forms of degradation. This increase will reduce recharge and hence impact on land salinisation. George (1990a) estimated recharge under the Ulva (sandplain) surface to average between 6 and 13 mm. As stated, this sandplain can be subject to acidity and compaction. Table 5.4 shows that, by applying lime and Mo to acid soils, wheat yields can increase by 350 kg ha^{-1}; Table 5.5 shows that liming acid subsoils increased yields by a massive 1120 kg ha^{-1}. Using even the potential water use efficiency of 20 kg ha^{-1} mm^{-1}, these yield increases mean that the crops used an extra 17.5 and 56 mm of water, respectively, certainly enough to have a significant impact on recharge.

Similarly, the average yield response of 500 kg ha^{-1} to deep ripping of compacted soils, as reported by Jarvis et al. (1985), would translate into 25 mm of extra water use; and the average first year yield response to ripping of 545 kg ha^{-1}, as shown in Table 5.10, translates into 27 mm of additional water use; again, enough to have an impact on recharge.

Farming to live today and to conserve for the future requires farmers to make very complex decisions regarding their land use practices. Their decisions about which crop or pasture species, which herbicide, which fertiliser, how much and when, what tillage technique, whether to deep rip or not, whether to apply gypsum, whether to apply lime, or whether to use a wetting agent or not, etc., are all factors which will impact not only on their viability, but also on the state of the soil from which they and future generations will hopefully continue to produce food and fibre.

References

Abbott, I., Parker, C.A., and Sills, I.D. (1979) Changes in the abundance of large soil animals and physical properties of soils following cultivation. *Australian Journal of Soil Research* 17:343–353.

Bond, R.D. (1964) The influence of microflora on physical properties of soil. II. Field studies on water repellent sands. *Australian Journal of Soil Research.* 2:123–131.

Bond, R.D. (1968) Water repellent sands. *Transactions 9th International Congress of Soil Science, Adelaide* 1:339–347.

Bond, R.D. (1972) Germination and yield of barley when grown in a water-repellent sand. *Agronomy Journal* 64:402–403.

Bond, R.D. and Harris, J.R. (1964) The influence of microflora on physical properties of soil. I. Effects associated with filamentous algae and fungi. *Australian Journal of Soil Research* 2:111–122.

Carter, D.J. (1990) The amount of stubble needed to reduce wind erosion. *Western Australian Department of Agriculture Farmnote* 40/90.

Chepil, W.S. and Woodruff, N.P. (1963) The physics of wind erosion and its control. *Advances in Agronomy* 15:211–302.

Clothier, B.E. and White, I. (1981) Measurement of sorptivity and soil water diffusivity in the field. *Soil Science Society of America Journal* 45:241–249.

Crabtree, B. and McGhie, D. (1990) Managing water repellence in sandy soils. *Western Australian Department of Agriculture Farmnote* 55/90.

De Bano, L.F. and Krames, T.S. (1966) Water-repellent soils and their relation to wildfire temperatures. *Bulletin of the International Association Science and Hydrology* 11:14–19.

Dickerson, B.P. (1976) Soil compaction after tree-length skidding in northern Mississippi. *Soil Science Society of America Journal* 40:956–966.

French, R.J. and Shultz, J.E. (1984a) Water use efficiency of wheat in a Mediterranean-type environment. I. The relation between yield, water use, and climate. *Australian Journal of Agricultural Research* 35:743–764.

French, R.J. and Schultz, J.E. (1984b) Water use efficiency of wheat in a Mediterranean-type environment. II. Some limitations to efficiency. *Australian Journal of Agricultural Research* 35:765–775.

George, R.J. (1990a) Reclaiming sandplain seeps by intercepting perched ground water with eucalypts. *Land Degradation and Rehabilitation* 2:13–25.

George, R.J. (1990b) The 1989 salt land survey. *Journal of Agriculture–Western Australia* 31:159–166.

Gill, W.R. (1968) Influence of compaction hardening of soil on penetration resistance. *Transactions of the American Society of Agricultural Engineers* 11:741–745.

Gill, W.R. (1971) Economic assessment of soil compaction. In K.K. Barnes, W.M. Carleton, H.M. Taylor, R.I. Throckmorton, and G.E. van den Berg (eds.) *Compaction of Agricultural Soils*. American Society of Agricultural Engineers Monograph. St Joseph, Michigan. pp. 431–458.

Glencross, R.N. and Clarke, M.G. (1984) Soil acidity and liming in the lower Great Southern. *Journal of Agriculture–Western Australia* 25:142–144.

Greacen, E.L. and Sands, R. (1980) Compaction of forest soils. A review. *Australian Journal of Soil Research* 18:163–189.

Hamblin, A.P. (1982) Soil water behaviour in response to changes in soil structure. *Journal of Soil Science* 33:375–386.

Hamblin, A.P. and Tennant, D. (1979) Interactions between soil type and tillage level in a dryland situation. *Australian Journal of Soil Research* 17:177–189.

Hamblin, A.P., Tennant, D., and Cochrane, H. (1982) Tillage and the growth of a wheat crop in a loamy sand. *Australian Journal of Agricultural Research* 33:887–897.

Hamilton, G.J., Carter, D.J., and Porritt, S. (1987) Wind erosion and soil fertility. *Western Australian Department of Agriculture Division of Resource Management Technical Report* 95:60–73.

Helyar, K.R. (1976) Nitrogen cycling and soil acidification. *Journal of the Australian Institute of Agricultural Science* 42:217–221.

Henderson, C., Levett, A., and Lisle, D. (1988) The effects of soil water content and bulk density on the compactability and soil penetration resistance of some Western Australian sandy soils. *Australian Journal of Soil Research* 26:391–400.

Howieson, J.G. and Ewing, M.A. (1984) Soil acidity and legume nodulation. *Journal of Agriculture–Western Australia* 25:125–127.

Jarvis, R.J. (1985) Subsoil acidity prevents deep ripping response. *Australasian Field Crops Newsletter* 20:126–127.

Jarvis, R.J. (1988) Tillage systems and soil stability. *Journal of Agriculture–Western Australia* 29:54–57.

Jarvis, R.J. and Porritt, S.E. (1985) Soil compaction and its effects on wheat yield. *Australasian Field Crops Newsletter* 20:122–125.

Jarvis, R.J., Hamblin, A.P., and Delroy, N.D. (1986) Continuous cereal cropping with alternative tillage systems in Western Australia. *Western Australian Department of Agriculture Technical Bulletin* No. 71.

Jarvis, R.J., Riethmuller, G.P., Henderson, C., Hamblin, J., and Hillman, B.J. (1985) Deep ripping. *Western Australian Department of Agriculture Farmnote* 61/85.

Jarvis, S.C. and Robson, A.D. (1983) The effects of nitrogen nutrition of plants on the development of acidity in Western Australian soils. I. Effects with subterranean clover grown under leaching conditions. *Australian Journal of Agricultural Research* 34:341–353.

King, P.M. (1981) Comparison of methods for measuring severity of water repellence of sandy soils and assessment of some factors that affect its measurement. *Australian Journal of Soil Research* 19:275–285.

Makin, A.W. (1989) Soil compaction under single and dual rear tractor tyres. *Journal of the Agricultural Engineering Society (Australia)* 18:3–8.

Marsh, Ba'B. and Carter, D.J. (1983) Wind erosion. *Journal of Agriculture–Western Australia* 24:54–57.

McGarry, D. (1990) Soil compaction and cotton growth on a vertisol. *Australian Journal of Soil Research* 28:869–877.

McGhie, D.A. (1980a) The contribution of the mallet hill surface to runoff in the Narrogin region of Western Australia. *Australian Journal of Soil Research* 18:299–307.

McGhie, D.A. (1980b) *The Origins of Water Repellence in Some Western Australian Soils.* Ph.D. Thesis, University of Western Australia. 171 pp.

McGhie, D.A. and Posner, A.M. (1981) The effect of plant top material on the water repellence of fired sands and water repellent soils. *Australian Journal of Agricultural Research* 32:609–620.

McGhie, D.A. and Tipping, P.I. (1983) Beating the non-wetting soil problem. *Journal of Agriculture–Western Australia* 3:84–86.

Northcote, K.H. and Skene, J.K.M. (1972) Australian soils with saline and sodic properties. *CSIRO Australia Soil Publication* No. 27.

Nulsen, R.A. (1981) Critical depth to saline ground water in non-irrigated situations. *Australian Journal of Soil Research* 19:83–86.

Nulsen, R.A. and Baxter, I.N. (1982) The potential of agronomic manipulation for controlling salinity in Western Australia. *Journal of the Australian Institute of Agricultural Science* 48:222–226.

Payne, D. (1988) Soil structure, tilth, and mechanical behaviour. In A. Wild (ed.) *Russell's Soil Conditions and Plant Growth,* 11th Ed. Longman, London U.K. pp. 378–411.

Porter, W.M. and Wilson, I.R. (1984) Soil acidity in the eastern wheatbelt. *Journal of Agriculture–Western Australia* 25:132–135.

Porter, W.M. and Yeates, J.S. (1984) The effects on plant growth. *Journal of Agriculture–Western Australia* 25:123–124.

Quirk, J.P. (1978) Some physico-chemical aspects of soil structural stability. A review. In W.W. Emerson, R.D. Bond, and A.R. Dexter (eds.) *Modification of Soil Structure.* John Wiley and Sons, Chichester. pp. 3–16.

Reeves, T.G., Haines, P.J., and Coventry, D.K. (1984) Growth of wheat and subterranean clover on soil artificially compacted at various depths. *Plant and Soil* 80:135–138.

Roberts, F.J. (1966) The effects of sand type and fine particle amendment on the emergence and growth of subterranean clover (*Trifolium subterraneum* L.) with particular reference to water relations. *Australian Journal of Agricultural Research* 17:657–672.

Roberts, F.J. and Carbon, B.A. (1971) Water repellence in sandy soils of southwestern Australia. I. Some studies related to field occurrence. *CSIRO Division of Plant Industries Field Station Record* 10:13–20.

Roberts, F.J. and Carbon, B.A. (1972) Water repellence in sandy soils of southwestern Australia. II. Some chemical characteristics of the hydrophobic skins. *Australian Journal of Soil Research* 10:35–42.

Rosenberg, N.J. (1964) Response of plants to the physical effects of soil compaction. *Advances in Agronomy* 16:181–196.

Soane, B.D. (1970) The effects of traffic and implements on soil compaction. *Journal of the Proceedings of the Institute of Agricultural Engineers* 25:111–126.

Soane, B.D. (1981) The role of field traffic studies in soil management research. *Soil and Tillage Research* 1:205–206.

Soane, B.D., Dickson, J.W., and Campbell, D.J. (1982) Compaction by agricultural vehicles: A review. III. Incidence and control of compaction in crop production. *Soil and Tillage Research* 2:3–36.

Stoneman, T.C. (1962) Loss of structure in wheatbelt soils. *Journal Agriculture–Western Australia* 3:493–497.

Stoneman, T.C. (1973) Soil structure changes under wheatbelt farming systems. *Journal of Agriculture–Western Australia* 14:209–214.

Stoneman, T.C. (1990) An introduction to the soils of the Lake Grace advisory district. *Western Australian Department of Agriculture Bulletin* 4178.

Summers, R.N. (1987) *The Induction and Severity of Nonwetting in Soils of the South Coastal Sandplain of Western Australia.* M.Sc. Thesis. University of Western Australia.

Teakle, L.J.H. and Burvill, G.H. (1938) The movement of soluble salts in soils under light rainfall conditions. *Journal of Agriculture–Western Australia* 15:218–245.

Teakle, L.J.H. and Southern, B.L. (1937) The peat soils and related soils of Western Australia. A survey of Herdsman Lake. *Journal of Agriculture–Western Australia* 14:404–424.

Tisdall, J.M. and Oades, J.M. (1982) Organic matter and water-stable aggregates in soils. *Journal of Soil Science* 33:141–163.

Wetherby, K.G. (1984) The extent and significance of water repellent sands on the Eyre Peninsula. *South Australian Department of Agriculture Technical Report* No. 47.

Wilson, I.R. (1984) Soil testing for acidity. *Journal of Agriculture–Western Australia* 25:121–122.

van Eimern, J., Karshon, R., Razumova, L.A., and Robertson, G.W. (1964) Windbreaks and shelter belts. *World Meteorological Organisation Technical Note* 59.

Chapter 6

Changes in the Hydrologic Cycle

D.J. McFarlane, R.J. George, and P. Farrington

6.1 Introduction

The deep-rooted native vegetation that originally grew in the wheatbelt was well adapted to the rainfall regime and diverse soil types. As a result, there was little or no runoff or groundwater recharge prior to clearing (Table 6.1). After the vegetation was replaced by shallow-rooted annual crops and pastures, runoff and groundwater recharge increased (Table 6.1). This resulted in serious water erosion, waterlogging, flooding, and salinity. It is ironic that all of these problems are caused by excess water when agricultural production is limited by a lack of water at particular times of the year.

The loss of arable land cannot be sustained without a major loss in agricultural production. For example, in the Kellerberrin Shire, 5568 ha (3.5%) of previously arable land were affected by secondary salinity in 1989 (George 1990a) and, in the neighbouring Tammin Shire, 9507 ha (9.3%) were affected (Fig. 6.1). Between 1979 and 1989, the rates of salinisation in the Kellerberrin and Tammin Shires were 150 and 350 ha y^{-1}, respectively.

Agricultural development has probably irreversibly changed some aspects of the hydrologic cycle. Whereas the effects on the hydrology of the wheatbelt appear simple (i.e., increased runoff and groundwater recharge), the processes involved are complex. It is necessary to understand the changes to the cycle thoroughly if sustainable farming practices are to be developed and if viable areas of remnant vegetation are to be retained.

The extent of clearing and salinity can be seen in the LANDSAT TM image of an area north east of Kellerberrin (Fig. 6.2). The main valleys contain saline playas which existed prior to clearing. Erosion and secondary salinity now affect the tributary valleys. Most paddocks are rectangular and do not correspond with the distribution of soil types. Small areas of remnant vegetation can be seen along roadways and in isolated reserves.

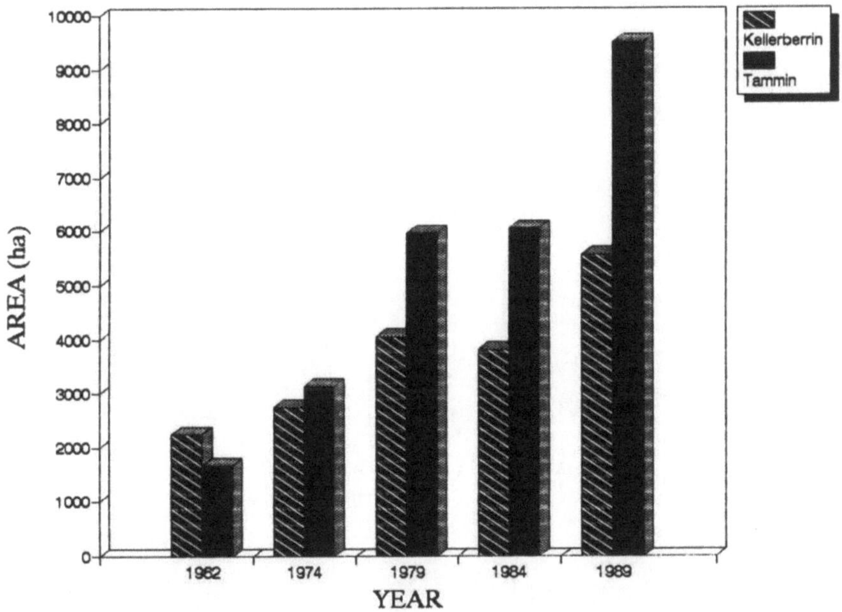

FIGURE 6.1. The area of secondary salinity in the Kellerberrin and Tammin Shires, as reported by farmers in surveys between 1962 and 1979. (Data from unpublished Australian Bureau of Statistics reports.)

TABLE 6.1. Annual average water balance for a native vegetated and agricultural catchment northeast of Newdegate (R.A. Nulsen unpublished data). All figures in mm.

	Rainfall	Runoff	Evaporation	Interception	Recharge
Native vegetation	370	0	359	11	0
Agriculture	370	18	319	7	26

Landscapes reflect their geological history (McArthur, Chapter 2). The lateritic soils (10 to 50 m deep) and alluvial-filled valleys (1 to 5 km wide) in the wheatbelt formed in a climate with a higher rainfall than at present. The river channels within the broad valleys are small, ill-defined, and braided (Fig. 6.2). These channels are inadequate for the runoff that is now taking place, resulting in increased water erosion, particularly in areas affected by secondary salinity.

Stranded playas and siliceous hardpans in the regolith provide evidence that the salt-affected areas have been larger than at present in the geological past. Clearing has increased the amount of rainfall that infiltrates the soil and reactivated some of these old discharges. How-

Figure 6.2. LANDSAT TM image of the Kellerberrin area showing saline playa lakes in the main valleys and tributary valleys affected by secondary salinity. The paddocks are laid out in rectangular blocks which do not relate to the soil types. Areas of remnant vegetation (dark blocks) are scattered throughout the area.

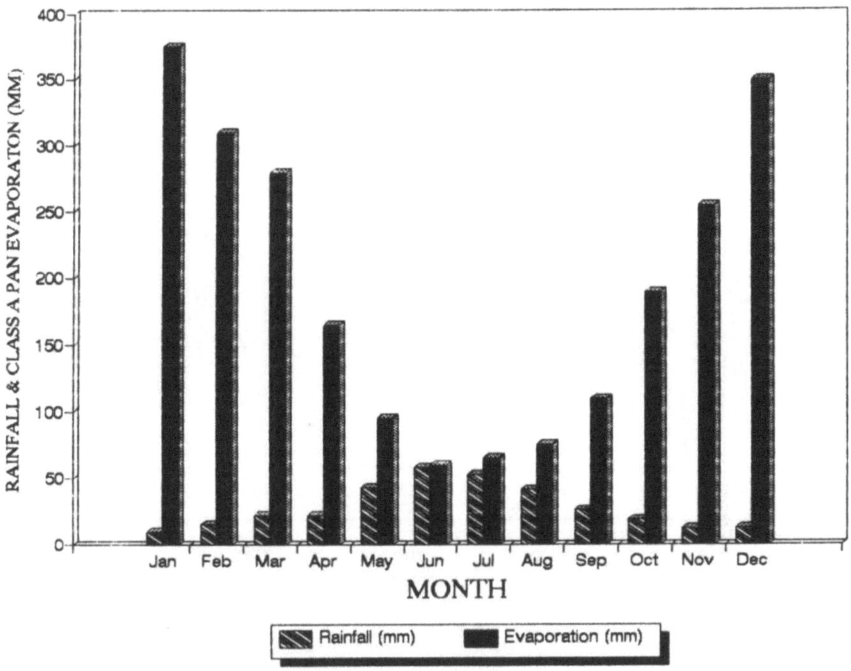

FIGURE 6.3. Average monthly rainfall and Class A pan evaporation for Kellerberrin (data from Australian Bureau of Meteorology).

ever, clearing has also altered the soil properties and changed the distribution of recharge and discharge in the landscape.

This chapter reviews the current knowledge of the hydrology of the central wheatbelt and indicates areas which require a better understanding if the current loss of arable land and native vegetation is to be arrested.

6.2 Rainfall

6.2.1 Annual and Seasonal Rainfall

The climate of the central wheatbelt is Mediterranean with cool, wet winters and hot, dry summers. At Kellerberrin, three-quarters of the annual rainfall of 334 mm and one-quarter of the annual potential evaporation (measured by a Class A pan evaporimeter) occurs between April and September (Fig. 6.3). Class A pan evaporation exceeds the rainfall in all months and by 2000 mm over a whole year. This results in a very short growing season (e.g., about four months at Kellerberrin;

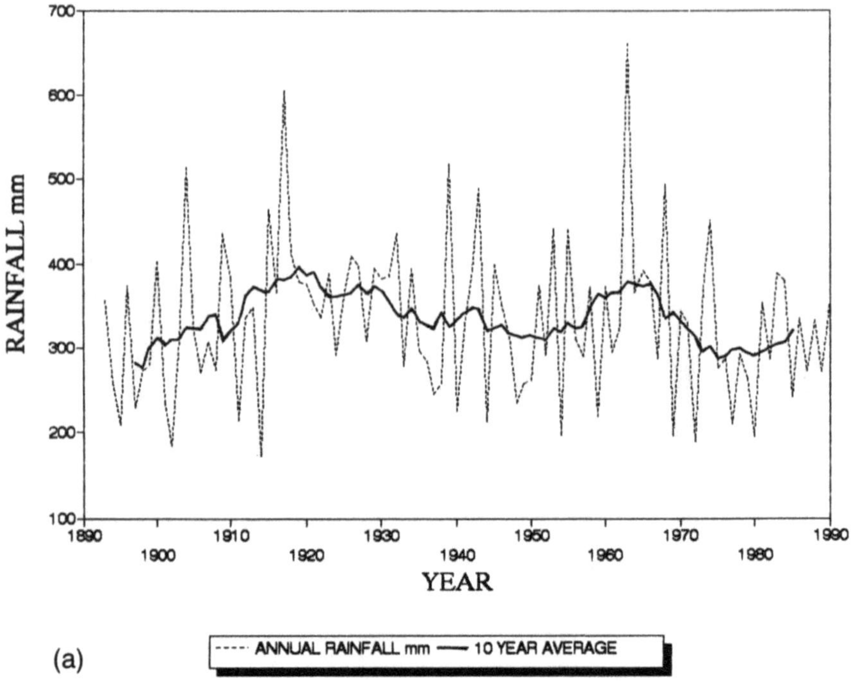

700
600
500
400
300
200
100
RAINFALL mm
1890
1900
1910
1920
1930
1940
1950
1960
1970
1980
1990
YEAR
ANNUAL RAINFALL mm
10 YEAR AVERAGE

(a)

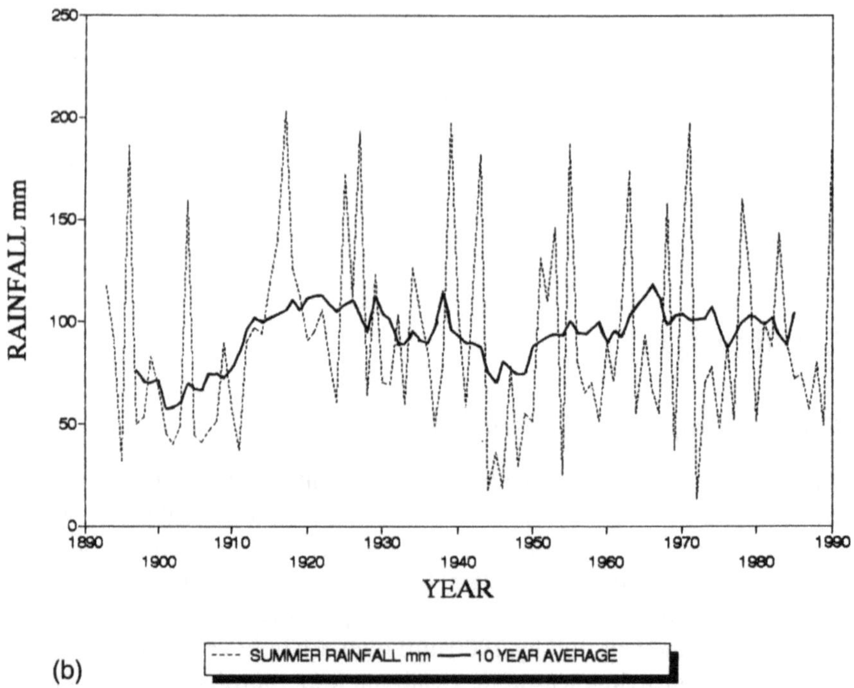

250
200
150
100
50
0
RAINFALL mm
1890
1900
1910
1920
1930
1940
1950
1960
1970
1980
1990
YEAR
SUMMER RAINFALL mm
10 YEAR AVERAGE

(b)

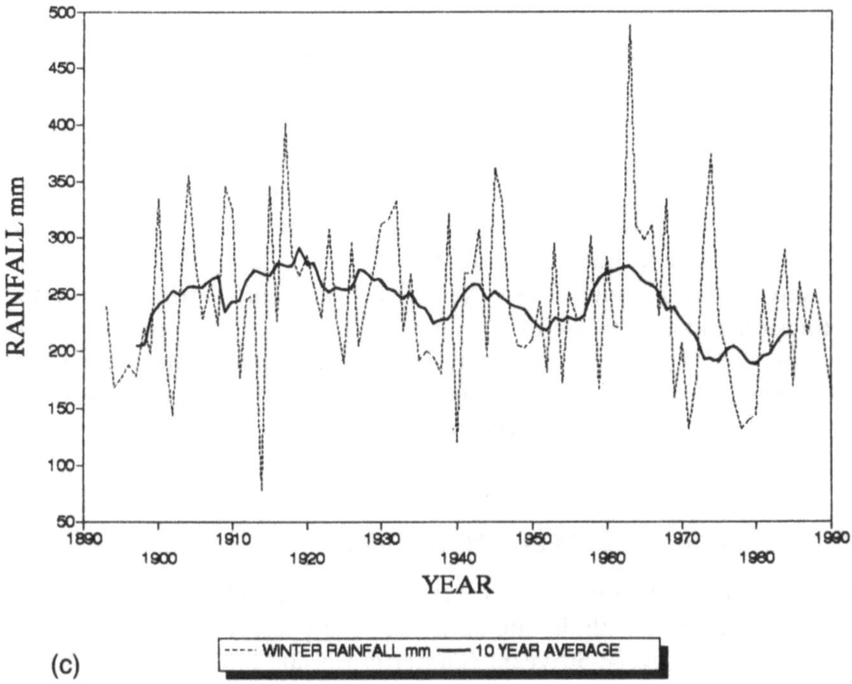

(c)

FIGURE 6.4. (a) Annual; (b) April to September (winter); and (c) October to March (summer) rainfall for Kellerberrin between 1893 and 1990, with ten-year moving averages. (Data from Australian Bureau of Meteorology.)

Bureau of Meteorology 1958). The water storage properties of some soils allow a longer growing season, as will be shown later.

Kellerberrin's rainfall is reliable, having a lower variability index than most of the cereal growing areas of the eastern states of Australia (Division of National Mapping 1986). However, periods of extreme rainfall are thought to be responsible for most erosion and recharge. Rainfall was above the long-term average for 16 of the 18 years between 1915 and 1932 (Fig. 6.4) and was also high in the early 1960s, including 662 mm in 1963. Fallowing was a common practice until the 1970s, and these high rainfalls resulted in widespread water erosion and raised the level of saline groundwaters.

The greenhouse effect is predicted to reduce winter rainfall by about 10% and increase surface temperatures by between 2 and 4°C in the wheatbelt (Pittock and Allen 1990). Since 1969, the average rainfall between April and September has been 14% below the long-term average at Kellerberrin. Between 1976 and 1985, the April to September rainfall was 23% below average, the driest ten years on record (Fig. 6.4). Pittock and Allen (1990) noted that the drying trend in the southwest since the 1960s has been much larger than had been predicted by the climate models of the greenhouse effect. The authors stressed that present climate models cannot accurately predict regional changes in the rainfall due to the greenhouse effect. Rainfall at Kellerberrin between October and March has remained close to the average since 1910 (Fig. 6.4). Whereas it is too early to say whether the changes predicted to occur due to the greenhouse effect are affecting the amount and distribution of Kellerberrin's rainfall, farmers have been urged to act as if this is the case (Chittleborough 1985).

Rain deposits about 25 kg NaCl ha^{-1} at Kellerberrin each year (Hingston and Galaitis 1976). This is a low rate (e.g., Perth receives over 200 kg NaCl $ha^{-1} y^{-1}$). However, as will be detailed later, the poor drainage from the wheatbelt landscape has led to the salts accumulating to form large stores in the regolith.

6.2.2 Rainfall Intensity and Erosivity

On average, Kellerberrin can expect 14 mm of rainfall in one hour once every two years, 22 mm in one hour once every ten years, and 35 mm in one hour once every 50 years (Canterford et al. 1987). It takes water about one hour to travel from the most remote point of a 2 km^2 wheatbelt catchment to its outlet (Flavell et al. 1987).

These are low intensities when compared to areas with tropical and subtropical climates. Geraldton in the northern wheatbelt can expect 22 mm of rainfall in one hour once every two years, which is the once in ten year intensity at Kellerberrin. Therefore, in addition to Kellerberrin's rainfall being reliable, it also has a relatively low intensity.

The amount of water erosion depends upon the erosivity and timing of the rainfall relative to the condition of the soil (e.g., whether it is bare or detached). Wischmeier and Smith (1958) found that of a number of rainfall variables, storm energy explained most of the variability in soil loss from runoff plots. The prediction of soil loss was improved when storm energy (E) was multiplied by the maximum amount of rain falling in 30 minutes (I_{30}) to produce the EI_{30} index.

The EI_{30} index has been calculated from eight years of pluviometer recordings at Kellerberrin and from eleven years of recordings at Nungarin, 60 km NW of Kellerberrin (McFarlane et al. 1986). On average, erosive rainfall was recorded in only three months of each year at both sites. In some years (e.g., 1980), no rain sufficient to cause erosion was recorded. The most erosive months were found to be February, April, May, June, and August (Fig. 6.5). There were large differences in the EI_{30} recordings from Kellerberrin and Nungarin in the summer months. Summer rain is often caused by isolated thunderstorms.

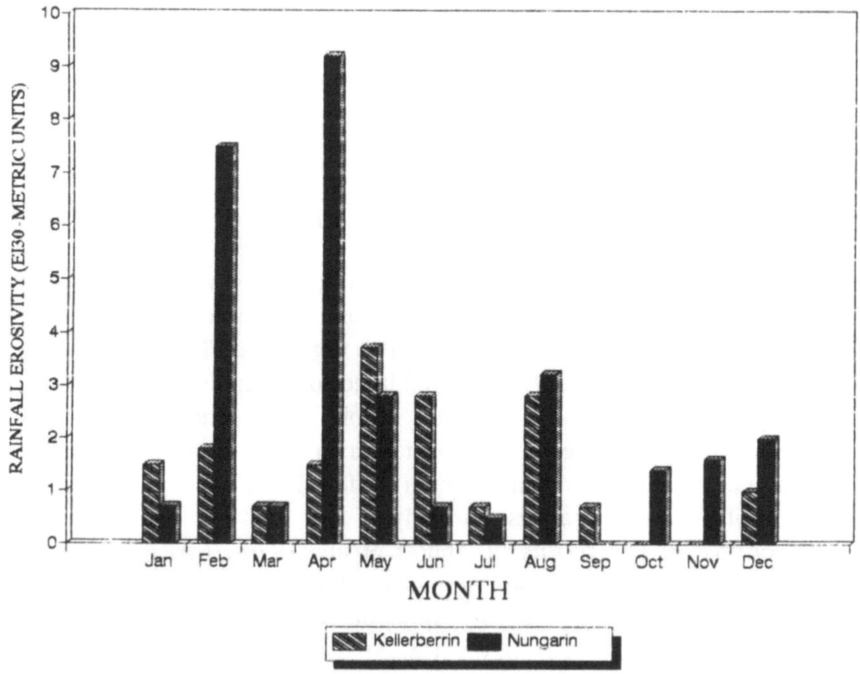

Figure 6.5. Average monthly rainfall erosivity (EI_{30} index) for Kellerberrin (1975–1982) and Nungarin (1973–1983).

Sheet erosion is common between February and April when the soils are bare and detached by sheep. When fallowing was a common practice, storms in these months caused widespread erosion. Cultivated soils are bare and detached in May and June, and rilling is common in cropping paddocks. Heavy rain in August causes little erosion because of the protective cover provided by pasture and crop canopies.

To the east of Kellerberrin, most of the erosive rain falls between November and April and, to the west, most occurs between May and October (McFarlane et al. 1986). This is because the influence of cold fronts in winter decreases with increasing distance from the coast.

In summary, Kellerberrin has a reliable, relatively low-intensity rainfall which contains low amounts of salt. However, these characteristics cannot explain the widespread occurrence of water erosion, waterlogging, flooding, and salinity in the wheatbelt.

6.3 The Effect of Clearing on Runoff and Erosion

To estimate the increase in runoff after natural vegetation is cleared, measurements can be made before and after clearing. Inferences can also be made from changes in soil properties after clearing.

6.3.1 Changes in Infiltration After Clearing

Abbott et al. (1979) reported that wheatbelt soils which had been regularly ploughed and stocked were more compacted and less permeable to water than soils under native vegetation. They found that regular cultivation virtually eliminated large soil animals which, in turn, reduced soil friability and infiltration. Hobbs et al. (Chapter 4) discuss the effects of agriculture on the soil fauna in more detail and indicate that faunal activity may remain high in farmland, but that the composition of the fauna changes markedly.

In contrast to the results of Abbott et al. (1979), McFarlane et al. 1992a reported lower sorptivities and infiltration rates in four uncleared wheatbelt soils with an undisturbed surface crust compared with nearby cleared soils. They surmised that an organic crust caused local redistribution of surface water within uncleared areas. In addition, water repellency was reduced in a gravelly soil after clearing.

Low ^{137}Cs contents have been reported in uncleared wheatbelt soils in comparison with cleared soils (McFarlane et al. 1992b). As ^{137}Cs is known to become rapidly attached to soils and act as a tracer of soil movement (Ritchie and McHenry 1975), it appears that the soils under the native vegetation had limited contact with rain containing ^{137}Cs. This is further evidence of local redistribution of surface water in areas with native vegetation.

The low infiltration rates in uncleared soils do not appear to produce significant runoff. Considerable redistribution of surface water was recorded inside the uncleared Newdegate catchment by Nulsen et al. (1986). Up to 7.7 mm of overland flow was collected from 4 m^2 plots when 30.9 mm of rain fell in one day. However, there was almost no runoff recorded from the uncleared catchment.

Nulsen et al. (1986) also found that rainfall was redistributed in the canopy of mallee vegetation. Stemflow was directed to the base of the trees, where it entered the soil along the soil-root interface. On a catchment basis, about 3% of the rainfall was lost to interception and 5% was redirected as stemflow.

Therefore, under native vegetation, rainfall is redirected both in the canopy and on the soil surface so that most of it infiltrates deeply into the soil profile at selected points. This reduces the loss of soil water by evaporation and enhances the uptake of water by roots over the summer (Nulsen et al. 1986). After clearing, infiltration is more uniform. During large storms, the runoff from uncleared areas is slowed by surface cover which results in the infiltration of more water. In cleared areas, there is less cover and the soil surface is not protected by an organic crust, so the risk of erosion is much greater.

6.3.2 Direct Measurements of Runoff

At Bakers Hill, 120 km west of Kellerberrin, streamflows were monitored from two uncleared catchments for six years before one of the catchments was cleared (Williamson and Bettenay 1979). In the four years after clearing, the annual streamflow from the cleared catchment was about eight times greater than estimated from a preclearing calibration with the uncleared catchment. The increased streamflow was not due to baseflow as the groundwater levels had not risen sufficiently after clearing to form seeps. The additional streamflow probably resulted from wetter soils caused by reduced interception and transpiration. The average annual rainfall over the monitoring period was 550 mm, much higher than Kellerberrin's average annual rainfall of 334 mm. Therefore, the effect of clearing on streamflow at Bakers Hill is only indicative of clearing in the western wheatbelt.

Runoff was monitored for twelve years from an uncleared catchment, with an annual average rainfall of 376 mm at Newdegate, 200 km SE of Kellerberrin (Nulsen et al. 1986). The catchment contained mallee and heath vegetation until it was partly cleared in 1985. Before clearing, streamflow from the catchment averaged only 63 $m^3 y^{-1}$. Since clearing, the annual streamflow increased to an average of 88 $m^3 y^{-1}$, despite the annual rainfall after clearing being 11% lower. In the twelve years prior to clearing, the peak runoff recorded on the catchment was 0.077 $m^3 s^{-1}$, whereas in the four years after clearing, the peak runoff has increased to 0.096 $m^3 s^{-1}$.

The effect of clearing on runoff rates can also be estimated from regional relationships between runoff and catchment characteristics. Flavell et al. (1987) developed the following relationship from runoff measurements on 20 wheatbelt catchments which had loamy and lateritic soils:

$$Q_5 = 0.304 * A^{0.60} * 10^{0.0052 C_L} \quad r = 0.97$$

Q_5 is the peak discharge ($m^3 s^{-1}$) for an average recurrence interval of five years, A is the catchment area (km^2), and C_L is the percentage of the catchment that is cleared.

As an example, clearing a 10 km^2 catchment is estimated to increase the runoff rates from 0.6 to 2.0 $m^3 s^{-1}$ for runoff with an average recurrence interval (ARI) of 2 years and from 2.1 to 7.1 $m^3 s^{-1}$ for runoff with an ARI of 10 years (Fig. 6.6).

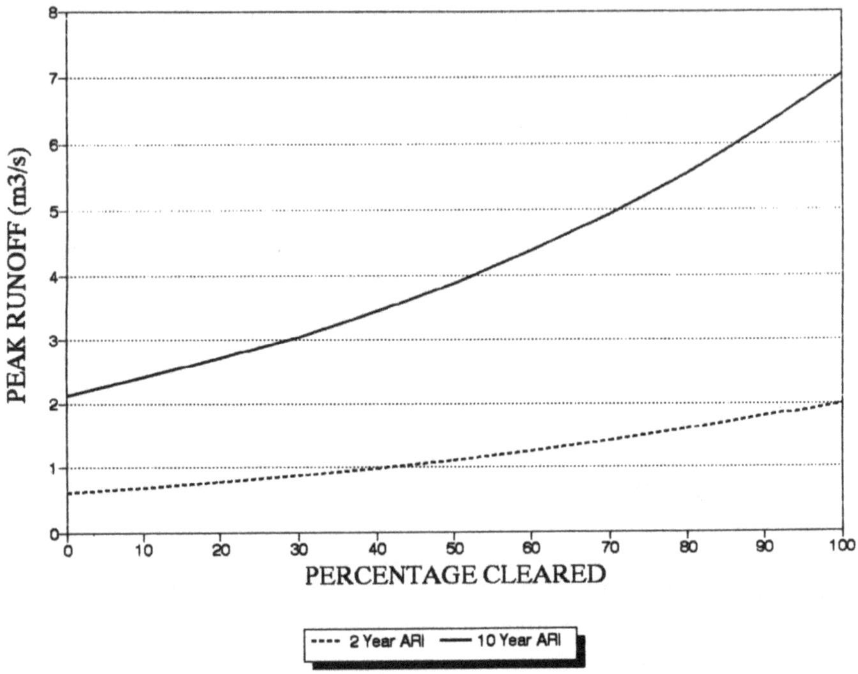

FIGURE 6.6. The effect of clearing a 10 km² catchment on the two-and ten-year average recurrence interval (ARI) peak runoff, as estimated by the method of Flavell et al. (1987).

In summary, clearing wheatbelt catchments is likely to result in an immediate increase in peak runoff rates and in annual streamflow. Runoff will further increase if the soil structure declines and when groundwater levels reach the surface and form saline seeps.

6.3.3 Runoff Mechanisms

There are two main processes which produce runoff on wheatbelt hillslopes, rainfall-excess (or Hortonian) overland flow (Topalidis and Curtis 1982) and saturation-excess overland flow (Betson 1964). Rainfall-excess overland flow occurs when the rainfall rate exceeds the infiltration rate of the soil and surface storages are exceeded. The main wheatbelt soils which produce this runoff are those with a poor structure which develop a surface seal under rainfall impact (Howell 1987), water-repellent soils (McGhie 1980), and soils compacted by vehicles and stock.

Saturation-excess overland flow occurs when the surface horizon is saturated and rain results in immediate ponding and runoff once surface storages are exceeded. The main wheatbelt soils which produce this runoff are those with shallow sandy topsoils overlying clayey subsoils (duplex soils) and those where groundwater is discharging (McFarlane and Davies 1988). Duplex soils are very common in the wheatbelt. The permeability of the subsoil is thought to have decreased markedly after clearing due to the loss of macropores, particularly those formed by tree roots (Hamblin and Howell 1987). The dispersion of clay can increase clay illuviation and further decrease the permeability of clayey subsoils. Therefore, runoff is expected to increase progressively after clearing as the permeability of both the topsoil and subsoil decreases. In addition, the expanding area of saline seeps after clearing results in saturation-excess overland flow of saline water.

6.3.4 Water Erosion of Agricultural Hillslopes

There must be both detached soil and runoff for erosion to take place. The main causes of soil detachment in the wheatbelt are raindrop impact, runoff, cultivation, and sheep hooves. Using rainfall erosivity as a guide to the ability of raindrops to detach soil, detachment will be highest in late summer and early autumn when the rainfall is most erosive (Fig. 6.5) and the soils are bare after summer. Soils are also detached by cultivation in late autumn and early winter.

Over the summer, many wheatbelt soils become crusted or hard set. One sheep can detach two-thirds of a tonne of sandy soil per week in summer and one-third of a tonne of clayey soil (Marsh 1982). Therefore, at most times of the year, water erosion is limited more by the availability of runoff than by the absence of detached soil. The short growing season in the wheatbelt results in little vegetative cover to protect the soil in summer and autumn. This explains why water erosion can be significant despite the relatively low erosivity of the rainfall.

Water erosion is evident on hillslopes below rock outcrops and lateritic breakaways. Soil losses on four hillslopes near Kellerberrin and Northam were estimated using the ^{137}Cs technique (McFarlane et al. 1992b). The losses were calculated from an equation relating the percentage of ^{137}Cs lost (relative to a reference value) to the loss of soil from long-term runoff plots (Elliott et al. 1990). Soil samples were taken downslope of two rock outcrops and two lateritic breakaways (McFarlane et al. 1992b). The amount of ^{137}Cs below the rock outcrops was much lower than below the breakaways, indicating a higher soil loss below the outcrops (Fig. 6.7). These findings were unexpected, as erosion below breakaways is usually more obvious than below rock outcrops. However, it is possible that there was more sheet erosion below the rock outcrops, which is less obvious than the gullies that can develop below lateritic breakaways (McGhie 1980).

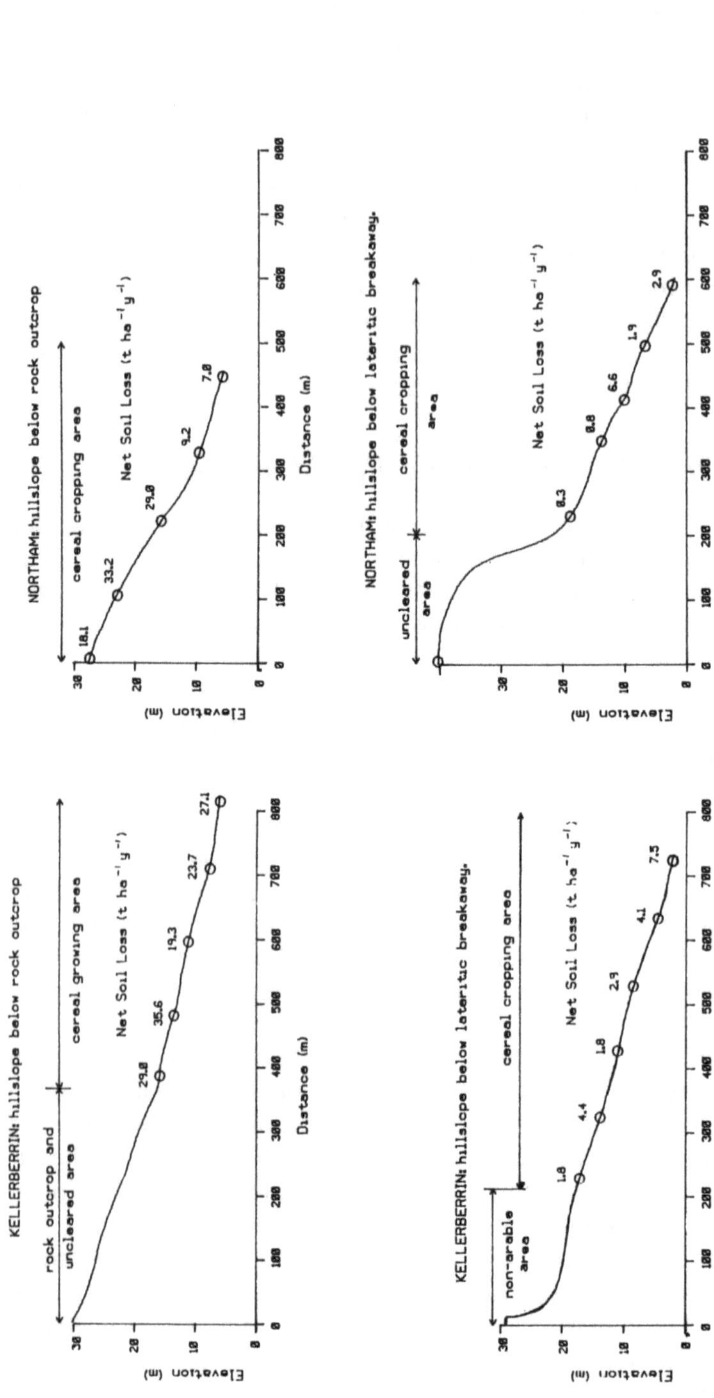

FIGURE 6.7. Estimated soil loss rates since 1954 on hillslopes below rock outcrops and below lateritic breakaways at Kellerberrin and at Northam. (After McFarlane et al. 1992b.)

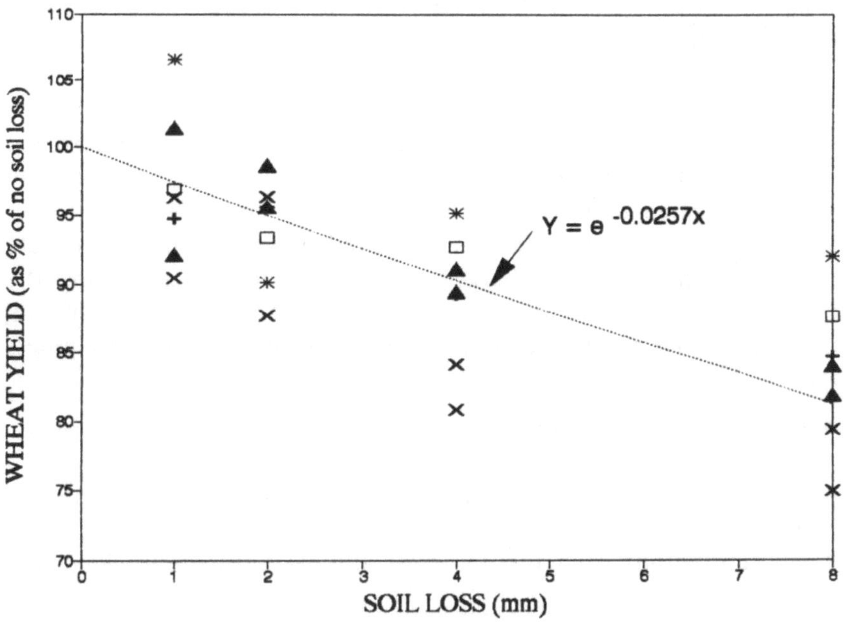

FIGURE 6.8. The effect of soil loss from seven pasture paddocks on subsequent wheat yield. (Data of Brian Marsh presented in McFarlane and Carter 1990.)

6.3.5 The Effect of Erosion on Soil Fertility

There is evidence that the erosion of wheatbelt soils results in a greater loss in crop yield than in many other parts of Australia (McFarlane and Carter 1990). An exponential decrease in wheat yield was recorded after the top eight millimetres were artificially removed from pasture soils over the summer (Fig. 6.8). The yield decline was correlated with the removal of organic nitrogen, which was concentrated at the top of the soil profile under pasture.

Wheat yields were depressed by 10 to 50%, 16 years after topsoil had been moved to infill gullies (McFarlane and Carter 1990). Some recovery in yields occurred after 27 years. However, where there is a high incidence of water and wind erosion, complete recovery is unlikely and the effect on soil fertility will be cumulative.

Average wheat yields in the wheatbelt have increased little in the past 25 years. Increased yields should have resulted from improved crop varieties, fertilisers, weed and insect control, and machinery. The expected increase may have been offset by reductions in soil fertility as a result of soil degradation.

6.4 The Effect of Clearing on Groundwater Levels and Salinity

The main type of aquifer responsible for secondary salinity in the wheatbelt is a semi-confined (sometimes confined), coarse-textured saprolite aquifer underlain by bedrock and overlain by the intensely weathered pallid zone of the laterite profile (George 1992a) (Fig. 6.9). There are also three types of unconfined aquifers which contribute significant amounts of recharge to the main aquifer. Managing the unconfined aquifers is often the most efficient method of restricting recharge to the main aquifer. The three types of unconfined aquifers are:

- ephemeral perched aquifers, which develop in shallow (<1 m) duplex soil profiles;
- perched aquifers, which develop in deep (1 to 8 m) sandplain soils; and
- aquifers within deep (up to 25 m) alluvial channels in floodplain sediments.

The unconfined aquifers are discussed below, before the features of the main aquifer are detailed.

6.4.1 Perched Aquifers in Duplex Soil Profiles

The majority of wheatbelt soils have a permeable sandy topsoil overlying a low-permeability clayey subsoil. In winter, an ephemeral, unconfined aquifer develops when water becomes perched on the clayey subsoil. Duplex soils conserve soil water because the sandy topsoil aids infiltration but reduces evaporation, while the clayey subsoil inhibits the soil water from draining below the root zone.

In wet winters, and in parts of the landscape where runoff and shallow seepage accumulate, the aquifer in duplex soils can cause waterlogging of the root zone. To the west of Kellerberrin, waterlogging causes millions of dollars of lost crop and pasture production each year (McFarlane and Wheaton 1990). Waterlogged areas contribute recharge to the main aquifer (Fig. 6.9) and limit what can be grown in saline discharge areas. As the macropores through the clay subsoil become blocked, waterlogging will become a more serious problem. Collgar (midslope) soils (McArthur, Chapter 2) are most noted for waterlogging due to a siliceous or ferruginous pan at the top of the clay subsoil. Perching is also common on Booraan (pediment), Danberrin (rocky hills), and Belka (flat valley) soils in wet years.

Recharge of the duplex soil aquifer is mainly from in situ rainfall and from runoff and shallow seepage from upslope areas. The amount of

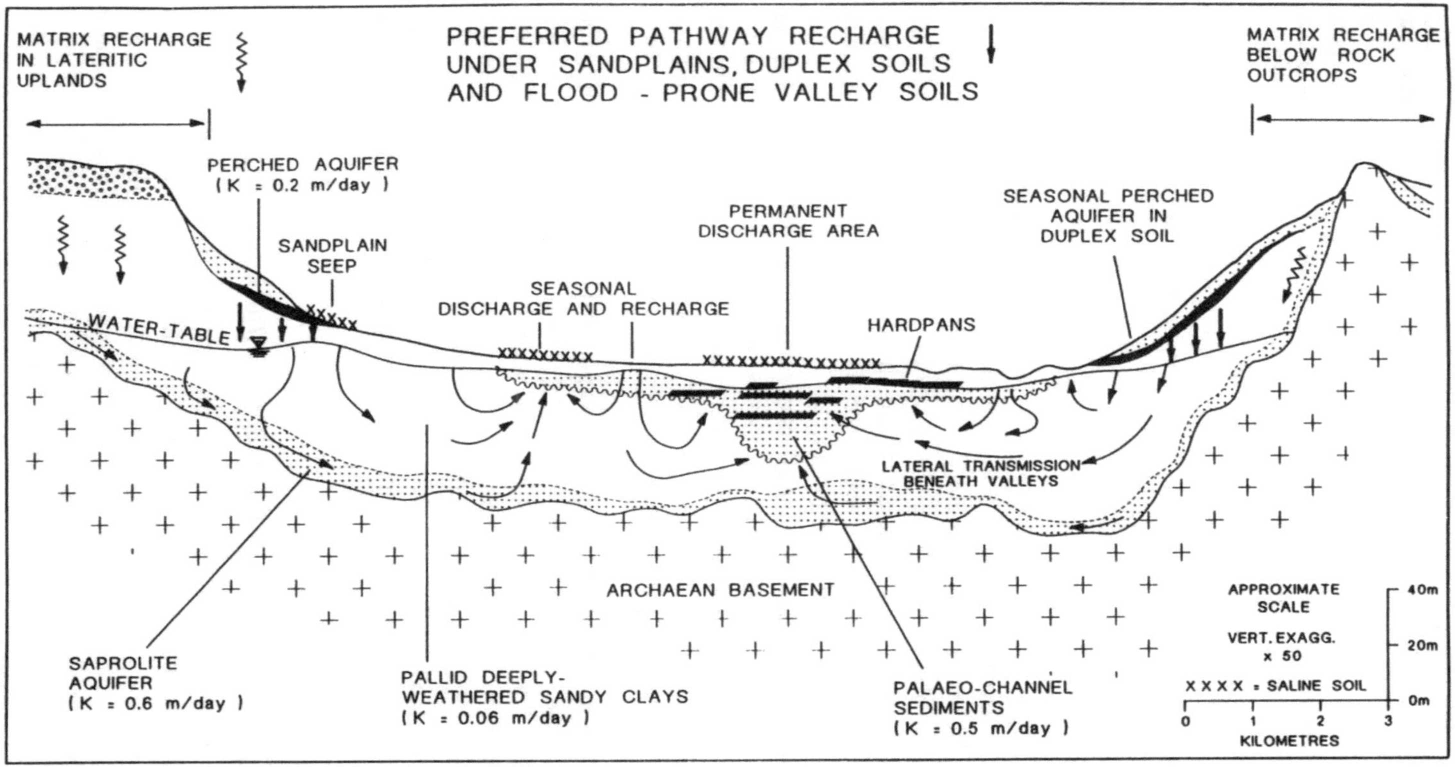

FIGURE 6.9. Diagrammatic cross-section across a wheatbelt valley showing the location of the three unconfined aquifers and main saprolite aquifer in relation to recharge (George et al. 1991).

recharge can be estimated from the amount of water discharging from seepage interceptor drains constructed with their channel into the subsoil clay. Between 1 and 7% of the annual rainfall (4 to 31 mm) has been collected by these drains at Narrogin and Cuballing, 140 km SSE of Kellerberrin (McFarlane and Cox 1990). Discharge is by evaporation, transpiration, discharge to streams, and leakage through the subsoil clay to recharge the main aquifer.

The salinity of the groundwater in the aquifer is low (<200 mg total soluble salts (TSS) l^{-1}), unless deeper groundwaters have discharged into the aquifer. The water can, therefore, be diverted to dams.

6.4.2 Sandplain Aquifer

In the upper parts of the landscape, areas of deep sandy loam (Norpa) are underlain by clays with a low permeability (Bettenay et al. 1964). The mean hydraulic conductivity of the sandy loam is about 0.15 m d^{-1} while that of the underlying clay is about 0.07 m d^{-1} (George 1990b). Seeps of brackish water occur where the depth of sand becomes less than about 2 m. Sandplain seeps are the most common form of secondary salinity east of Kellerberrin.

Recharge of the sandplain aquifers occurs over all of the sandplain but may be highest in areas which supported wodjil vegetation (*Acacia* spp.) before clearing. These soils have very acidic subsoils and aluminium toxicity restricts the depth to which roots can extract water. George (1992b) used flow net analyses to estimate recharge to be between 6 and 13 mm y^{-1}.

Sandplain aquifers discharge by evapotranspiration when uncleared and mainly by evaporation from seeps after clearing. There is also evidence that leakage through macropores in the underlying clay is significant beneath sandplain seeps (George 1992b). The leakage contributes recharge to the main saline aquifer (Fig. 6.9).

Salt storages in the sandplain are low (<50 t TSS ha^{-1}) and where they have not been concentrated by evaporation, groundwaters are usually < 5000 mg TSS l^{-1} (George 1992b).

6.4.3 Aquifer Within Alluvial Channels

Many wheatbelt valleys contain deep sandy channels within clayey floodplain sediments (Fig. 6.9). The unconfined aquifer which forms in the channels has a mean hydraulic conductivity of about 0.6 m d^{-1} (George 1992a). Some channels contain water fresh enough to water stock. However, the channels in the Wallatin Creek catchment near Kellerberrin (Chapter 3, Fig. 3.1) have groundwaters with a salinity of 34,000 mg TSS l^{-1}.

Recharge to the aquifer is thought to occur mainly during flood runoff on the catchments (George and Frantom 1990a) and by groundwater flow from adjacent aquifers. Runoff waters have been observed to soak into sandy parts of valleys. Discharge is mainly by evaporation and by leakage to the surrounding floodplain sediments and main aquifer, if there is a hydraulic continuity between them. Flow to playas is thought to be small due to low hydraulic gradients (<1:500).

6.4.4 Main Aquifer

This semi-confined (sometimes confined) aquifer has a mean hydraulic conductivity of about $0.6\,m\,d^{-1}$, which is about ten times higher than that of the overlying pallid zone (George 1992). Some flow is likely in fractures in the bedrock.

Recharge to the main aquifer was thought to be concentrated in small parts of the landscape, but work in the last 20 years has progressively increased the areas within the wheatbelt catchments that are considered to contribute recharge.

Smith (1962) identified sandy and lateritic soils high in the landscape as contributing recharge to the main aquifer. Bettenay et al. (1964) showed that arkosic sands below granite outcrops could contribute large amounts of recharge. Nulsen and Henschke (1981) proposed that differences in surface and subsurface infiltration rate could be used to define recharge areas. From such measurements, Nulsen and Baxter (1982) estimated that up to 25% of one wheatbelt catchment had the potential to contribute recharge.

Recharge through preferred pathways in the pallid zone was inferred from the distribution of salt in the Darling Range SSE of Perth by Peck et al. (1981). Johnston et al. (1983) and Johnston (1987) showed that these preferred pathways could be root channels, quartz veins, and areas adjacent to dolerite dykes. Preferred pathway recharge was identified in agricultural land near Narrogin, 140 km SSW of Kellerberrin (Engel et al. 1989). The areas thought to produce preferred pathway recharge had perched groundwater in the soil profile and higher salt storages than coarser-textured soils, in which matrix recharge was thought to occur. Water flowing in the preferred pathways probably by-passes the clayey areas storing the salt. From the same area, George and Conacher (1992) reported a shallow lens of low salinity groundwater underneath areas with waterlogged soils. This is further evidence of recharge from the perched aquifer in duplex soils to the main saline aquifer.

Recharge has also been reported in the floodplain of wheatbelt catchments (McFarlane et al. 1989; George and Frantom 1990a). Recharge during the wet years was considered to be considerable as plant water use was low because of floodwaters inundating the valleys for several months of the year. Some soil profiles in wheatbelt valleys have ferruginous and siliceous hardpans, which can restrict such recharge.

The area believed to contribute recharge now includes most of the landscape (George et al. 1991). Using responses in groundwater levels, Salama et al. (1991) identified recharge throughout the year under 37% of a wheatbelt catchment, seasonal recharge and discharge under 57% and discharge under only 6%. This has implications for management which will be discussed later.

The amount of recharge responsible for the rising groundwater levels in the Darling Range area, SSE of Perth, has been estimated to be between 23 and 65 mm by Peck and Hurle (1973). In the drier wheatbelt, the amount of recharge is less than 10 mm y^{-1} (George 1992c). Nulsen et al. (1986) estimated that recharge was negligible in an uncleared catchment near Newdegate. George (1992c) estimated that for the Wallatin Creek and six other eastern wheatbelt catchments, the pre-clearing recharge rates were about 0.05 mm y^{-1}. Post-clearing rates were about two orders of magnitude higher.

Discharge from the main aquifer is mainly by evaporation from hillside seeps and valley floors. The aquifers in most wheatbelt catchments have low hydraulic gradients and transmissivities resulting in low rates of groundwater flow (George 1992a). Therefore, groundwater outflow to the main valleys containing the playas is not considered a major form of groundwater discharge since clearing.

In many wheatbelt catchments, the main aquifer is already full of water and the pallid zone aquitard is now filling. The aquitard has a higher porosity (because of the many small pores in the clay) and contains more stored salt than the main aquifer. As most management practices (e.g., drains and trees) are carried out in the pallid zone aquitard and on the colluvial and alluvial sediments in the valleys, these are important parts of the groundwater system.

Many hillside seeps in the western wheatbelt are caused by dolerite dykes (Fig. 6.10), which weather to less permeable clay than the clay and quartz which are formed from weathering granite and granitic gneiss (Engel et al. 1987). The dykes are often not apparent in the valleys as they are covered by alluvial sediments. They can be located using magnetometers as dolerite has a higher magnetic susceptibility than granite.

In the Kellerberrin area, hillside seeps mainly form behind ridges of basement rock called bedrock highs (McFarlane and George 1992). These ridges are generally broad (>100 m), which makes them difficult to breach. The dykes and bedrock highs break up large catchments into smaller units of groundwater flow. Other causes of hillside seeps are reductions in the thickness of the aquifer above the bedrock (George and Frantom 1990b) and decreases in slope (Fig. 6.10).

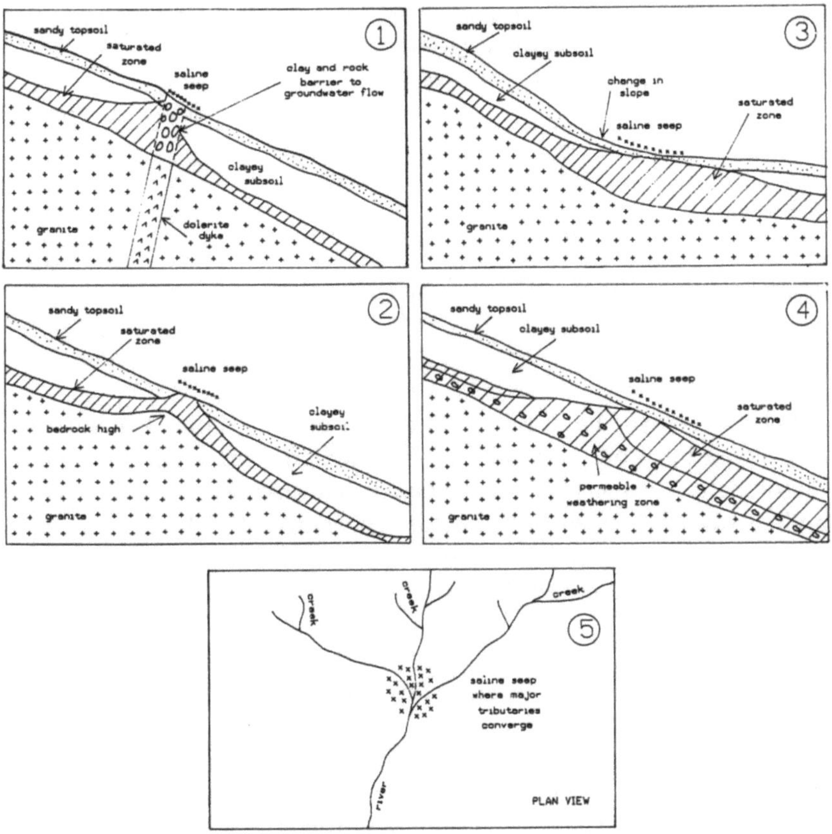

FIGURE 6.10. Types of discharge from the main saline aquifer: (1) dolerite dyke; (2) bedrock high; (3) reduced slope; (4) change in aquifer thickness; and (5) groundwater convergence where valleys meet.

Valley salinity can also occur behind restrictions to groundwater flow (Lewis 1991), but the presence of alluvial sediments in the valleys makes the reason for discharge more difficult to determine. Large areas of salinity occur at the confluence of streams and may result from an overloading of the aquifer where groundwater flows converge (McFarlane and George 1992). Lewis (1991) postulated that shear and fault zones may preferrentially carry groundwater, and saline seeps develop where these zones intersect barriers to flow (e.g., dolerite dykes and basement highs) or other shear zones. Groundwater flow in many wheatbelt valleys is likely to be mainly vertical, as the valleys are very flat and groundwater gradients are low.

Discharge of groundwater from saline seeps is low in the absence of vegetation. In a bare hillside seep north of Kellerberrin, only about 35 mm of groundwater was discharged each year (George and Frantom

Table 6.2. Average total soluble salts (TSS) in storage in soils in the Wallatin Creek and North Baandee catchments (after McFarlane and George 1991). Soil types are described by McArthur, Chapter 2.

Soil Type	Average TSS Storage ($t\,ha^{-1}$)	Range TSS Storage ($t\,ha^{-1}$)	Number of Profiles	Average Depth (m)
Danberrin	247	7–657	7	5.9
Ulva	289	139–422	5	19.7
Booraan	802	43–1,798	10	13.5
Collgar	1,056	109–2,231	12	16.5
Belka	2,571	44–6,206	14	20.4
Baandee	13,533	5,752–21,314	2	51.0

1990b). Greenwood and Beresford (1980) measured only 0.4 mm day^{-1} of evaporation from bare saltland near Kellerberrin, while from piezometric data, George (1992a) estimated that discharge was between 50 and 230 mm y^{-1} in wheatbelt areas. The reason for the low discharge from saline seeps may be due to the low hydraulic conductivity of the soils (Thorburn et al. 1991) and deeply weathered materials (George 1992a).

The amount of salt stored in the lateritic profile increases markedly from the upslope rocky (Danberrin) and lateritic (Ulva) soils to the heavier-textured soils (Belka and Baandee) in the valley (Table 6.2). The depth of the regolith also increases towards the valleys. The amount of stored salt can easily be accounted for by salt in the rainfall. It would take only 40,000 years to accumulate 1000 t NaCl ha^{-1} if none of the 25 kg NaCl y^{-1} of salt in the rainfall was drained from wheatbelt catchments. Drainage is so poor that it is feasible that almost no salt is exported from the main valleys containing saline playas.

6.5 Decreasing Runoff and Erosion

Practices which improve soil structure (e.g., applying gypsum, adopting minimum tillage, and retaining stubble) have been shown to increase infiltration and therefore reduce runoff (Howell 1987; Nulsen Chapter 5). Many of these practices are being adopted by farmers in the wheatbelt as they increase crop yields, save time at seeding, and reduce the need for earthworks to control runoff.

Most wheatbelt farms have some earthworks for controlling water erosion and flooding, although many areas remain untreated. After severe erosion in the 1960s, small grade banks were installed on many farms to break up long slope lengths and to carry runoff at nonerosive velocities to waterways (usually natural drainage lines). In areas where the drainages became gullied, some farmers installed large level banks which stored the water in their channel rather than carrying it to drainage lines.

Whereas the level banks reduced the need for rebuilding the banks and reduced the gullying of drainage lines, they have several disadvantages:

- the banks cost about five times more to build and occupy about three times more land than the smaller, grader-built banks. An economic study showed that the greatest cost of earthworks was the lost production from the land occupied by the earthworks (Salerian and McFarlane 1987);
- the banks can increase salinity by diverting fresh runoff and shallow seepage waters through the subsoil, where they pick up salt and raise groundwater pressures in the main aquifer (McFarlane et al. 1990b); and
- by storing water above ground, the banks can cause severe erosion when they fail.

Level banks are used for controlling floodwaters as well as for mitigating water erosion. Flooding in the wheatbelt damages fences, roads, and railways. In addition, inundation reduces access to paddocks. Davies et al. (1988) showed that level banks can reduce the peak flow at critical road and rail crossings if they are strategically placed in a catchment. However, the banks have limited use in preventing runoff from major storms from entering the broad valleys as the volumes of runoff greatly exceed the capacity of the banks' channels. Therefore, it may be more economical to accept that the valleys will be periodically flooded and install relief drains, which remove the ponded waters as quickly as possible (Davies et al. 1988). Removing the water has short term benefits (reduces waterlogging and improves vehicular access) as well as long term benefits (reduces groundwater recharge). However, there can be problems in finding suitable areas to discharge the drained water.

Levee banks have been installed in some wheatbelt catchments to prevent floodwaters spreading out over the floodplain. These schemes have been effective when there is widespread adoption by landholders. Unfortunately, many schemes have not been properly designed or constructed and there have been difficulties in taking the channelled waters across roads and railway lines (Davies et al. 1988).

A major problem in the wheatbelt is the ponding of water behind inadequate culverts at road and rail crossings. These areas often become bare of vegetation as they are affected by both waterlogging and salinity. Barrett-Lennard (1986) showed that many plants are unable to cope with these combined problems. Once bare, the areas erode, which silts the culverts under roads and railways. Floodways, which are at the same level as the valley, reduce the problem of water ponding behind the crossings. However, they result in traffic hazards due to water flowing over the road.

In summary, improving soil structure can reduce runoff and the need for earthworks in wheatbelt catchments. Properly designed and maintained earthworks can be very effective in reducing water erosion and flood damage. However, poorly designed or maintained earthworks can cause more problems that they solve.

6.6 Decreasing Groundwater Levels and Salinity

Revegetating cleared catchments with plants with a high water use can lower groundwater levels by reducing the amount of recharge and increasing the amount of discharge through greater foliar interception and extracting more soil water. The two most important plant characteristics which promote high water use are the depth of rooting and the length of time that the plant carries transpiring leaves. Discharge is increased by salt-tolerant plants extracting groundwater from saline seeps.

Vegetation which can decrease recharge in the wheatbelt includes:

- Annual crops such as cereals and lupins, rather than annual pastures;
- Trees, mainly *Eucalyptus* species;
- Perennial fodder shrubs such as tagasaste (*Chamaecytisus palmensis*); and
- Remnants of uncleared or regenerating native vegetation.

Vegetation which can increase discharge in the wheatbelt includes:

- Salt-tolerant trees, mainly *Eucalyptus* species;
- Salt-tolerant shrubs, mainly *Atriplex* and *Maireana* species; and
- Salt-tolerant grasses such as *Puccinellia ciliata* and *Paspalum distichum*.

6.6.1 Annual Crops and Pastures on Recharge Areas

The evapotranspiration of annual crops and pasture has been measured using the ventilated chamber method (Greenwood et al. 1985) at Kondut (130 km NW of Kellerberrin) and Cunderdin (45 km W of Kellerberrin) by Nulsen and Baxter (1982) and 30 km N of Kellerberrin by Farrington et al. (1992). The crops transpired more water than the annual pasture in each of the studies (Table 6.3). The distribution of rainfall affected the transpiration of the crops, particularly lupins. At Cunderdin and North Kellerberrin, autumn rain enabled the lupins to grow early and have a higher transpiration rate than the cereals. The failure of the lupins in dry years (e.g., at Kondut in 1980) can result in little water being transpired.

Table 6.3. Mean daily evapotranspiration rates over the growing season for agricultural plants at three locations in the wheatbelt.

Location	Mean Daily Evapotranspiration Rate for: Wheat	Barley (mm day^{-1})	Lupins	Pasture	Growing Season Rainfall (mm)
Kondut*	0.74	0.99	0.51	0.24	162
	(155)***	(155)	(162)	(155)	
Cunderdin*	0.73	1.08	1.22	0.66	258
	(162)	(162)	(162)	(145)	
North	1.41	–	1.50	1.12	199
Kellerberrin**	(136)	–	(165)	(160)	

*Nulsen and Baxter (1982).
**Farrington et al. (1992).
***Values in brackets are the length of growing period in days for each agricultural plant.

The crops used more water because they maintained higher evapotranspiration rates than pasture for longer periods in spring (Fig. 6.11). The differences in water use have been attributed to the differences in rooting depth, leaf area index, and biomass. Clover roots penetrate to depths of less than 0.8 m in loamy sands (Sedgely et al. 1981), whereas wheat, barley, and lupin roots penetrate to 2.5 m (Hamblin and Tennant 1987). At Cunderdin, barley attained the highest leaf area index, followed by lupins, wheat, and clover (Nulsen 1984). At North Kellerberrin, the much higher transpiration rate of lupins over wheat and pasture in late winter was due to greater biomass. Doubling the recommended plant density of wheat and lupins did not significantly increase the water use of crops over the growing season (Farrington et al. 1992).

Nulsen and Baxter (1982) used a simple water balance model to estimate the mean annual recharge under alternative crop rotation systems. Much less recharge occurred in rotations where lupins replaced clover. A barley-lupin rotation reduced annual recharge by 50% compared with a wheat-clover rotation. Nulsen (1983, 1984) concluded that changes in crop rotation could reduce the spread of salinisation but would be unable to reverse the trend.

The capacity to increase water use depends on the distribution of rainfall during the growing season and the soil's ability to store water. If the growing season commences early, crop growth is enhanced and high water use can be achieved. A major weakness with winter growing annual species is that they are unable to use most of the rain which falls outside the growing season.

Overcoming nonsaline forms of soil degradation will improve crop and pasture productivity and reduce recharge. Nulsen (1991) identified deep ripping to remove hardpans and reducing soil acidity as methods

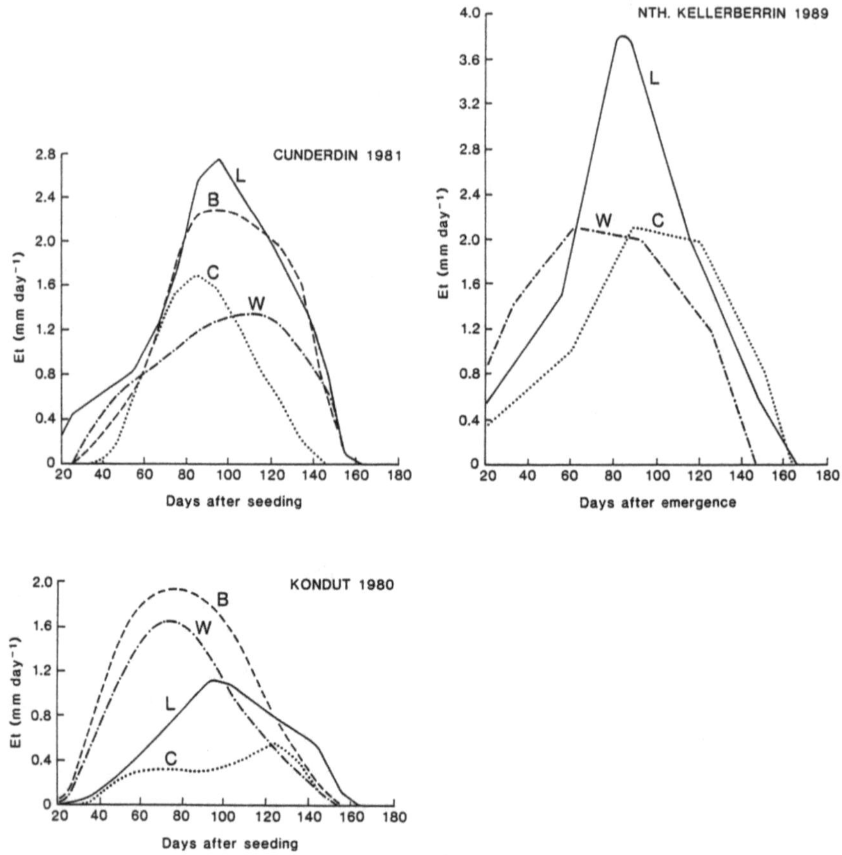

Figure 6.11. Seasonal evapotranspiration rates (Et) for wheat (W), barley (B), lupins (L), and clover-based pasture (C) at Cunderdin, North Kellerberrin, and Kondut.

of improving soil water extraction from deep sandy soils. Similarly, draining waterlogged duplex soils and clay flats will increase rooting depth and soil water extraction in these soils. Improving soil structure enhances infiltration and the deep drainage of soil water. Provided that the improved soil structure increases plant growth, the increased infiltration in well-structured soils should not result in increased recharge.

6.6.2 The Economics of Manipulating Recharge by Agronomic Means

A dynamic, nonlinear programming model was developed to examine the long-term (50 years) economic effects of managing recharge using crops and pastures. In one case study, the recharge and discharge areas were on the same farm (Salerian et al., in preparation) and, in another, the discharge was on the same farm and on a neighbouring farm (Salerian 1991). The single farm model tested whether farmers were suboptimally managing their recharge areas due to an imperfect knowledge of the hydrologic and economic effects of their actions; the two farm model tested whether suboptimal management was due to failing to account for the effect on their neighbour's farm.

The single farm model showed that there was little difference in the amount of salinity after 50 years between the long-term strategy (i.e., managing the recharge area with regard to the effect on salinity) and the short-term strategy (i.e., growing the most economic crop irrespective of the effect on salinity; Salerian et al., in preparation). This was because, in many cases, the most economic crops also were the highest water users. In addition, there is little economic incentive to adopt land uses which are less profitable on large recharge areas in order to have an effect on the comparatively small discharge area. Growing saltbushes on the saline area reduced the marginal cost of salinity and was a profitable means of dealing with soil salinity, as there was a deficiency of sheep feed in autumn, which is the best time for saltbushes to be grazed.

The two farm model showed that there was little incentive for the farmers to negotiate or bargain in order to reduce the area of salinity through modifications in land use. This is because the recharge area is generally much larger than the likely difference in the size of the discharge area (Salerian 1991). The cost of obtaining the necessary information and negotiating a settlement would exceed the benefits that could be achieved. In addition, there is uncertainty regarding the outcomes of any changes due to modified land use.

In conclusion, farmers could improve their profitability while, at the same time, reducing the rate of salinisation by adopting the optimum landuse for each soil on their farm, independent of the effect on salinity. Matching crops, pastures, and trees to the soil types improves plant growth and maximises profits. The plants also use water which would otherwise run off or become recharge. Whereas this may not be the long-term optimum land use, it is much closer to the long-term optimum than the current land use and would be a significant improvement. Emphasis should be placed on encouraging farmers to maximise short-term profits as most improvement in salinity control will be made by such a change, and there is a strong financial incentive for farmers to adopt the improved practices.

6.6.3 Native Vegetation on Recharge Areas

A study which measured evapotranspiration of the native vegetation in the Durokoppin Nature Reserve showed that the vegetation transpired

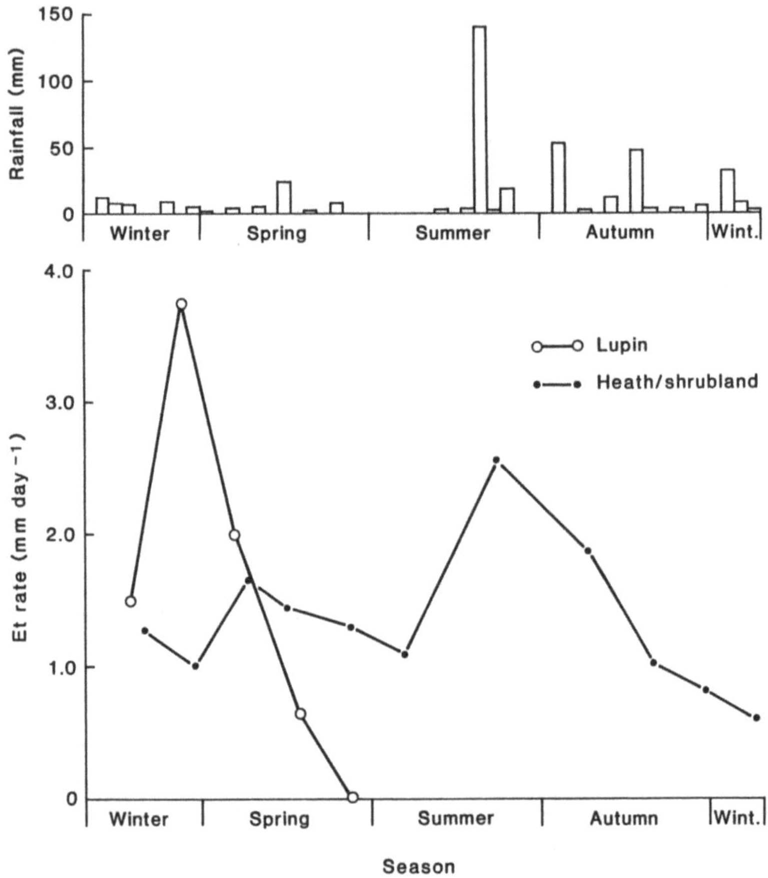

FIGURE 6.12. Rainfall and evapotranspiration rates for native heath/shrubland and lupins in the Wallatin Creek catchment.

rain which fell outside of the growing season for annual crops and pastures (Farrington et al. 1992). During the growing season, evapotranspiration rates of the native communities were lower than the agricultural plants until midspring, when the annuals started to mature (Fig. 6.12). During late spring and early summer, when rainfall was low, the daily evapotranspiration rate from the native vegetation declined. A further decline in evapotranspiration rates during summer would have been expected in an average year when little or no rain falls. However, the area received above average rain in summer and autumn (Fig. 6.12). Following the rain, there was a rapid rise in daily evapotranspiration by the native communities to above 2 mm in early February. In early autumn, the rate was maintained above the rates recorded in the previous winter and spring (Fig. 6.12).

Thus, the perennial plants were able to transpire a significant amount of rainfall in summer and autumn, when most of the rain would have run off, increased soil water storage, or become recharge under annual crops and pastures. Nulsen et al. (1986) found that rainfall was balanced by evapotranspiration in an uncleared catchment and so there was generally no runoff or net change in soil moisture storage. Annual evapotranspiration from annual pasture was about half that of regenerating *E. wandoo* near Collie (Greenwood and Beresford 1982) and about one-seventh that of *Eucalyptus* species in a five-year-old plantation at Bannister (Greenwood et al. 1985).

Heath and shrublands play an important role in the contribution to the evapotranspiration of native plant communities. In the jarrah forest near Dwellingup (195 km SW of Kellerberrin), 37% of the annual rainfall was evaporated by the understorey, litter, and soil components (Greenwood et al. 1985). In the banksia woodland near Perth, evapotranspiration from the same components accounted for about two-thirds of the annual rainfall (Farrington et al. 1989).

Most studies on trees which grow in the wheatbelt have been carried out in areas with a higher annual rainfall than Kellerberrin. Greenwood et al. (1985) found that interception losses from *Eucalyptus* species in five-year-old plantations ranged from 16% to 37%. Sharma (1984) estimated that in a jarrah-marri (*E. marginata, E. calophylla*) forest in a 1100 mm annual rainfall area, interception losses were up to 40% of annual rainfall.

The amount of water transpired by trees depends on species, age, soil type, seasonal conditions, and depth and salinity of the groundwater. Greenwood et al. (1985) compared the evapotranspiration rates of three *Eucalyptus* species in a five-year-old plantation in an area with an average annual rainfall of 750 mm. The annual evapotranspiration of the trees ranged from 2300 mm for *E. maculata* to 2700 mm for *E. cladocalyx* and *E. globulus*, well in excess of the 648 mm rain recorded during the study period. Excavations showed that tree roots had reached the watertable, which was 5 to 7 m deep.

Daily transpiration of mature *E. wandoo* and *E. salmonophloia* were monitored by Farrington et al. (1992) in Durokoppin Nature Reserve using the heat pulse technique (Edwards and Warwick 1984). Evapotranspiration was consistently higher in *E. wandoo* than *E. salmonophloia* during spring and summer (Fig. 6.13). Rates in *E. wandoo* rose steadily in late spring to reach 70 l day^{-1}. During summer, the rates declined to below 50 l day^{-1} as soil water became limiting. Heavy rainfall in late January resulted in increased transpiration (Fig. 6.13). Later in summer and in autumn, the rates declined steadily in response to either reduced soil water supply or the lower evaporative demand of the atmosphere. *E. salmonophloia* tended to be more conservative in water use, with transpiration rates ranging from 20 to 40 l day^{-1}.

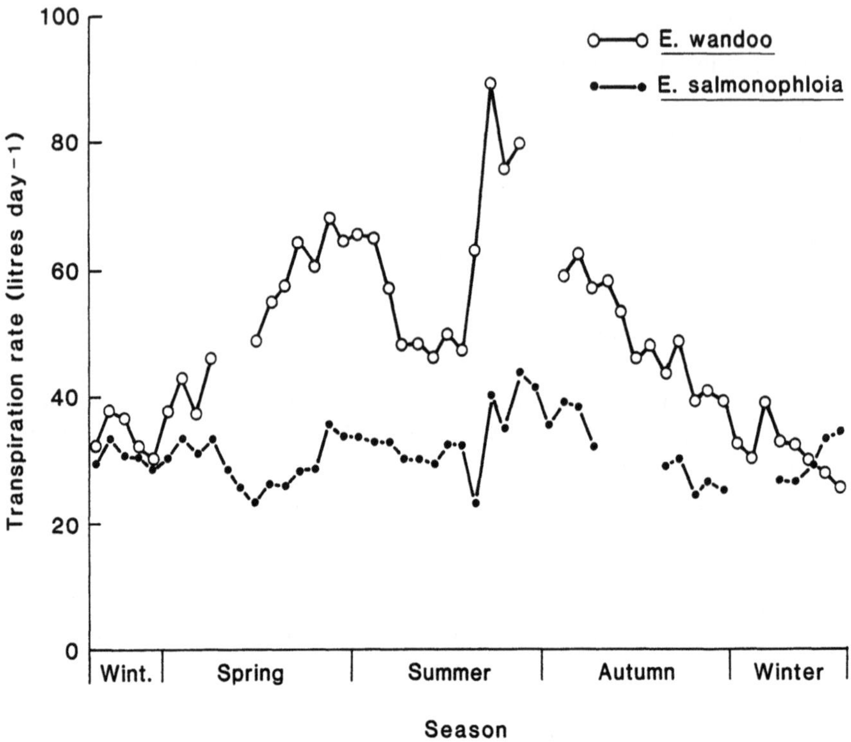

FIGURE 6.13. Seasonal transpiration rates for *E. wandoo* and *E. salmonophloia* in the Durokoppin Nature Reserve, Wallatin Creek catchment.

To determine the effect of the areas of remnant vegetation on groundwater levels, transects of bores were drilled through the Durokoppin and Kodj Kodjin Nature Reserves (McFarlane and George 1992). Groundwater levels were up to 7 m lower under the vegetation than under nearby cleared areas (Fig. 6.14). Comparable landscape positions in an adjoining catchment were saline where the native vegetation had been cleared. The catchment containing the reserves remains 22% vegetated and is less than 0.1% salt-affected. In contrast, the adjoining catchment with only 4% vegetation has over 2.8% of its area salt-affected (McFarlane and George 1992).

6.6.4 Fodder Shrubs on Recharge Areas

A leguminous shrub, tagasaste (*Chamaecytisus palmensis*), is a promising perennial fodder shrub for the wheatbelt, which is suitable for deep sandy soils but not for areas prone to waterlogging or salinity. Engel and Scott (unpublished data) found that, at a site with an annual rainfall of 700 mm, annual evapotranspiration from ungrazed tagasaste shrubs

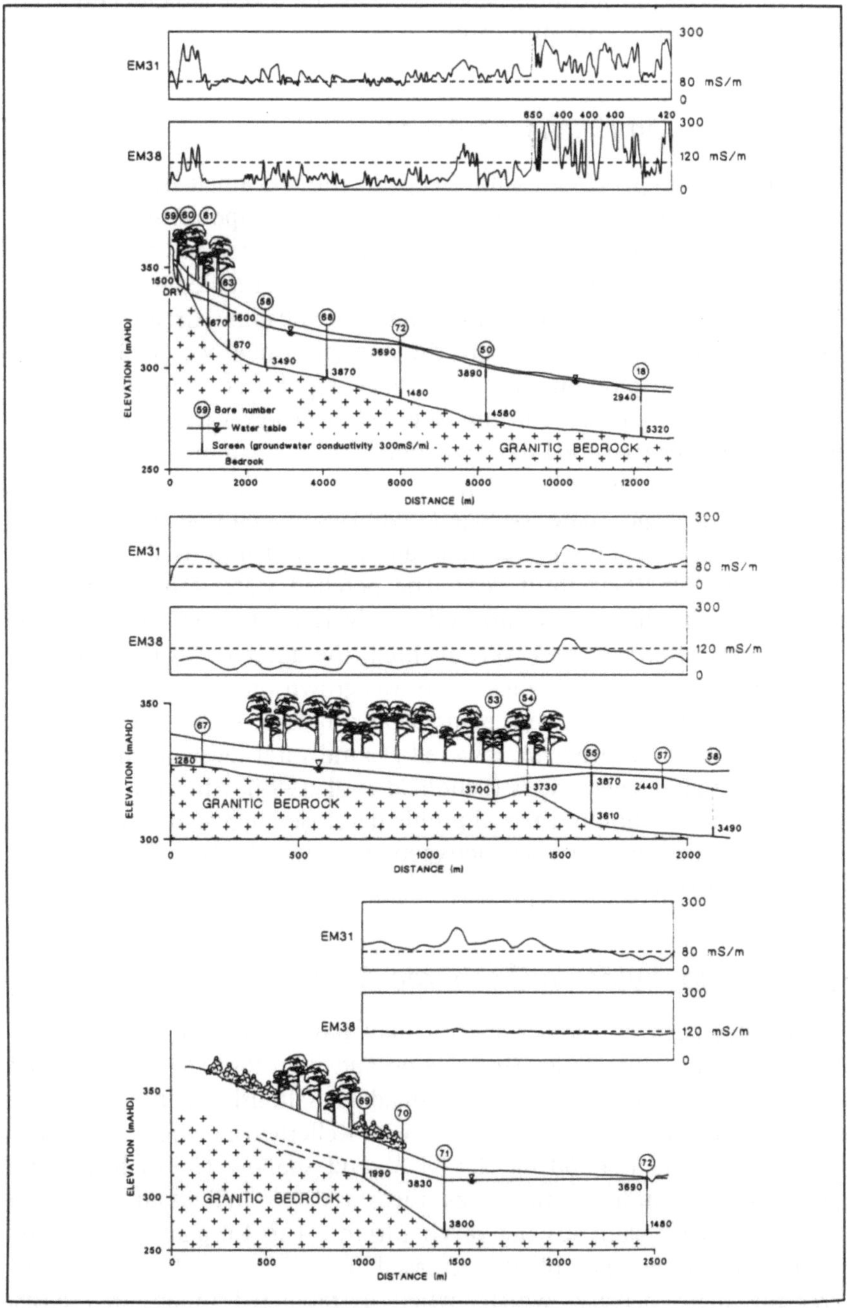

FIGURE 6.14. The effect of native vegetation on groundwater levels, as shown by transects of bores in the Wallatin Creek catchment (McFarlane and George 1992). EM31 and EM38 measurements are electromagnetic terrain conductivity in the top six and one metres, respectively.

with a density of 600 plants per hectare was 400 mm. Near New Norcia (annual rainfall of 500 mm), tagasaste has dried out a sandplain seep (R. Speed, personal communication).

6.6.5 Trees on Discharge Areas

Trees in plantations above saline seeps in the wheatbelt transpire large amounts of water during summer. In March, the evapotranspiration rate from several species of young *Eucalyptus* trees averaged 27 l day^{-1} $tree^{-1}$ (equivalent to 2.2 mm day^{-1}) at a site with a water table between 2 and 5 m deep and a mean annual rainfall of 500 mm (Greenwood and Beresford 1979). At a site where the annual rainfall was 420 mm, the mean evapotranspiration rate was 19.5 l day^{-1} $tree^{-1}$ (equivalent to 1.6 mm day^{-1}). At the 500 mm site, *E. cladocalyx* var. *nana* had the highest rate per tree (30 l day^{-1}) while, at the 420 mm site, *E. wandoo* had the highest rate (26 l day^{-1}).

Using hydrological data, George (1990b) estimated that trees growing upslope of a sandplain seep (annual rainfall of 330 mm) used at least 10 l day^{-1} $tree^{-1}$ in their third year. The perched water table under the plantation was shallow (<2 m) and relatively fresh (5000 mg TSS l^{-1}). Three years after the trees were planted, the aquifer was dry over the summer and a successful wheat crop was grown on the seepage site.

Strips of salt-tolerant trees planted on a hillside seep near Narrogin, 150 km SSW of Kellerberrin, lowered groundwater levels in summer by up to 2.8 m after seven years (Schofield et al. 1989). After reforestation, the salt concentration of the soil profile decreased between 0 and 0.75 m below the surface, increased between 0.75 and 2.25 m, and was unchanged below 2.25 m.

6.6.6 Halophytes on Discharge Areas

Halophytes such as *Atriplex* species may have a role in discharge enhancement in saline areas. Greenwood and Beresford (1980) measured evapotranspiration of *A. vesicaria* (bladder saltbush) using ventilated chambers during summer at a site near Kellerberrin where the water table was saline (24,000 mg NaCl l^{-1}) and shallow (1.2 m below the ground). Evapotranspiration was measured on individual plants in treatments with plant spacing varying from 0.1 to 1.0 plants m^{-2}. Daily evapotranspiration per plant decreased from 3.3 to 1.3 mm with increasing plant density, which, when extrapolated to an area basis, became an increase from 0.7 mm to 1.3 mm day^{-1}. Evaporation from bare soil was only 0.4 mm day^{-1}. The saltbush stand was estimated to evaporate 250 mm of water each summer, which is a large proportion of the average annual rainfall of 334 mm.

6.6.7 Revegetation Strategies

It is unclear what proportion of wheatbelt catchments needs to be planted to high water using plants and the best location for these plants. Deep-rooted plants are required when matrix recharge takes place in coarse-textured soils. However, when rapid recharge takes place down preferred pathways, drains need to be considered (see Section 6.6.8).

Bell et al. (1990) investigated several reforestation strategies for recharge areas in the Darling Range area SSE of Perth. They found that, while groundwater levels under reforested areas decreased relative to those under pasture, the reduction was related to the proportion of the cleared area and the crown cover and not to the position of the reforestation in the catchment. The reforestation had to cover more than 15% of the cleared landscape to lower the water tables significantly in this area, which had an average annual rainfall of about 700 mm. The salinity of the groundwater under the trees fell by 12%. Whereas there may be no clear strategy for reforestation of recharge areas, Schofield et al. (1989) found that lower slope and seep reforestation was required if saline discharge into streams was to be eliminated in the Darling Range area.

In summary, preferred strategies for reforesting recharge areas have not been defined, possibly because recharge is such a widespread process and groundwater flow rates are low. The adoption of high water using crops and pastures is likely to be beneficial over such extensive areas (see Section 6.6.1). Reforestation which intercepts groundwaters that are fresh enough for root uptake appear to be effective, as is planting salt-tolerant species on saline seeps. Mounding is necessary to reduce the amount of waterlogging on salt-affected areas so that trees can become established (Schofield et al. 1989). By lowering the groundwater level, trees reduce soil evaporation and the salinity of the upper soil profile. However, the rate of salt leaching from poorly-structured soils can be slow.

6.6.8 Engineering Strategies

Seepage interceptor drains have been recommended to reduce the amount of recharge taking place via preferred pathways on waterlogged duplex soils (McFarlane and Cox 1990). In addition to reducing the duration of ponding on the clay subsoil, the drains enable plant roots to grow more deeply and further reduce recharge. George (1991) used buried pipe to lower watertables and soil salinities in a sandplain seep in the Wallatin Creek catchment. Crops were grown on the reclaimed land.

As recharge is thought to be important when wheatbelt valleys are inundated, drains are required to remove this water as quickly as possible (McFarlane et al. 1990a). However, it may not be possible to

Table 6.4. The cost of controlling surface, shallow subsurface, and deep ground water using banks and drains.

Main Problem	Solution	Cost ($\$ m^{-1}$)	Spacing (m)	Cost ($\$ ha^{-1}$)
Water erosion	Grade banks	0.30	100	30
	Level banks	1.20	100	120
Waterlogging	Seepage interceptor drains	0.45	60	75
Salinity	Tube drains	5.00	20	2500

remove the water to a place where no recharge will take place. In these cases, extra efforts are needed to reduce the amount of water entering the valleys and to improve the infiltration of water in poorly-structured soils (e.g., using gypsum and minimum tillage).

Deep, open drains have been installed by many farmers to increase groundwater discharge in wheatbelt valleys. Significant drawdowns around the drains have only been recorded when the drains have been dug into regoliths formed from sediments west of the wheatbelt (P.R. George, personal communication). Any improvements in plant growth near the drains in the wheatbelt are probably caused by reduced water-logging and inundation. Drawdown has also been minimal around subsurface tube and mole drains in most wheatbelt soils.

Groundwater pumping has been used to reduce groundwater levels in saline areas (George and Nulsen 1985, George and Frantom 1990b). As pumping is very expensive, it can only be recommended where large areas are to be reclaimed or where a valuable asset is to be protected. A major problem with deep drains and pumping is safely disposing of the saline effluent.

6.7 Integration of Hydrologic Solutions

When considering solutions to hydrologic problems, it is necessary to understand both the main effects and the side effects of the solutions on water flows. Water flowing over the soil surface can be controlled by grade banks costing $\$30\,ha^{-1}$ or, if stable sites for discharging the water are not available, be level banks costing $\$120\,ha^{-1}$ (Table 6.4). Water-logging caused by ponding on the clay subsoil of duplex soils can be controlled by seepage interceptor drains costing $\$75\,ha^{-1}$, whereas draining saline seeps in clayey soils costs $\$2500\,ha^{-1}$. There is, therefore, a strong financial incentive to prevent surface waters from becoming saline groundwaters. However, some wheatbelt farmers adopt level banks to control surface waters because they need less maintenance and because of the problems finding stable waterways to accept the bank discharge.

There are two incentives for farmers to define and fence waterways on their farms: being able to use grade banks instead of level banks; and growing trees beside the waterway to lower groundwater levels in a part of the landscape that is most susceptible to salinisation. Adequate provision should be made for surface waters to enter and flow down the drainage lines. Vegetated waterways can cause storm waters to slow and drop sediment, causing the water to flow outside the drainage line and cause erosion. Drainage lines may need to be improved by excavating the main channel and constructing trainer banks on each side which can be overtopped safely during major storms.

Given the lack of a preferred strategy for reforesting recharge areas, there is considerable scope for farmers to plant trees wherever there are other benefits (e.g., wind erosion control, wildlife refuges and corridors, and stock and crop shelters) while aiding the salinity problem at the same time. Planting salt-tolerant trees and shrubs where they can access the groundwater near discharge areas is essential if the area of salt is to be minimised.

The barriers to groundwater flow created by dolerite dykes and bedrock highs divide surface catchments into a number of groundwater cells or subcatchments (George et al. 1991). Controlling recharge on one groundwater cell may only affect the discharge of a small area. In many cases, the barriers to groundwater flow are not apparent at the surface. Breaching one barrier may result in worsening salinity in the next down-gradient barrier.

The possibility that groundwater may move along shear zones between catchments also has management implications (Lewis 1991). Tree planting may need to be concentrated on the shear zones, which could be outside the surface water catchment containing the salinity. Pumping stock-quality water from the zones may also be an option (Lewis 1991).

In flat areas, the lateral flow of groundwater may be negligible because of low hydraulic gradients and conductivities (George et al. 1991). In these cases, farmers may unilaterally lower their groundwater and not experience significant groundwater inflow from surrounding areas. However, surface water flows do need to be managed on a catchment basis.

6.8 Conclusions

The central wheatbelt has a reliable, low-intensity rainfall which is strongly winter dominant. The potential evaporation rate is about seven times higher than annual rainfall.

The native vegetation redistributes rainfall both within the canopy and on the soil surface so that the water infiltrates deeply into the soil profile at selected points (e.g., around tree roots). In remnant vegetation, water use continued after winter rainfall ceased, until all of the water was transpired. Thus, the small amount of salt in the rainfall accumulated in the soil profile over many years prior to clearing.

After clearing, infiltration was more uniform. Runoff amounts and rates increased due to the wetter soils and loss of soil structure in the topsoil and the closure of the macropores in the clayey subsoil. Rising groundwater levels formed saline seepages which further increased runoff. The low amount of vegetative cover (resulting from the short growing season) and soil detachment by raindrop impact, cultivation, and sheep hooves greatly increased water erosion. Soil fertility appears to decrease rapidly after erosion. Recovery from erosion events is slow and the effects may be cumulative.

Runoff and erosion can be reduced by minimum tillage, applications of gypsum, and stubble retention which increase soil structure and, therefore, infiltration. Earthworks can also be effective in reducing erosion when they are properly designed and maintained. However, level banks, which pond water in their channels, can increase recharge and result in severe erosion if they fail.

The main aquifer responsible for wheatbelt salinity is located between the deeply-weathered clays in the laterite profile and bedrock. Recharge is by matrix flow in coarse-textured soils and by preferred pathway flow through subsoil clays. Three unconfined aquifers contribute water to the preferred pathways: (1) perched water in duplex soils; (2) perched water in deep sandplain soils; and (3) sandy channels in wheatbelt valleys. It is now thought that recharge takes place over 90% of the landscape.

Saline seeps form at hydrologic discontinuities in the aquifer, behind dolerite dykes and bedrock highs, at changes in slope and decreases in aquifer thickness, and where groundwaters converge. Groundwaters may also be preferentially carried along shear zones and form seeps where these meet barriers to flow or other carriers.

Cereal and lupin crops use significantly more water than annual pastures and therefore have the potential to reduce the rate of salinisation. Overcoming other forms of soil degradation, particularly those which affect rooting depth (e.g., subsoil compaction, subsoil acidity, and waterlogging), can increase water use and therefore reduce recharge. Selecting the right rotation and land use for each soil type will maximise short-term profits and greatly reduce recharge, compared with present land uses on many wheatbelt farms.

Native vegetation, in particular *Eucalyptus* trees, use much more water than annual crops and pastures. Groundwater levels under areas of remnant vegetation in the wheatbelt are up to 7 m lower than adjacent areas. The understorey component contributes substantially to the high water use of areas of native vegetation and therefore should be included in any revegetation plan.

There do not appear to be strategic places for locating high water-use crops, fodder shrubs, and trees on recharge areas as recharge is so widespread on wheatbelt catchments. Trees should be placed into the landscape wherever possible, particularly if they have other roles such as the protection of soils and stock from wind and the conservation of flora and fauna.

In the absence of vegetation, saltland discharges little groundwater. Salt-tolerant shrubs and trees are essential to lower groundwater levels and lower the marginal cost of salinity. Saltbushes help fill the autumn feed gap on many wheatbelt farms. Trees in discharge areas, or where they can access groundwater, can be incorporated in double-fenced waterways to reduce both erosion and salinity. Where recharge to the main saline aquifer is via preferred pathways which are fed by groundwaters in unconfined aquifers, drains are necessary to remove the perched water. Deep drains and pumping to remove saline groundwater are not usually cost effective in the wheatbelt. There are also major problems with disposal of the saline effluent.

In addition to locating the right plant in the right soil type and including well-designed earthworks to cope with extreme events, consideration needs to be given by farmers to catchment planning. Unfortunately, surface water catchments do not often coincide with groundwater catchments because of the barriers and carriers which affect groundwater flow. Recent developments in geophysics may help to define groundwater catchments.

Acknowledgments. We would like to gratefully acknowledge the following help. Useful comments on a draft of this chapter were made by Mr. E. Hauck, Ms. M.F. Lewis, Dr. M.J. McFarlane, Dr. R.A. Nulsen, Dr. J.S. Salerian, and Ms. K. Veltrop. A.T. Ryder drafted Figs. 6.1, 6.3 to 6.8, and 6.10.

References

Abbott, I., Parker, C.A., and Sills, I.D. (1979) Changes in the abundance of large soil animals and physical properties of soils following cultivation. *Australian Journal of Soil Research* 17:343–353.

Barrett-Lennard, E.G. (1986) Effects of waterlogging on the growth and sodium chloride uptake by vascular plants under saline conditions. *Reclamation and Revegetation Research* 5:245–261.

Bell, R.W., Schofield, N.J., Loh, I.C., and Bari, M.A. (1990) Groundwater response to reforestation in the Darling Range of Western Australia. *Journal of Hydrology* 115:297–317.

Betson, R.T. (1964) What is watershed runoff? *Journal of Geophysical Research* 69:541–551.

Bettenay, E., Blackmore, A.V., and Hingston, F.J. (1964) Aspects of the hydrologic cycle and related salinity in the Belka Valley, Western Australia. *Australian Journal of Soil Research* 2:187–210.

Bureau of Meteorology (1958) *Climatological Survey: Eastern Wheat Belt West Australia*. Commonwealth of Australia.

Canterford, R.P., Presod, N.R., Pearce, H.J., and Turner, L.H. (1987) Design intensity-frequency-duration rainfall. In D.H. Pilgrim (ed.) *Australian Rainfall and Runoff. A Guide to Flood Estimation, Vol.* 1. The Institution of Engineers, Australia. pp. 31–40.

Chittleborough, R.G. (1985) Towards a state conservation strategy. 1. Planning to meet climatic change. *Department of Conservation and Environment Bulletin* 207.

Davies, J.R., McFarlane, D.J., and Ferdowsian, R. (1988) The effect of small earth structures and channel improvements on the flooding of agricultural land in southwestern Australia. *Western Australian Department of Agriculture Division of Resource Management Technical Report* 77.

Division of National Mapping (1986) *Atlas of Australian Resources. Vol. 4. Climate*. Commonwealth Government Printer, Canberra.

Edwards, W.R.N. and Warwick, N.W.M. (1984) Transpiration from a kiwifruit vine as estimated by the heat pulse technique and the Penman-Monteith equation. *New Zealand Journal of Agricultural Research* 27:537–543.

Elliott, G.L., Campbell, B.L., and Loughran, R.J. (1990) The correlation of erosion measurement and soil caesium-137 content. *Applied Radiation and Isotopes* 41:713–717.

Engel, R., McFarlane, D.J., and Street, G. (1987) The influence of dolerite dykes on saline seeps in southwestern Australia. *Australian Journal of Soil Research* 25:125–136.

Engel, R., McFarlane, D.J., and Street, G. (1989) Using geophysics to define recharge and discharge areas associated with saline seeps in southwestern Australia. In M.L. Sharma (ed.) *Groundwater Recharge*. A.A. Balkema Publishing Co., Rotterdam. pp. 25–39.

Farrington, P., Greenwood, E.A.N., Bartle, G.A., Beresford, J.D., and Watson, G.D. (1989) Evaporation from *Banksia* woodland on a groundwater mound. *Journal of Hydrology* 105:173–186.

Farrington, P., Salama, R.B., Bartle, G.A., and Watson, G.D. (1992) Water use of agricultural and native plants in a Western Australian wheatbelt catchment. *Agricultural Water Management* (in press).

Flavell, D.J., Martin, D.K., and Belstead, B.S. (1987) Flood estimation procedures for Western Australia. *Western Australian Main Roads Department Technical Report* 50T.

George, P.R. and Nulsen, R.A. (1985) Saltland drainage: Case studies. *Journal of Agriculture – Western Australia* 26:115–118.

George, R.J. (1990a) The 1989 saltland survey. *Journal of Agriculture – Western Australia* 31:159–166.

George, R.J. (1990b) Reclaiming sandplain seeps by intercepting perched ground water with Eucalypts. *Land Degradation and Rehabilitation* 2:13–25.

George, R.J. (1991) Management of sandplain seeps in the wheatbelt of Western Australia. *Agricultural Water Management* 19:85–104.

George, R.J. (1992a) Hydraulic properties of groundwater systems in the saprolite and sediments of the wheatbelt, Western Australia. *Journal of Hydrology* 130:251–278.

George, R.J. (1992b) Groundwater processes, sandplain seeps, and interactions with perched groundwater systems. *Journal of Hydrology* (in press).

George, R.J. (1992c) Estimating and modifying the effects of agricultural development on the groundwater balance of large wheatbelt catchments, Western Australia. *Applied Hydrogeology* (in press).

George, R.J. and Conacher, A.J. (1992) Interactions between perched and saprolite aquifers on a small salt-affected and deeply-weathered hillslope. *Earth Surface Processes and Landforms* (in press).

George, R.J. and Frantom, P.W.C. (1990a) Preliminary groundwater and salinity investigations in the eastern wheatbelt. 3. Welbungin and Beacon catchments. *Western Australian Department of Agriculture Division of Resource Management Technical Report* 90.

George, R.J. and Frantom, P.W.C. (1990b) Using pumps and siphons to control salinity at a saline seep in the Wallatin Creek Catchment. *Western Australian Department of Agriculture Division of Resource Management Technical Report* 91.

George, R.J., McFarlane, D.J., and Lewis, M.F. (1991) A review of recharge and implications for management in saline agricultural catchments, Western Australia. *Proceedings International Hydrology and Water Resources Symposium*, Perth. pp. 193–197.

Greenwood, E.A.N. and Beresford, J.D. (1979) Evaporation from vegetation in landscapes developing secondary salinity using the ventilated chamber technique. 1. Comparative transpiration from juvenile *Eucalyptus* above saline groundwater seeps. *Journal of Hydrology* 42:369–382.

Greenwood, E.A.N. and Beresford, J.D. (1980) Evaporation from vegetation in landscapes developing secondary salinity using the ventilated chamber technique. 2. Evaporation from *Atriplex* plantations over a shallow saline water table. *Journal of Hydrology* 45:313–319.

Greenwood, E.A.N. and Beresford, J.D. (1982) Evaporation from vegetation in landscapes developing secondary salinity using the ventilated chamber technique. 4. Evaporation from a regenerating forest of *Eucalyptus wandoo* on land formerly cleared for agriculture. *Journal of Hydrology* 58:357–366.

Greenwood, E.A.N., Klein, L., Beresford, J.D., and Watson, G.D. (1985) Differences in annual evaporation between grazed pasture and *Eucalyptus* species in plantations on a saline farm catchment. *Journal of Hydrology* 78: 261–278.

Hamblin, A.P. and Howell, M.R. (1987) Maintenance and improvement of soil structure. *Western Australian Department of Agriculture Division of Resource Management Technical Report* 95:16–32.

Hamblin, A.P. and Tennant, D. (1987) Root length density and water uptake in cereals and grain legumes: How well are they correlated? *Australian Journal of Agricultural Research* 38:513–527.

Hingston, F.J. and Galaitis, V. (1976) The geographic variation of salt precipitated over Western Australia. *Australian Journal of Soil Research* 14:319–335.

Howell, M.R. (1987) The measurement of surface soil structure using a tension infiltrometer under a range of dryland cropping systems. In K.J. Coughlan and P.N. Truong (eds.) *Effects of Management Practices on Soil Physical Properties.* Department of Primary Industries, Queensland. pp. 48–51.

Johnston, C.D. (1987) Preferred waterflow and localised recharge in a variable regolith. *Journal of Hydrology* 94:129–142.

Johnston, C.D., Hurle, D.H., Hudson, D.R., and Height, M.I. (1983) Water movement through preferred pathways in lateritic profiles in the Darling Plateau, Western Australia. *CSIRO Groundwater Research Technical Paper* No. 1.

Lewis, M.F. (1991) Lineaments and salinity in Western Australia – carriers or barriers? *Proceedings International Hydrology and Water Resources Symposium,* Perth. pp. 202–209.

Marsh, B. (1982) Summer-grazing wind erosion. *Western Australian Department of Agriculture Farmnote* No. 7/82.

McFarlane, D.J. and Carter, D.J. (1990) The effect of erosion on soil productivity in southwestern Australia. In *Proceedings Erosion/Productivity and Erosion Prediction Workshop.* 5th Australian Soil Conservation Conference, Perth. pp. 3–14.

McFarlane, D.J. and Cox, J.W. (1990) Seepage interceptor drains for reducing waterlogging and salinity. *Journal of Agriculture – Western Australia* 31:66–69.

McFarlane, D.J. and Davies, J.R. (1988) Soil factors affecting flood runoff on agriculture catchments in Western Australia. *Western Australian Department of Agriculture Division of Resource Management Technical Report* 18.

McFarlane, D.J. and George, R.J. (1992) Factors affecting dryland salinity in two wheatbelt catchments in Western Australia. *Australian Journal of Soil Research* 30:85–100.

McFarlane, D.J. and Wheaton, G.A. (1990) The extent and cost of waterlogging. *Journal of Agriculture – Western Australia* 31:44–47.

McFarlane, D.J., Davies, J.R., and Westcott, T. (1986) Rainfall erosivity in Western Australia. In *Proceedings of the Hydrology and Water Resources Symposium,* Brisbane, The Institution of Engineers, Australia. pp. 350–354.

McFarlane, D.J., Engel, R., and Ryder, A.T. (1989) The location of recharge areas responsible for valley salinity in the Lake Toolibin Catchment, Western Australia. In M.L. Sharma (ed.) *Groundwater Recharge.* A.A. Balkema Publishing Co., Rotterdam. pp. 255–268.

McFarlane, D.J., Howell, M.R., Ryder, A.T., and Orr, G. (1992a) The effect of agricultural development on the physical and hydraulic properties of four Western Australian soils. *Australian Journal of Soil Research* 30:in press.

McFarlane, D.J., Loughran, R.J., and Campbell, B.L. (1992b) Soil erosion of agricultural land in Western Australia estimated by Caesium-137. *Australian Journal of Soil Research* 30:in press.

McFarlane, D.J., Negus, T.R., and Ryder, A.T. (1990a) Shallow drains for reducing waterlogging and salinity on clay flats. *Journal of Agriculture – Western Australia* 31:70–73.

McFarlane, D.J., Engel, R., Ryder, A.T., and Eales, M. (1990b) Level banks used to decrease waterlogging can increase salinity. *Journal of Agriculture – Western Australia* 31:74–77.

McGhie, D.A. (1980) The contribution of the mallett hill surface to runoff and water erosion in the Narrogin region of Western Australia. *Australian Journal of Soil Research* 18:299–307.

Nulsen, R.A. (1983) Manipulation of recharge by agronomic techniques. In *International Conference on Groundwater and Man, Vol. 2: Groundwater and the Environment.* Australian Water Resources Council, Sydney. pp. 317–325.

Nulsen, R.A. (1984) Evapotranspiration of four major agricultural plant communities in the southwest of Western Australia measured with large ventilated chambers. *Agricultural Water Management* 8:191–202.

Nulsen, R.A. (1991) Land degradation – The hydrological issues. *Proceedings International Hydrology and Water Resources Symposium,* Perth. pp. 35–41.

Nulsen, R.A. and Baxter, I.N. (1982) The potential of agronomic manipulation for controlling salinity in Western Australia. *Journal of Australian Institute of Agricultural Science* 48:222–226.

Nulsen, R.A. and Henschke, C.J. (1981) Groundwater systems associated with secondary salinity in Western Australia. *Agricultural Water Management* 4: 173–186.

Nulsen, R.A., Bligh, K.J., Baxter, I.N., Solin, E.J., and Imrie, D.H. (1986) The fate of rainfall in a mallee and heath vegetated catchment in southern Western Australia. *Australian Journal of Ecology* 11:361–371.

Peck, A.J. and Hurle, D.H. (1973) Chloride balance of some farmed and forested catchments in southwestern Australia. *Water Resources Research* 9:648–657.

Peck, A.J., Johnston, C.D., and Williamson, D.R. (1981) Analyses of solute distributions in deeply weathered soils. *Agricultural Water Management* 4: 83–101.

Pittock, A.B. and Allen, R.J. (eds.) (1990) *The Greenhouse Effect. Regional Implications for Western Australia, 1st Interim Report 1989–90.* CSIRO Division of Atmospheric Research, Melbourne.

Ritchie, J.C. and McHenry, J.R. (1975) Fallout Cs^{137}: A tool in conservation research. *Journal of Soil and Water Conservation* 30:283–286.

Salama, R.B., Farrington, P., Bartle, G.A., and Watson, G.D. (1991) Identification of recharge and discharge areas in the wheatbelt of Western Australia using water level patterns in relation to basin geomorphology. *Proceedings International Hydrology and Water Resources Symposium,* Perth. pp. 841–846.

Salerian, J.S. (1991) Application of an economic model to dryland soil salinity in Western Australia. *International Economics Postgraduate Research Conference Volume,* 97–108 *Economic Record.*

Salerian, J.S. and McFarlane, D.J. (1987) BANKS: A method of financially assessing banks used to mitigate water erosion in southwestern Australia. *Western Australian Department of Agriculture Division of Resource Management Technical Report* 55.

Schofield, N.J., Loh, I.C., Scott, P.R., Bartle, J.R., Ritson, P., Bell, R.W., Borg, H., Anson, B., and Moore, R. (1989) Vegetation strategies to reduce stream salinity of water resource catchments in southwest Western Australia. *Water Authority of Western Australia Report* WS 33.

Sedgley, R.H., Smith, R.E., and Tennant, D. (1981) Management of soil water budgets of recharge areas for control of salinity in southwestern Australia. *Agricultural Water Management* 4:313–334.

Sharma, M.L. (1984) Evapotranspiration from a *Eucalyptus* community. *Agricultural Water Management* 8:41–56.

Smith, S.T. (1962) *Some Aspects of Soil Salinity in Western Australia.* M.Sc. Thesis, University of Western Australia.

Thorburn, P.J., Walker, G.R., and Graecen, E.L. (1991) Diffuse discharge from shallow water tables in arid and semi-arid areas: Examination of past results. *Proceedings International Hydrology and Water Resources Symposium,* Perth. pp. 103–109.

Topalidis, S. and Curtis, A.A. (1982) The effect of antecedent soil water conditions and rainfall variations on runoff generation in a small Eucalypt catchment. *First National Symposium on Forest Hydrology,* The Institution of Engineers, Australia. pp. 43–49.

Williamson, D.R. and Bettenay, E. (1979) Agricultural land use and its effect on catchment output of salt and water – Evidence from southern Australia. *Progress in Water Technology* 11:463–480.

Wischmeier, W.H. and Smith, D.S. (1958) Rainfall energy and its relationship to soil loss. *Transactions of the American Geophysical Union* 39:285–291.

Section III

Landscape Reintegration

Landscape reintegration. Clearing for agriculture has left native vegetation only in relatively small patches, connected to varying degrees by roadside and fenceline vegetation. Can agricultural production and nature conservation coexist in this altered landscape? The photograph shows North Bungulla Reserve, in Tammin Shire.

(Photo: D. Steven)

Chapter 7

Landscape Reintegration: Problem Definition

A.R. MAIN

7.1 Introduction

It is apparent from the previous chapters that settlement by Europeans has been an ecological disturbance on the landscape scale. Many of the ecological functions of the system such as maintaining soil structure, nutrient recycling and transport, and the hydrological cycle have been impaired. It is because of the loss of these functions that the landscape has disintegrated. Any attempt to reverse the landscape trends now apparent require that it be managed in such a way that natural areas, revegetated and farmed land together form a landscape in which the impaired or lost functions are restored. In this chapter, I set out the principles to be followed so that functional systems are established in the reintegrated landscape. These principles are developed from what is known to be needed to maintain life, ensure regeneration, and maintain species richness. I then interpret species richness as filling a functional role in a system by relating nutrient retention and recycling to the need to have organisms which will perform these functions under all conditions likely to be experienced in a fluctuating environment, thereby tending to keep the nutrient pool intact. I then compare agricultural and natural ecosystems to provide a common framework before discussing reintegration and the boundary conditions for the natural persistence of systems.

7.2 Conditions for Life

It is a matter of observation that plants require space, light, water, and nutrients before they can grow successfully. Moreover, unless there are plants, there is neither food nor shelter for herbivores and these in turn are food for carnivores. Finally, dead plants and animals provide nutrients for a vast number of invertebrates, protists, and bacteria which are referred to as detritivores. The soil is the medium in which these organisms function and it maximises the availability of the inherent level of chemical elements which, in turn, determines the biological activity. Together these functions can be referred to as communities or ecosystems.

The question of how these systems are maintained is at the core of reintegrating the landscape. In particular, the problem of how space, light, water, and nutrients are made available so that a system can be renewed needs to be addressed. The problem is highlighted by the observation that all plant species do not establish equally or readily or grow as rapidly or as large as other species; that is, they have specific biological characteristics and needs. However, it appears that, in nature, these diverse needs can be met since there may be a great number of species with different longevities, sizes, and biological needs in any one area. Furthermore, the richness may be equalled in nearby localities but composed of quite different species. An understanding of how this species richness is achieved or maintained, along with any functional role that it may have, is a prerequisite for adequate reintegration and management.

7.3 Regeneration

In any locality or system, resources are limiting when they are utilised by all organisms, so that space is occupied or light, water, and nutrients are used to the maximum. Under these circumstances, resources will only become available when an organism dies, perhaps because of age, light (being over-grown), crowding (shortage of space), drought, shortage of nutrients, or being eaten (herbivory or carnivory) by other organisms. These events can be considered as disturbances of a small scale, affecting the individuals of an assemblage. However, there is, in reality, a continuum from individual deaths to the vast numbers killed by wide-spread lethal diseases or wind storms, tornados, floods, drought, or fire. All of these catastrophes make resources available for the organisms which survive the mortality or can colonise the depopulated area because their particular conditions or biological needs are met.

Various scales of disturbances lead to the availability of postdisturbance areas of various sizes. Pickett and White (1985) and Pickett et al. (1989) interpret these disturbance areas as initiating patch dynamics, which is a requirement for the maintenance of a regional biota. Patch dynamics studied as secondary succession has, in the past, been seen as being a consideration of the conditions which would lead to the reestablishment of the former assemblage or its replacement by a different one. Early interpretations were based on an assumption of a balance of nature expressed as an assemblage typical of a locality. It was believed that this typical assemblage tended to be stable in the sense that, following disturbance, there was believed to be a tendency for the former assemblage to be restored via a characteristic successional pathway, so that the ideal or balance was restored. In this view, departures of populations or systems from the ideal were treated as oscillations.

In contrast with the deterministic interpretation given above, there is the alternative that populations or systems are subject to chance or stochastic events. This interpretation views any apparent equilibrium or stability exhibited as due to the shortness of time over which observations have been made, i.e., the observations have not included the rare, or infrequent, yet significant, events which can permanently influence the numbers in a population or relative abundance of species in a system.

The corollary of this interpretation is that any statement about the constancy of population numbers or species richness is a probabilistic one, depending on the chance that the significant stochastic event, which could influence the number of individuals present or species richness, will occur or has occurred.

The interpretation of the systems as being deterministic or dynamic may be reconciled by relating the apparent stability to local short-term, as contrasted with long-term, observations over large areas.

The outcome of disturbance in terms of patch dynamics depends on the biota surviving or being present nearby and being able to invade or colonise, establish, and reproduce. All of these depend on the kind, size, intensity, or frequency of disturbance. The foregoing can be illustrated by taking extreme examples, such as extensive disturbances of great intensity which may leave the only possible colonisers at great distances from the now available resources while, at the other extreme, low-intensity, high-frequency disturbances may give only short-lived organisms the opportunity to complete a life cycle. Those species which can take advantage of patch dynamics occur as metapopulations composed of demes of various size, each occupying patches following localised disturbance. Thus, the effects of disturbance can account for the regional maintenance of a biota by a process of regeneration in patches following disturbances of different sizes and intensities at irregular intervals. The effects of each of these sorts of events, even when wide-spread, are not uniform, perhaps because of local topographical relief, and, so, there is often a local minor variability of survivors which can account for initial species richness in postdisturbance situations. That is, those potential colonisers survived the disturbance by chance and may be quite unrepresentative of the species composition of the assemblage prior to the disturbance. Moreover, the postdisturbance conditions may not favour the survivors or many of the potential colonisers, with the result that the ultimate postdisturbance array of species need not, in any way, resemble the prior distribution. Thus, a model of regeneration based on disturbances and patch dynamics influenced by chance events is not inevitably set to return to its predisturbance composition.

Whether a system is viewed as deterministic or stochastic and dynamic has markedly different implications for management. An interpretation based on an assumption of inherent stability implies that management must be directed to assisting, or at least not impeding, the return of populations, species, or systems to their characteristic natural state, represented by the predisturbance condition. With this interpretation, the maintenance of species richness, which is a goal of management, is seen as almost inevitable. This is not the case with an interpretation based on chance or stochastic processes, which pose for management quite difficult questions in a postdisturbance situation. These questions relate to the following: Will the final product of the postdisturbance regeneration be of the same species richness, composed of the same or different species than formerly? If the latter is the likely outcome, then where will the species, which were formerly present in the patch but are not after regeneration from disturbances, be found, so as to be retained within the conservation estate?

The foregoing outline of how biotas are regenerated emphasises the role of chance in several ways:

1. The chance that an organism able to exploit the resources made available by disturbance will be present;
2. That favourable environmental conditions will persist, allowing continued exploitation and the development of assemblages into systems which give the appearance of stability;
3. That environmental conditions will provide, by chance, the necessary disturbances of the appropriate kind (fire, storm, flood, drought, etc.), interval (frequency), and size (area affected), to retain the pattern of patch dynamics within a landscape, thus providing the habitat diversity on which local and regional species richness depend; and
4. Resources are in limited supply and are made available for reuse following death.

7.4 Species Richness

Management of conserved areas can have several goals, such as the retention of examples of a characteristic habitat, particular examples of a species-rich habitat, and/or rare and endangered species. Each of these goals poses problems because of the differences in the biology and life histories of organisms which vary widely: some are quick-growing and early maturing while others are the converse; some are apparently rare and others always common; and some are tall-growing shady species (trees) which might place low, slow-growing shade-intolerant species at a disadvantage. Yet, it frequently happens that just such a combination of organisms occur together in a natural assemblage. Why should this be so? Why should one or a few species dominate species-poor assemblages so infrequently? A possible explanation lies in the experimental demonstration by Paine (1980) of strongly interacting species which, when removed, lead to the expansion of prey species and crowding out of other species which are now disadvantaged, with the final outcome a reduction in species richness. Thus, at this level, biological activity which interferes with the potential growth of organisms capable of numerically or physically dominating the environment of a locality acts as a disturbance, just as physical disturbance does on another scale. Biological, like physical, disturbance varies, in effect, in step with the frequency of the intensity of predation. Little or no predation leaves a species-poor assemblage of long-lived species, while heavy or sustained predation leaves a species-poor assemblage composed of predation-tolerant species.

The results of physical and biological disturbances are shown, in terms of control of species richness, in Table 7.1 which shows that biological disturbances in the form of herbivory or predation produce many similarities with the species richness resulting from different regimes of physical disturbances. However, physical and biological disturbances may act synergetically or antagonistically and so their effects may be enhanced or modified in some way. The possibilities are given in Table 7.2 and the combinations shown are likely to produce differences in species richness depending on the disturbance regimes and can be ranked in the following order: 1, 3, 7, 9 < 2, 4, 6, 8 < 5.

In contrast with the foregoing dynamic interpretation which relates species richness and disturbance regimes, there have been attempts to relate species richness simply to area available. Main (1979) related richness to reduced areas available as refugia on the soil mosaics of the wheatbelt following presettlement wildfires, i.e., simple physical disturbance. Kitchener et al. (1982) related species richness to the area of the insular remnant contained within present reservation. Certainly both these are possible explanations that may be mediated directly by the size of the available habitat or indirectly by way of the area needed to maintain a population of organisms of the higher trophic level associated with high species richness. This possibility has serious consequences for management attempting to maintain species richness. Moreover, the loss of species may not yet be apparent because the longevity of conspicuous plants gives the appearance of a "stable climax" (Frank 1968).

To this point, the argument developed suggests that biotic regeneration follows disturbance, which produces a mosaic of patches of different sizes and ages, thus maintaining a regional species richness. Species

Table 7.1. Species richness related to disturbance.

Kind of disturbance	Frequency/Intensity/Size of Disturbance: Rare, infrequent, or slight	Moderate	Frequent, intense, and extensive
Physical, e.g., death from windstorm, drought, flood, or fire.	Low: dominated by long-lived plants 1	High: patches of different size and postdisturbance ages; great differences between patches 2	Low: wide-spread occurrence of a few disturbance-tolerant, e.g., grasses, in fire-induced grassland 3
Biological, e.g., herbivory predation	Low: dominated by rapidly growing, space occupying organisms 4	High: within area richness high 5	Low: grazed or predated species infrequent; mostly unpalatable species present. 6

Table 7.2. Likely species richness resulting from interactions between physical and biological disturbances.

Biological	Intensity of Disturbance Physical: Infrequent or slight	Moderate	Intense or high frequency
Rare or slight	few dominant long-lived species 1	species richness determined by those species maintained by disturbance 2	few species and those tolerant of physical stress 3
Moderate	species rich: species intolerant of physical but tolerant of biological disturbances 4	highest species richness as physical and biological factors complement each other 5	species presence determined by tolerance of physical stress, richness maintained by biological disturbance. 6 .
Very intense	species poor: dominated by species tolerant of grazing or predation 7	species richness determined by response to moderate physical disturbance yet tolerant of grazing or predation. 8	low: species present determined by tolerance to physical and biological factors 9

richness within patches results from the cascading effect of higher trophic levels within the patch. This conclusion raises the question of whether species richness is merely the consequence of biological or physical factors or fills a role or function within the system also.

7.5 A Possible Role – A Function for Species Richness

Environments which are rich in species, such as tropical rainforests, have always fascinated biologists. Interpretation of the reason for this richness, and any possible advantage the richness may confer on the system, has been coloured by the prevailing belief in deterministic theories and interpretations based on the balance of nature and inherent stability within populations and systems. MacArthur (1955) argued that a qualitative condition for stability was the number of pathways that energy could take up through the system. He concluded that, where there is a small number of species, stability is hard to achieve but, where there is a large number of species, stability can be achieved. This paper has stimulated a large research effort directed towards calculating a single appropriate measure of species richness or diversity, making observations, or performing experiments to test whether species richness makes for stability or resilience within a system subject to disturbances.

The earlier parts of this chapter have argued for a nondeterministic interpretation; in particular, I have argued that disturbance has a role in preserving the opportunities for regeneration of a system, while predator-prey relationships produce the cascading relationships which result in species diversity. In effect, the role of predators and herbivores in the system is to redistribute resources, i.e., prevent the accumulation of resources by one or a few common species or in one or a few localities. This is of particular importance in the wheatbelt where mineral nutrients, especially phosphorus, are very low. Under such conditions, it is particularly difficult to replace nutrients lost from the system by leaching or in other ways. Moreover, the environment is very variable and disturbances common. As argued earlier, these events give opportunities to many organisms and, if there is a variety of organisms present and able to function under the various conditions, there is the possibility that the loss of difficult-to-replace nutrients will be minimised. In this view, the roles of nitrogen fixation, carbon fixation, nutrient abstraction from soil, and recycling and transport within the system by means of herbivores, predators, decomposers, the soil biota, fungi, and mycorrhizae all increase the retention times of nutrients within the system. At any one time, the species present that are capable of filling a particular role may be very numerous but the species may not be equally abundant. However, the abundance may change under different environmental conditions when another species capable of thriving under the changed circumstances fills the functional role. For example, nitrogen fixation may be by cyanobacteria on bare soil, a rhizobial symbiont associated with Fabaceae or Mimosaceae in postfire situations, by symbiotic *Frankia* associated with Casuarinacea, by free-living actinomycetes, or by aerobic or anaerobic bacteria deriving their energy from plant material in the soil.

Similarly, a great variety of mycorrhizae and fungi scavenge phosphorus and other soluble nutrients which would otherwise be lost from the system. In essence, the diversity present can be conceived of as an inbuilt redundancy which ensures that all functions will be operating under all experienced environmental conditions. A rich variety of organisms, capable of filling essential roles under all experienced circumstances, endows a system with robust qualities, so that the probability of the system persisting approaches one. Main (1981) discussed species richness in terms of maximising the retention time of biologically important material within the system and diversifying the modes or routes by which the biologically useful pool of nutrients is introduced into, retained by, and redistributed within the system. Thus, the supply and retention of nutrients is ensured under all possible or conceivable circumstances. This contributes to stability or persistence in the context of managed systems.

7.6 Comparison of Natural and Agricultural Systems

The foregoing sections have shown how systems may regenerate, how species or richness may be retained, and what function species richness may have. Taken together, the three approaches show why a management goal to retain species richness is an appropriate goal and what mechanisms are necessary in order to achieve that goal. In the context of landscape reintegration, three quite different systems will need to be managed together. In the agricultural ecosystem, the intention is to maintain pure stands of crop or pasture species, i.e., eliminate weeds (nonbeneficial species) and reduce consumption by pests. In contrast, natural ecosystems do not aim to have single species stands without herbivores or predators. Rather, the aim is to have a suite of plants or animals with life histories and biologies which complement each other. The two ecosystems are compared in Table 7.3. This comparison shows that, in an ecological sense, agriculture establishes and maintains a system of low species richness by ensuring that, through cultivation, there is an appropriate disturbance regime which ensures that the favoured species approximates to a monoculture. In the reintegrated landscape, the minimum requirements are that natural vegetation should provide the maximum species richness, require minimal expenditure on maintenance, and provide landscape amenity

Table 7.3. A comparison of agricultural and natural ecosystems.

Focus	Agricultural	Natural
Goal:		
short-term	Profit	No irreplaceable losses
mid-term	Profit margin for maintenance of system	Mimimal external input
long-term overall	Persistence of system, economic viability	Persistence of system, self-perpetuation
Provision of basic requirements		
Plants		
Space	Appropriate husbandry e.g., cultivation	Appropriate disturbance regime
Light	Cultivation and weed control	An array of species each capable of taking advantage of light regimes at various stages after disturbance
Water	Management practices, conserving water, weed control, maintaining groundwater table	Water harvesting by some species, soil structure maintained by animals, groundwater table regulated by evapo-transpiration
Nutrients	Applied or by husbandry and crop rotation	Naturally occurring nutrients retained by recycling within the system but with a net loss
Animals		
Water	Husbandry	From plants as natural occurrence
Nutrients	Forage or provided by appropriate husbandry	Plants of appropriate postdisturbance stage or by eating herbivores or detritivores
Shelter/space	Appropriate husbandry	Distribution of plants of appropriate post-disturbance stage
Soil structure	Husbandry and cultivation	Determined by above ground burrowing animals, soil-dwelling animals, plant roots, fungi, bacteria and plant material buried in soil by spoil produced by burrowing animals

(*Continued*)

TABLE 7.3. *Continued*

Focus	Agricultural	Natural
Tolerances		
Nutrients:		
General	Low levels result in poor growth	Low levels tolerated, high levels toxic
Major nutrients	N by husbandry or applied	N fixed naturally
	P applied	P naturally occurring and recycled
Trace elements	Applied	Low levels tolerated
Losses from system	N lost as gas or taken off in harvest	Lost as gas or transported by mobile animals
	P lost by leaching, taken off in harvest, bound to soil fraction	Lost by leaching or transported by mobile animals
Replacement of losses	Applied or replaced by husbandry as appropriate	In unfragmented system on average is redistributed through transport by animals
Maintenance	Management practices which retain soil, soil structure, and productive status of soil	Appropriate patch dynamics, determined by natural disturbance regime and size of area
Failure to persist	Loss of soil structure	Loss of soil structure
Physical	Changes in soil salinity, acidity, nonwetting soils, waterlogging	Changes in salinity or acidity
Environment	Changes in rainfall regime, erosion by wind and water	Changes in rainfall regime, erosion by wind and water
Management practices	Inappropriate husbandry	Patch dynamics of inappropriate scale

in the form of control for wind and water erosion and control or amelioration of the groundwater table and salinity. Moreover, the reintegrated system has a disturbance which is qualitatively different from that of presettlement times. The problems of managing and monitoring such a system within a landscape dominated by agricultural practices have been listed by Main (1987 Tables 2, 3, and 4) and in the following section.

7.7 The Reintegrated Landscape

Within the reintegrated landscape, there should be a consciously maintained mosaic of areas, differing in terms of naturalness, species richness, size, and purpose as follows:

(a) little-disturbed natural vegetation;
(b) disturbed or degraded natural vegetation, e.g., grazed by stock;
(c) revegetated land; and
(d) agricultural land.

For the present, only (a) to (c) above are of concern. Depending on size and presence of strongly interacting species, areas classified under (a) may be able to maintain their species richness. Areas classified under (b) will have to have the desired level of species richness restored. Revegetated land (c) will need a decision as to what minimal diversity is to be developed so that the area will maintain both nature conservation and landscape amenity functions, as discussed in the previous section (7.6). Ultimately, this reduces to a matter of perceiving the money spent as an investment rather than an outright cost.

In earlier sections (7.3, 7.4), it was noted how disturbances and subsequent patch dynamics led to the persistence of a system while the cascading effects of trophic interaction led to the maintenance of species richness. Can these processes produce the desired outcome in a reintegrated, much-modified landscape compared with farm agricultural conditions? There is a temptation to say that it cannot be done, or not done readily, because the extinctions and changes discussed by Hobbs et al., (Chapter 4) almost certainly mean that the upper trophic levels which might be expected to maintain species richness are no longer present. Moreover, the presence of the rabbit (*Oryctolagus cuniculatus*) and the fox (*Vulpes vulpes*) will tend to lower the species richness and move it towards cells 7, 8, and 9 of Table 7.2. At present, it is possible but costly to regulate the number of foxes and rabbits. Ultimately, it would be desirable to eliminate them or reduce their numbers so much that their effect on species abundance is insignificant. Until this is possible, continued control is the only option.

In the recent geological past, a very significant change in the fauna occurred with the extinction of the so-called megafauna (Murray 1984). The loss of the large grazing and browsing herbivores must have produced significant changes in the postfire grazing and, consequently, affected plant regeneration. Moreover, the loss of these large animals removed the minor disturbing effects of their feeding on browse and breaking and damaging growing shrubs and trees. Additionally, the loss of predators such as *Thylacoleo* and *Thylacinus* and their replacement by the dingo (*Canis familiaris dingo*) must have had important repercussions on the remaining biota. Yet, we do not now readily recognise in what way the preagricultural system differed from that which preceded the extinction of the megafauna. So does this give hope in the present situation where we are trying to reintegrate the landscape? The question may be irrelevant because of the fragmented nature of the remaining natural vegetation. However, the continued existence of a species-rich biota, even after the loss of the megafauna, suggests that once one set of strongly interacting species was lost, the systems were reorganised, and those which persisted, rich in species, were those in which a formerly weakly interacting set of species became a strongly interacting one in the changed circumstances after the loss of the megafauna. In the current circumstances, the fragmented landscape no longer contains the former suite of strongly interacting species such as the vertebrate herbivores and predators. The most likely candidates for strongly interacting herbivores are the herbivorous arthropods, especially insects, while the strongly interacting species of predators could then be birds, reptiles, other insects, arachnids, or bacterial or viral disease.

A tendency to dismiss the ability of arthropods to act as strongly interacting species should be moderated by the effect of *Cactoblastis* on reducing the population of the cactus *Opuntia* or the effect of the gall-forming *Apiomorpha*, in the absence of a hymenopteran parasite, on the growth of the tuart *Eucalyptus gomphocephalus* on Rottnest Island (Short 1947). Additionally, many eucalypts and shrubby Australian species which are neither weedy nor abundant in Australia become aggressive invaders in Africa and America when freed from their arthropod herbivores (Groves 1986). In Australia, these may range from hepialid and cossid moths to leaf-mining forms or gall-forming Homoptera and lerps. It is known that both these latter are eaten by parrots. What degree of control this predation has is not known but, should it be significant, the loss of parrot populations because of the absence of tree-nesting hollows could have repercussions as serious as the absence of predators of *Apiomorpha* had on tuart growth on Rottnest Island. It is idle to speculate further but, if strongly interacting invertebrate species which will maintain species richness are to emerge, it is clear that our present lack of knowledge precludes even a tentative identification of those which might have potential for strong interactions. The absence of knowledge to suggest whether the persistence of potential regulators is assured in the transition period while the landscape is reintegrated is even more critical.

A further significant shortcoming is our lack of knowledge relating to the maintenance of soil structure and the incorporation of carbon into the upper soil horizons. In the past, the litter-disturbing and burrowing vertebrates were both abundant and varied. The losses through local extinction have been discussed by Hobbs et al., (Chapter 4). With the extinction of the mallee fowl (*Leipoa ocellata*), extensive litter disturbance is no longer generated by birds. Quail (*Turnix* sp.) scratchings are still important in some areas, but disturbance is less extensive than that of mallee fowl. All litter-disturbing marsupials are now locally extinct and the only vertebrates persisting are the echidna (*Tachyglossus aculeata*), the lizard (*Varanus gouldi*), some burrowing agamids and skinks, and the frogs (*Heleioporus albopunctatus* or *Neobatrachus* sp.). The effect of these vertebrates is unlikely to produce the quantities of burrow spoil formerly produced. Thus, the soil turnover, soil structure, and incorporation of litter into the carbon pool of the soil is likely to decline. The only possible compensation could arise from the increased activity of macroinvertebrates which, in the absence of marsupial predators, might be expected to increase. Candidates for this role are the centipedes (*Cormocephalus* and *Scolpendra*), the scorpion (*Urodacus* sp.), or beetles such as scarabids and carabs, which produce deep burrows as adults (Scaritinae). Other burrowing insects are to be found among spider wasps (Pompilidae), fossorial wasps (Sphecoidea), solitary bees (Colletidae) when they construct their ground nest, ants (Formicidae), and cicadas. Mygalomorph, lycosid, and zodarid spiders are significant in undisturbed places. More commonly, soil is disturbed by burrowing insect larvae, including some, such as the hepialid moth [*Trictena argentata* (H. Sch.)], which produce deep burrows. This species tunnels to about one metre in soil where it feeds upon the roots of *Eucalyptus* species.

Other beetles whose larvae live in burrows are to be found in the Geotrupidae. All Australian species as far as known are dung feeders. Adults dig deep burrows and bury large quantities of dung on which eggs are laid. Since dung-producing herbivores such as macropods (kangaroos and wallabies) no longer occur in many remnants of native vegetation, this very efficient contribution to soil structure and turnover will be lost. Larvae of Scarabaeidae feed on dung and decaying vegetable matter, while the chafers (Melolonthinae) pupate in a cell which may be deep in the soil; click beetles (Elateridae) are phytophagus, feeding on roots. All these must, in some way, contribute to soil turnover and structure.

It is likely that, in the future, the most significant contribution to soil disturbance will be termites and ants. Termites harvest large amounts of plant material and the undigested part is incorporated in the gallery walls. However, even when the nests or mounds are abandoned, they are not readily reinvaded or colonised by plants. On the other hand, both termites and ants make an immense contribution to the recycling of nutrients, either through the workers being predated while foraging or during the mating flights when many vertebrate predators gorge themselves. The danger in the latter is that nutrients may be exported out of the system.

There are, of course, many insect larvae, especially of Lepidoptera, which break down fallen leaves and litter, but the present state of knowledge suggests that a dearth of soil-burrowing animals and a decline in organic material incorporated into the soil are likely to impair seriously the productivity of areas of natural vegetation. The amount of soil turnover in the area is presented by Hobbs et al., (Chapter 4). It is unlikely that the effect of changes in the soil structure will be seen immediately because the already established shrubs and trees may be long lived. It will be early death and failure to regenerate that will be the first, obvious, and perhaps only, evidence.

The decline in activity of soil-disturbing animals will have an immediate influence on the procedures adopted when restoring degraded or disturbed natural areas. Such areas lack burrowing vertebrates and are unlikely to have a large number of invertebrates capable of deep soil disturbance, so root penetration by the seedlings of native plants will be restricted. Additionally, such areas are likely to be invaded by weeds, especially annual grasses such as barley grass (*Hordeum* sp.) or wild oat (*Avena fatua*), which effectively further restrict the establishment of the seedlings of shrubs and trees. Under such circumstances, the establishment of native plants may only be possible following mechanical disturbance of the soil to allow water and roots to penetrate more readily.

7.8 Boundary Conditions

Boundary conditions are the limits (nutrient status of the soil, disturbance regimes, species interactions, water availability, or physical environment) to be met in order to satisfy the conditions within which the native biota can reproduce and maintain species assemblages.

Species richness is only one of the several goals that have been proposed for management of natural areas; others have been the maintenance or preservation of typical or characteristic habitats or the retention of "rare and endangered" species. The former is based on a belief that, in an ideal or deterministic concept of nature, there is, in any locality, a typical or characteristic assemblage or system. In the argument developed in this chapter, the biotic assemblage present at any locality is the result of chance events, such as disturbance, climate, weather, and the presence of other organisms. It follows that, should it be desired to retain typical or characteristic habitats, active intervention to circumvent the effects of chance events will be necessary. A somewhat different perspective arises in the case of retaining "rare and endangered species." It is likely that such species will be favoured by certain postdisturbance developmental stages, and their current rarity reflects the fact that, with the present fragmented landscape where natural vegetation only occurs as remnants, there is no opportunity for disturbance and patch dynamics to provide a continuous supply of habitat patches at a stage favouring the species of concern. When the species' biology is known, it may be possible to manipulate the environment to favour it.

The question of maintaining species richness has been the central theme of this chapter. The argument developed suggests that boundary conditions for a system containing a species-rich assemblage will be determined by:

1. The provision of the conditions for life (space, light, water, nutrients, and shelter). Nutrients will be bound in the standing crop unless it is consumed by herbivores or predators and so redistributed or recycled.
2. The conditions for regeneration provided by disturbance and patch dynamics. These are stochastic in frequency, intensity, or size of area affected. Predictions of occurrences and effect are probabilistic statements.
3. The conditions which will maintain species richness. This implies that, should strongly interacting species be lost from a system, then the cascading effects of trophic interaction will need to be maintained, perhaps by a former weakly interacting species taking a role as a strongly interacting species.
4. The ability to maintain a species-rich assemblage, which will be very much determined by the success of measures to control or eliminate the two common exotics, namely, the rabbit and the fox.

Taking the effects of physical disturbance and the biological effects of trophic interactions together, as in Table 7.2, it is clear that cells 1, 3, 7, and 9 of the table fail to satisfy the boundary conditions for the retention of a species-rich assemblage. Cells 2, 4, 6, and 8 provide an indication of what will satisfy the minimal conditions for species richness. Cell 5 indicates the conditions under which maximum species richness might be achieved. This last case is the optimal or upper boundary.

The foregoing set out the general principles but the mechanisms are not obvious in an actual situation. Even the effects of rabbits and foxes are not unequivocally obvious without experimental evidence, such as the exclusion of rabbits or control of the number of foxes. In other cases, causal relationships can be or have been established or inferred. For example, interpretations providing a mechanism for rising water tables or salinisation establish a causal relationship with clearing (in effect, an extreme physical disturbance), which effectively upsets the hydrological balance by changing or reducing evapotranspiration. Further, increased salinity will change soil structures while rising water tables lead to waterlogging of soils.

Other possible changes in soil, such as increased acidity (low pH), have not been investigated. Increased acidity is known to occur in agricultural soils, but it is also known that some soils of natural areas are acid when first cleared; for example, the soils of kwongan or wodjil shrub lands. These soils carry a vegetation dominated by *Acacia* and *Allocasuarina*. Species in both genera have the ability to fix nitrogen, the former by rhizobial symbionts and the latter by the symbiotic actinomycete *Frankia*. Actinomycetes are able to function in soils of lower pH than are rhizobial symbionts. It is a matter of observation that, in many nature reserves, *Allocasuarina* is coming to dominate or be the sole species in what were formerly stands composed of many species. Moreover, with its winged seed and good dispersal capacity, it is now actively invading other vegetation types from which it was formerly absent. A possible explanation of this observation lies in the absence of a strongly interacting upper trophic level, i.e., a grazing or browsing herbivore to control *Allocasuarina*. The result is that systems lacking such interactions are moving towards cell 1 of Table 7.2. Other relevant observations in this regard are as follows.

Some 30 years ago, while studying populations of the tammar wallaby (*Macropus eugenii*) at the Tutanning Reserve (south of the present area), it was common to see seedling *Allocasuarina* cropped and frequently killed by the grazing of the wallaby. At Tutanning, the wallaby is now much reduced in abundance due to the presence of the introduced fox. In the Kellerberrin area, wallabies have not been present for decades, and the grey kangaroo (*Macropus fuliginosus*) is restricted to larger areas and does not occur in the density necessary to exert the grazing pressure of a population of tammar. There is an interesting possibility that the harvesting by the tammar, whose metabolic needs by way of nitrogen was about 1 g per day, returned most of this to the atmosphere by way of bacterial metabolism of the urine and faeces; however, in the absence of harvesting, fixation has continued but nitrogen is now accumulating and contributing to soil acidity. *Allocasuarina* can grow in acid conditions to its advantage relative to other shrubs and so contributes to their decline in abundance. This would be an example of the mechanism producing the outcome predicted in cell 4 of Table 7.1 or of cell 1 of Table 7.2. A way to reverse this trend is not readily apparent. Now that the *Allocasuarina* are mature, the introduction of a terrestrial herbivore is futile. *Allocasuarina* and other species with a soil seedbank will regenerate following disturbance such as fire, but an introduced native herbivore, such as the tammar, could not survive in the presence of the fox. It would appear that, once *Allocasuarina* becomes dominant, the almost inevitable result will be an almost single species stand.

Another problem relates to the role of mycorrhizae in retaining and recycling nutrients within the system. The changes in abundance and significance of the role of mycorrhizal species as the population of the host species matures is not documented. However, many mycorrhizal species have underground fruiting bodies which were formerly eaten, and the spores dispersed in the faeces of the species which included them in their diet. In other parts of Western Australia, bettongs (*Bettongia penicillata*) ate the underground sporocarps (fruiting bodies) and dispersed the spores in their faeces (Christensen 1980). Christensen described how the dung of bettongs containing spores is buried by scarabid beetles of the genus *Onthophagus*. It has already been mentioned that scarabid beetles depending on dung of macropods cannot persist in the absence of kangaroos and wallabies; so it would seem that, in the absence of species of *Bettongia*, sporocarps will not be eaten and the spores dispersed. The absence of dung-burying beetles also means that the spores cannot be introduced to the appropriate depth in the soil for the establishment and maturity of the fungi. Once dispersal is no longer possible, it means that, should the hypogean fungi be lost from a locality, it is not likely to be reintroduced. It is unknown what possible effect such an extinction would have on the species richness of mycorrhizae and the retention and recycling of scarce nutrients in a system.

Finally, another unfavourable situation may arise unless management intervenes and this is the failure to control rabbits and foxes. In their presence, and with management favouring frequent low-intensity hazard reduction burns, it will lead to cell 9 of Table 7.2; even with their absence, cell 3 of Table 7.2 will be favoured. It may take some time but a reduction in species richness is the inevitable consequence.

The boundaries defined above for maintaining species richness indicate that a management goal of maximum species richness in natural or rehabilitated areas will not be achieved cheaply. Successful management will depend on well-executed biological studies directed towards an understanding of the biology of the species which are possibly capable of strong interactions and are a test of the mechanisms proposed above. Management will thus become experimental, and its success will depend on the nature of the records kept, i.e., monitoring over several generations of managers. Success will depend on truly long-term studies.

The other problem of managing for the retention of rare and endangered species has been discussed by Main (1982, 1984). Should an increase in abundance be deemed the only way to retain the species, then it will be necessary to manipulate the environment so that the rare species is favoured relative to some other species which are dominating the function that the rare species could perform. However, should the rare species be a strongly interacting one, then its abundance cannot be readily manipulated, because the best hope for a rare, strongly interacting species is to maintain the species richness of the assemblage within which it occurs. It will still be rare, but its future is dependent on the continuation of the system within which it fills a functional role that is independent of commonness. An insurance that the rare and endangered species will be retained would be provided by the replication of occurrence at other sites.

7.9 Discussion

In order to achieve an integrated landscape, not only must the goal or vision of the end product be clear, but the mechanisms involved must also be clear. In this chapter, I have tried to indicate that simple principles regulate or influence the outcome of ecological situations.

An interpretation of the sensitivity to disturbance and the rapidity with which wheatbelt ecosystems have collapsed is that they are "bootstrapping systems" (Perry et al. 1989), able to persist despite the poor nutritional status of the soils because they have developed a system of close mutual interactions and so have been able to rise above the limitations of the environment. The redundancy natural to the system and the close positive feedback maintained the system in the face of disturbances characteristic of the natural environment but not foreign disturbances.

The maintenance of the mutual interactions in the natural part of the reintegrated landscape and establishment of a new interacting system in the revegetated parts of the landscape might pose difficulties. Mycorrhizae provide significant nutrient pathways in soils of low nutrient status, but phosphorus can be toxic (Ozanne and Specht 1981). Moreover, phosphorus depresses mycorrhizae development and plants with mycorrhizae are often more resistant to disease than those without (Malajczuk and Lamont 1981; Mosse et al. 1981). Thus, the following are possibilities: aerially applied phosphate fertiliser for crops or pastures may drift into natural vegetation, or native plants reintroduced to formerly cultivated land being revegetated may be exposed to sufficient residual phosphorus to inhibit mycorrhizae formation. In either case, the net result may be that the native vegetation exposed to phosphorus may die, become prone to disease, or fail to establish.

When a reintegrated landscape is achieved, it will be a landscape that is a mosaic of agricultural, natural, and seminatural systems which together maximise biodiversity and economic returns by maintaining the landscape amenity function (minimising the loss of landscape qualities through soil salinisation or water and wind erosion) and so make possible a sustainable agriculture and a functionally diverse natural system. The redundancy in the natural system reflects the functional substitution of species under changed circumstances such as environmental change, disturbance, and patch dynamics. In effect, we create a landscape with multiple uses. The nonagricultural part of the landscape could consist of:

1. relatively large natural areas retained as samples of the biota in a self-sustaining system;
2. smaller natural areas retaining a subset of biota capable of surviving in a self-sustaining manner within the limits of the system;
3. samples of self-sustaining native biota capable of persisting by exploiting the array of areas under 1 and 2; or
4. assemblages of selected elements of the local biota (local provenance), native but not local, or exotic species, established to provide a

landscape amenity function or perform an ecosystem function and be self-sustaining.

The principles underlying the persistance of each of the above are simple. The problem is one of least costly implementation and maintenance. The principles outlined allow for predictions of likely outcomes arising from actions taken; thus, management will become experimental, which is a sure way of establishing a useful, i.e., validated or tested knowledge, base. At present, lack of knowledge is the most serious deficiency in nature conservation, as this chapter indicates.

Acknowledgments. I thank W. McArthur for comments on the manuscript.

References

Christensen, P.E.S. (1980) The biology of *Bettongia penicillata* Gray, 1837 and *Macropus eugenii* (Desmarest, 1817) in relation to fire. *Forests Department of Western Australia Bulletin* 91.

Frank, P.W. (1968) Life histories and community stability. *Ecology* 49:355–357.

Groves, R.H. (1986) Invasion of Mediterranean ecosystems by weeds. In B. Dell, A.J.M. Hopkins, and B.B. Lamont (eds.) *Resilience in Mediterranean-type Ecosystems*. Junk, Dordrecht. The Netherlands. pp. 129–145.

Kitchener, D.J., Dell, J., and Muir, B.G. (1982) Birds in Western Australian wheatbelt reserves – Implications for conservation. *Biological Conservation* 22:127–163.

MacArthur, R. (1955) Fluctuations of animal populations and a measure of community stability. *Ecology* 36:533–536.

Main, A.R. (1979) The fauna. In B.J. O'Brien (ed.) *Environment and Science*. University of Western Australia Press, Nedlands. pp. 77–99.

Main, A.R. (1981) Ecosystem theory and management. *Journal of the Royal Society of Western Australia* 64:1–4.

Main, A.R. (1982) Rare species: Precious or dross? In R.H. Groves and W.D.L. Ride (eds.) *Species at Risk: Research in Australia*. Australian Academy of Science, Canberra. pp. 163–174.

Main, A.R. (1984) Rare species: Problems of conservation. *Search* 15:93–97.

Main, A.R. (1987) Management of remnants of native vegetation – A review of the problems and the development of an approach with reference to the wheatbelt of Western Australia. In D.A. Saunders, G.W. Arnold, A. Burbidge, and A.J.M. Hopkins (eds.) *Nature Conservation: The Role of Remnants of Native Vegetation*. Surrey Beatty and Sons, Chipping Norton, NSW. pp. 1–13.

Malajczuk, N. and Lamont, B.B. (1981) Specialised roots of symbiotic origin in heathlands. In R.L. Specht (ed.) *Heathlands and Related Shrublands*. Elsevier, Amsterdam. pp. 165–182.

Mosse, B., Stribely, D.P., and LeTacon, F. (1981) Ecology of mycorrhizae and mycorrhizae fungi. *Advances in Microbial Ecology* 5:137–210.

Murray, P. (1984) Extinctions down under: A bestiary of extinct late Pleistocene monotremes and marsupials. In P.S. Martin and R.G. Klein (eds.) *Quarternary Extinctions A Prehistoric Revolution*. University of Arizona Press, Tucson. pp. 600–627.

Ozanne, P.G. and Specht, R.L. (1981) Mineral nutrition of heathlands: Phosphorus toxicity. In R.L. Specht (ed.) *Heathlands and Related Shrublands*. Elsevier, Amsterdam. pp. 209–213.

Paine, R.T. (1980) Food webs: Linkage interaction strength and community infrastructure. *Journal of Animal Ecology* 49:667–685.

Perry, D.A., Amaranthus, M.P., Borchers, S.L., and Brainard, R.E. (1989) Bootstrapping in ecosystems. *BioScience* 39:230–236.

Pickett, S.T.A. and White, P.S. (1985) *The Ecology of Natural Disturbance and Patch Dynamics*. Academic Press, London.

Pickett, S.T.A., Kolasa, J., Armesto, J.J., and Collins, S.L. (1989) The ecological concept of disturbance and its expression at various hierarchical levels. *Oikos* 54:129–136.

Short, J.R.T. (1947) Description and life history of a new Western Australian coccid. *Proceedings of the Linnean Society of New South Wales* 71:259–268.

Chapter 8

Integrating Economic and Ecological Considerations: A Theoretical Framework

E.C. Lefroy, J. Salerian, and R.J. Hobbs

> *. . . if the farm is to last and remain in health, the wilderness must survive within the farm. That is what agricultural fertility is: the survival of natural process in the human order.*
>
> W. Berry
> The Unsettling of America: Culture and Agriculture (1977)

8.1 Introduction – Disintegration as a Crisis of Values

Previous chapters have described the fragmentation of the Western Australian wheatbelt environment over the past one hundred years. Ecological disintegration has been apparent in the form of the extinction of plants and animals (Hobbs et al., Chapter 4), the loss of topsoil and decline in soil properties (Nulsen, Chapter 5), and an annual increase in the area of land affected by salinity (McFarlane et al., Chapter 6). Social disintegration has been apparent in the last third of that period, with a decline in populations in small towns to the point of collapse of the social infrastructure (B.Y. Main, Chapter 3). In economic terms, however, agricultural land use in the region as a whole has remained viable over this period. This viability has been maintained partly at the expense of social values through the amalgamation of family farm businesses. The trend toward further amalgamation, which saw an eightfold increase in the amount of farmland controlled per farmer in the 50 years up to 1971 in Australia generally, continues (Mauldon and Schapper 1974; Australian Bureau of Statistics 1990). The primary cause of the social disruption of farm amalgamation has been the long-term decline in the terms of trade in agriculture; the prices received for farm produce have been increasing at a slower rate than the costs of inputs.

This chapter puts forward a model that attempts to integrate ecological, economic, and social values. Essential to this model is the development of a set of indicators that can be used to reflect the condition of four fundamental ecological processes: the cycling of water, the cycling of nutrients, the flow of energy, and biodiversity (as an expression of the process of evolution). A suggested framework for such a set of indicators is given in Tables 8.1, 8.2, 8.3, and 8.4.

An integrated approach to the management of land must take ecological values, social values, and economic values into consideration. Furthermore, it must acknowledge that, as humanity is a function of nature and can only exist within nature, the patterns of human land use must ultimately respect the tolerances within which natural physical, chemical, and biological processes operate. Ecological values must ultimately take precedence over other values as the boundary conditions or tolerances within which natural processes operate will always be outside human influence and cannot be compromised without compromising human existence. By contrast, economic and social values are partly a function of human demands and expectations and can be influenced by human attitudes and actions. In other words, ecological tolerances are relatively fixed compared to economic and social tolerances. As Williams (1990) pointed out, this difference is to be seen in the origins of the words economy and ecology. The Greek root *oikos*, meaning house, is common to both words. Whereas *logos* implies knowledge of the house, or the inherent properties of the house, *nomos* implies management of the household, or, more specifically, the distribution of its resources. The process of integration then involves respecting those

Table 8.1. Suggested ecological parameters for assessment of land use on farms in the Western Australian wheatbelt. I. The water cycle

Parameter	Indicator(s)	Measure	Level		
			Present	Transitional	Desirable
1. Water use potential	Perennial plant density	Trees ha^{-1} Percent cover	<1	10	30–40
	Annual plant density by season	Percent cover			
2. Groundwater quality	Conductivity	mS m^{-1}	3450		<0–800
	Total phosphorous	mg l^{-1}			0.05–0.15
	Total nitrogen	mg l^{-1}			1.0–1.4
	Suspended solids	mg l^{-1}			
3. Surface water quality	Conductivity	mS m^{-1}			0–400
	Total phosphorus	mg l^{-1}			
	Total nitrogen	mg l^{-1}			
	Suspended solids	mg l^{-1}			
4. Groundwater accession	Change in depth to water table	m			

Table 8.2. Suggested ecological parameters for assessment of land use on farms in the Western Australian wheatbelt. II. The nutrient cycle

Parameter	Indicator	Measure	Level		
			Present	Transitional	Desirable
1. Nutrient loss through wind and water erosion	Retention of surface cover	Percent stable vegetation cover			
	Average distance between permanent shelterbelts	Units of [shelterbelt] height	> ×200	×25	×10
2. Leaching of applied nutrients	Coincidence between soil types and management units	Percent			100
3. Soil biological activity	Population density of key key species	Number cm^{-3}			
	Level of metabolic activity in soil	Respiration rate			
	Rate of OM and surface litter breakdown.	kg ha^{-1} $year^{-1}$			
	Acidity/alkalinity	Δ pH $year^{-1}$			
4. Nitrogen fixation	Density of plant species with N-fixing associations	Percent cover			

Table 8.3. Suggested ecological parameters for assessment of land use on farms in the Western Australian wheatbelt III. Energy

Parameter	Indicator	Unit of measure	Level		
			Present	Transitional	Desirable
1. Energy inputs in grain production	Calorie equivalent herbicide per calorie grain	Ratio			
	Calorie equivalent fertilizer per calorie grain	Ratio			
	Calorie equivalent liquid fuel per calorie grain	Ratio			
2. Solar energy interception	Weighted average sowing date for crops	Date			
	Weighted average maturity date for crops	Date			
	Annual pasture density	Percent ground cover	0–20	10	15
	May –		1–30	15	30
	July –		5–60	40	60
	September –				

Table 8.4. Suggested ecological parameters for assessment of land use on farms in the Western Australian wheatbelt IV. Diversity

Parameter	Indicator	Unit of measure	Level		
			Present	Transitional	Desirable
1. Diversity of plant species and forms (native vegetation)	Species count by plant form:	Proportion of species originally present			
	Trees				
	Shrubs				
	Herbaceous perennials				
	Perennial grasses				
	Annual grasses				
	Annual herbs				
2. Representation of plant functional groups	Proportion of area occupied by	Percent			
	Trees*		3	5	10–20
	Shrubs*		5	10	20
	Herbaceous perennials*		2	10	20
	Annual pasture		45	40	30
	Crops		45	45	40
3. Diversity of native fauna	Key species count:	Proportion of species originally present			
	Birds –				
	Mammals –				
	Other –				
4. Soil biological activity	(See Table 8.2)				
5. Conservation status	Congruence between protected remnants and minimum subset		20	50	100
6. Connectivity	Percent of remnants connected by corridors		0	50	100

* Percent areas are not additive as they may co-occur.

aspects of natural systems upon which we are ultimately dependent and changing those aspects of the human order over which we have control.

Crucial to this approach is a definition of the tolerances or boundary conditions within which natural processes operate. In this regard, we are little better equipped than the Europeans who came and settled in the wheatbelt at the end of the last century. Their anthropocentric view of the world meant that they saw nature as providing opportunities for human use without recognizing any inherent limits. Today we are aware of the consequences of this attitude but still lack a system of values by which to correct it. As McHarg (1969) stated, ever since the origin of society and exchange "... the valuations attributed to commodities have increased in range and precision and the understanding of the operation of the limited sphere of economics has increased dramatically. This imperfect view of world as commodity fails to evaluate and incorporate physical and biological processes.... We are now unable to attribute value to indispensable natural processes but we have developed an astonishing precision for ephemera." This is still largely true today. We still lack a defined set of ecological boundary conditions against which we can measure both current land use practice and the value of any changes that are proposed.

The model we present for the integration of ecological and economic values is intended for use in the context of farm planning. The development of whole-farm and catchment planning has provided an excellent framework for rational land use decision making. The process, described more fully in the following chapter, utilises computer-based geographic information systems and analyzes the outcomes of alternative practices. Similar methods have been applied to other types of land use, for instance, fire management in conservation areas (e.g., Kessel 1979; Good 1985). Until now, however, farm-planning models have almost exclusively considered the economic implications of altered management and had no systematic way of accounting for ecological parameters. Only with the full integration of ecological, economic, and social values can truly sustainable agriculture exist. In doing so, it must be acknowledged that, despite the ultimate precedence of ecological considerations, economic and social criteria must be satisfied if the restoration and maintenance of ecological values is to be an active process to the mutual benefit of the landscape and those dependent on it. In a landscape such as this, presently dominated by agriculture, without profitable farms, there is unlikely to be active integration. Without respect for natural processes and attempts to understand the limits within which they operate, there can be only further disintegration.

8.2 Sustainability – Reintegration as a Problem of Scale

Sustainability is a term increasingly used but seldom effectively defined. This problem of definition has aroused considerable debate. Brown et al. (1987) asked whether sustainability was "rapidly becoming one of those transcendent terms . . . which are cornerstones of environmental policy and research, but difficult to measure and rarely defined explicitly." Similarly, Tisdell (1988) asked why ecologists have not defined and measured sustainability more carefully since they consider it so important. Within the agricultural context, it is clear that sustainability has been used in an imprecise sense, and there have been no attempts to quantify what is meant by it (Lockeretz 1988; National Research Council 1989; Gleissman 1990a; Parr et al. 1990).

The word sustainable implies *the ability to support over time.* Its use remains imprecise unless it is defined by boundaries of both space and time. In other words, the ability to support at what physical scale and for how long. Lowrance et al. (1986) suggested that a useful way of looking at agricultural systems is as a series of levels in a hierarchy, with temporal and spatial scales increasing at each level (Fig. 8.1). Looking at agriculture in this way helps remove some of the confusion caused by the use of the term sustainability, as it becomes apparent that different values and constraints (agronomic, economic, ecological, and social) are seen, at present, to dominate at different levels. For instance, at the scale of the individual field, where most agricultural research and extension have been directed up until now, the dominant constraints are agronomic and the dominant goal is the productivity of crops and pastures over several seasons. At the farm level, the dominant constraints are microeconomic and the overriding goal is the survival of the farm business, with a time scale often expressed in units of several generations of the farm family. At the landscape level, the dominant constraints are ecological and the goal is the maintenance of the life-supporting capacity of the catchment or landscape unit over a time scale considerably longer than units of several human generations. At the regional or national level, the dominant constraints are macroeconomic, the goals a mix of domestic food supply, generation of export income, and support of a rural population, while the time scale is determined by the planning horizons of politics and economics.

The physical boundaries mentioned above, and shown diagramatically in Figure 8.1, can be seen on maps and aerial photographs as the grid of human activity made up of field boundaries, property boundaries, and roads on one hand, while less obvious are the irregular shapes of catchments and habitats on the other. Ecological constraints dominate within the naturally defined boundaries of the watershed and the

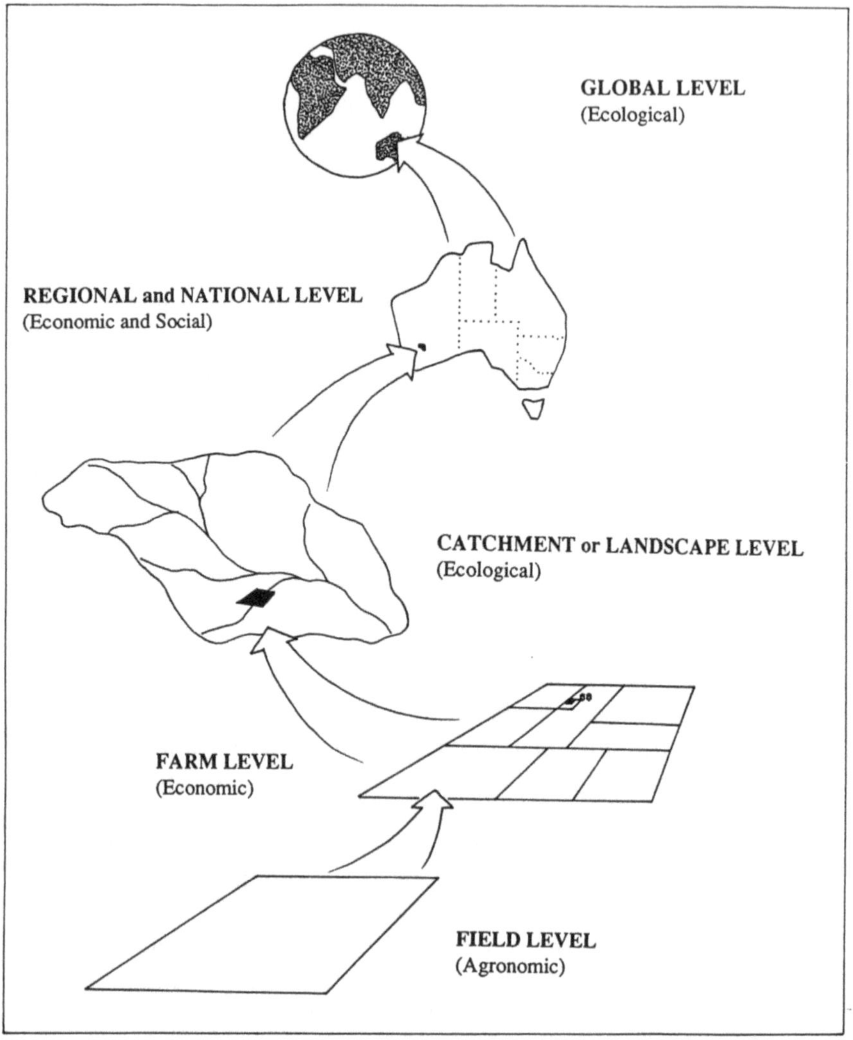

FIGURE 8.1. A hierarchical view of current agricultural systems showing different constraints dominating at different levels. (Modified from Lowrance et al. 1986.)

biosphere, while economic and social constraints presently dominate within the man-made boundaries of field, farm, shire, state, and nation. Seeing these different sets of boundaries in this way graphically highlights the problems inherent in trying to integrate values and constraints that apply at different levels. A consequence of integration, and a measure of its success, is the physical blurring of that grid typical of modern agriculture as field boundaries and vegetation are realigned according to ecological criteria.

Clearly, attempting to integrate ecological and economic values involves operating at two levels simultaneously. The ecological parameters, as they are applied to the basic unit of management, the farm, must themselves be determined at the catchment or landscape level, the smallest naturally defined land area within which the farm resides. Increasingly in Australia, the catchment area (i.e., that area defined by a watershed) is being recognised by farmers as a logical division of the landscape at which to begin addressing the problems of land degradation. Farmers within regionally based Land Conservation Districts and Land Care Groups are becoming involved in planning exercises at the catchment level (Robertson 1988). In that case, to apply ecological parameters directly at the catchment level would seem more logical than attempting to apply them at the individual farm level. However, no one person or group ultimately takes responsibility for what happens within a catchment. They first have to answer to the economic constraints of their farm business to ensure their own survival. For that reason, reintegration needs to begin at the basic level of human responsibility, the farm, and incorporate the economic constraints that apply at that level with the ecological constraints that are determined initially at the landscape level and ultimately at the global level.

A characteristic of hierarchical systems that becomes important when attempting to integrate between levels is their response to disturbance. According to Allen and Starr (1982), a disturbance at one level may be seen as a stabilizing force at another level. A.R. Main (Chapter 7) discussed the phenomenon of localised habitat disturbance contributing to the larger scale maintenance of a biota. In a different sense, this phenomenon can also be seen with farm amalgamation, where agriculture remains viable at a regional and national level partly due to the amalgamation that results from business failure at the individual level. Clearly, in the latter case social values are compromised. This highlights the need to consider the social implications and incorporate those values at the appropriate regional level in order to define fully the sustainability of a proposed land use. In the example that follows we present a model for integrating ecological and economic measures of sustainability at the farm level (Fig. 8.2), acknowledging that the logical extension of this requires the inclusion of social values.

8.3 A Theoretical Framework

The diagram in Figure 8.2 illustrates a theoretical decision-making process that incorporates ecological and economic considerations. The ecological parameters, specific to the catchment in which the farm is located, are given precedence over the economic parameters for the reasons discussed above.

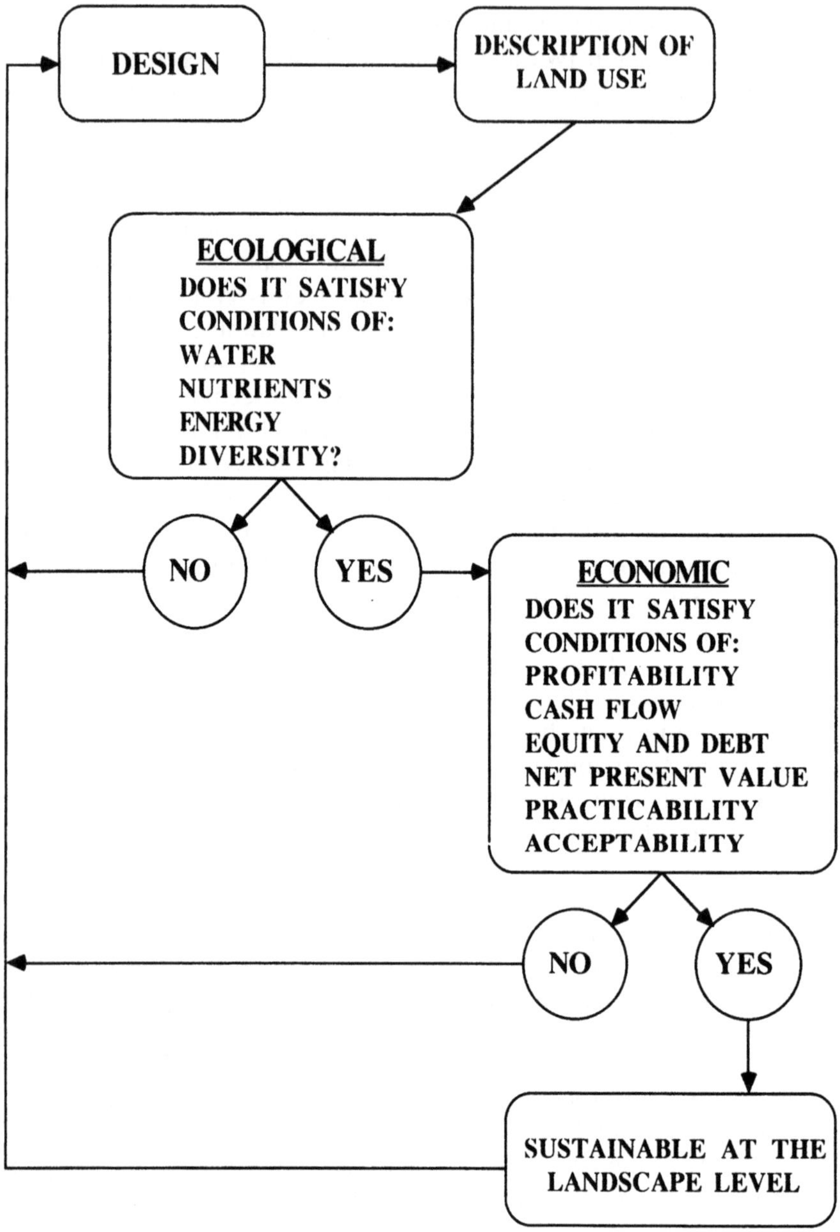

FIGURE 8.2. A model for the integration of ecological and economic values in agriculture.

The four quantitative economic parameters (profit, cash flow, equity and debt, and net present value) are generally accepted measures of the viability of businesses in general and farms in particular. The relative emphasis on these is a function of the individual manager's attitude to risk and will vary in each case. The two qualitative parameters (practicality and acceptability) emphasise that not all management decisions are based on rational economic principles. If a proposed change in farm practice or layout is considered unacceptable or impractical, despite its soundness by all other criteria, it is modified or discarded. There is no justification in advocating land use practices that will not be applied by the managers at whom they are directed.

The ecological parameters which need to be considered are more difficult to define and assign boundary conditions to. There is general agreement on the broad aspects of ecological sustainability, and these relate to the cycling of matter in the form of water and nutrients, the flux of energy, and the role of species richness in the dynamics of the biotic component (Savory 1988; Gleissman 1990b; A.R. Main, Chapter 7). We take these four processes (the water cycle, the nutrient cycle, energy flow, and diversity) as our starting point. The emphasis on an ecological process, rather than states, is important (Walker 1989), but it is this emphasis that makes the definition of boundary conditions difficult. While it is relatively simple to determine whether a particular land use will result in net economic loss or gain, it is much more complex to determine whether it will produce a net benefit in terms of nutrient or water cycling. However, the inclusion of parameters relating to each of these processes demands that some objective measures be attempted for each.

An important aspect of this model is that it represents a continual process. Given that meeting many of the desired ecological objectives is unlikely to be economic in the short term, we have suggested that parameters be expressed at three levels: present, desirable, and transitional. The transitional represents a realistic compromise, while the desirable remains as a target to focus whatever flexibility exists in the farm business in the future. The continual return to design in order to satisfy the desirable condition indicates the dynamic nature of this process. As knowledge increases, so will targets increase in number and degree, reinforcing this dynamic nature and making the point that, as we will never have perfect knowledge of ecological processes, sustainability will remain a moving target. This is not a new idea, but is encapsulated in the concept of "active adaptive management," as reviewed by Walters and Hilborn (1978).

8.4 Ecological Parameters – The Problems of Measurement

Indicators are needed to assess readily changes from the desirable condition. An example of where ecological status can be assessed using readily obtained indicators is in river and stream ecosystems, where pollution levels can be determined using simple chemical and biological indicators; as water quality declines, the suite of aquatic animals encountered also declines (Hynes 1960; Mills 1972). The success of restoration procedures can thus be quickly assessed with reference to the recolonising biota. For terrestrial ecosystems, such simple measures are distinctly lacking, particularly because of our lack of understanding of the processes occurring in soil and the inherent difficulties in studying them (see also Inhaler 1976; Paoletti et al. 1991). The problem is compounded by a lack of scientific consensus on many ecological issues. For instance, the concept of succession is central to ecology, but there is still debate concerning the mechanisms by which succession operates (Pickett et al. 1987; Walker and Chapin 1987; Tilman 1988, 1990). If we cannot provide a clear understanding of the major ecological processes, are we likely to be in a position to provide detailed guidelines which will assist in developing land management practices which are ecologically sound? The answer is that we *must* do so, since we do not have time to refine our understanding before acting on it. We must use the information we have available at present to provide guidelines that can be refined as more information becomes available. This process itself helps focus attention on gaps in our understanding and to direct investigation. It is in this light that we suggest the following set of ecological indicators, summarized in Tables 8.1, 8.2, 8.3, and 8.4. Most of the indicators represent only a snapshot view, whereas long-term sustainability can be assessed only in terms of changes or deviations from the desirable status. Also, some of these indicators have a fairly indirect relationship with the parameter under consideration, whereas they should ideally have a direct and definable relationship with them. Tables 8.1, 8.2, 8.3, and 8.4 provide estimates of the present, transitional, and desirable levels of parameters, where these are available; and indicate areas where parameter estimation is still required.

8.4.1 Water

There are two aspects of the water cycle that we are concerned with: attempting to restore balance between the various pathways by which water moves through the system and attempting to improve the quality of that water.

Considering the quantitative aspect, the widespread replacement of the original deep-rooted perennial vegetation with shallow-rooted, short-lived annual crops and pastures in southern Australia has caused disruption of the hydrologic balance, resulting in excess water in certain parts of the landscape and the subsequent problems of waterlogging, salinity, erosion, and flooding. This problem has been discussed frequently elsewhere in the world, with the removal of plant cover resulting in a less-buffered hydrologic system (Borman and Likens 1979; Uhl and Jordan 1984; Schofield et al. 1988). McFarlane et al. (Chapter 6) discussed in detail the hydrologic problems in the Western Australian wheatbelt and their possible solutions. The problem is to specify the desired outcomes and clear prescriptions for achieving these at the landscape level. The objective is the increased utilisation of water. This appears contradictory in a semiarid environment. However, we are dealing with a subtle redistribution that results in excess water at depth and in certain parts of the landscape, while water remains a limiting factor to plant growth at critical times during the growing season. The challenge is to use the excess water as productively as possible so that water tables are stabilised at a suitable depth and saline areas are stabilised or preferably returned to productive use. The key to this is to increase evapotranspiration by increasing the amount of actively transpiring plant cover present in the landscape throughout the year. There are two options available for the increased use of water. The first is the more effective use of annual plants while the second involves the increased use of perennial plants.

8.4.1.1 Increasing the Water Use of Annual Plants

Potential ways of increasing water use by annual plants includes increasing the area and duration of plant cover by the early planting of crops and by reducing the grazing pressure during germination and establishment of pastures (Fels 1990). Improvements can also be achieved by increasing water retention and decreasing runoff through contour cultivation, deep ripping, and minimum tillage techniques (Unger 1990) and through increased biomass production from genetic improvement and improved nutrient use efficiency. While all these approaches can refine water use during the growing season, the fact that 25% of the annual rainfall occurs outside this period (McFarlane et al., Chapter 6), combined with the poor moisture holding capacity of the soils of this region (Nulsen, Chapter 5), limits the effectiveness of using annual plants alone.

8.4.1.2 Increasing Perennial Plant Cover

The question here becomes one of defining the goal to be reached in terms of revegetation; what density of perennial plant cover is required and where in the landscape should it go? The answer to this will vary with the type of problem being tackled (Greenwood 1988; McFarlane et al., Chapter 6), but a general prescription can be provided by considering the amount of excess water which has to be used. An estimated 5 to 10% of the rainfall is currently unused and contributes to problems of salinity and waterlogging (Peck and Hurle 1973). With an annual rainfall of 350 mm per annum, this amounts to an average of $1000\,l\,ha^{-1}$ of excess water per day, assuming 10% of the rainfall is unused. From the limited information on water use by trees in this environment, we could assume that an individual eucalyptus tree would use about 25 to 30 l day^{-1} (Greenwood and Beresford 1979). This would amount to a target population of 30 to 40 trees ha^{-1} to account for the presently unused rainfall. This figure is dependent on age, health, species, and planting configuration but, for the purposes of this discussion, we will use this as a target figure.

Current analysis suggests that vegetation strategies for increasing water use in recharge areas are not clear, unlike strategies for localised areas of discharge (McFarlane et al., Chapter 6). The pattern by which this revegetation is laid down could then be dictated by other ecological requirements while minimising the impact on agronomic constraints. As an example, by concentrating this planting into belts with an average width of 15 m and a density of 400 trees ha^{-1} (25 m^2 per tree) and spacing those belts at an average distance of 200 m apart, the desired target of 30 to 40 trees ha^{-1} overall can be achieved while occupying less than 10% of the total area. The specific orientation of such planting can then be determined by the relative importance of environmental and agronomic factors including slope, runoff, wind erosion hazard, multiples of machinery widths, and the potential nature conservation value of such planting. For example, arranging tree belts in association with earthworks designed to intercept runoff water on sloping land would effectively bring the water to the trees and allow the continued production of annual crops and pastures in the broad-contoured areas between these wide-spaced belts. On areas where slope and soil type preclude the harvesting of water and increase the wind erosion hazard following the removal of native vegetation, orientation is more logically determined by prevailing wind direction. While, in the first case, this broadacre "alley-farming" approach has similarities with the Keyline System (Yeomans 1958, 1964) and, in the second case, is similar to the Forest Net System employed in China (Brandle et al. 1988), effective design solutions will most likely contain elements of both.

A further function of these vegetation belts is to reunite isolated patches of remnant native vegetation to act as corridors for endemic fauna and refuges for flora (Hobbs et al., Chapter 10). For their long-term survival, these vegetation belts would benefit from the presence of a mixed understorey, including some indigenous nitrogen-fixing species such as *Acacia* and *Allocasaurina*. An understorey of shrubs and herbaceous plants would also improve their effectiveness in preventing runoff, in acting as wildlife corridors, and in limiting weed infestation and the subsequent risk of frequent fires.

In attempting to satisfy initially the conditions of water use, it is therefore possible to envisage a network of perennial vegetation forming the framework for an irregular mosaic of agricultural, natural, and seminatural systems, as A.R. Main described (Chapter 7). The emergence of such a network, and the elements that go to make it up, are illustrated in Figure 8.3. Such a pattern could simultaneously satisfy some of the conditions of species diversity, nutrient cycling, and energy use efficiency (as discussed in subsequent sections) as well as water use. However, the widespread use of this approach and its effectiveness in restoring the active cycling of water is dependent on its direct and indirect economic benefit to the farm. Revegetation on such a scale would need to be economically driven. With about 16 million hectares of cleared land in the agricultural area of Western Australia, the target of 30 to 40 trees per hectare would require about 500 million trees. To put this into context, some seven million trees for noncommercial purposes were planted by farmers in Western Australia in 1989 (based on a survey conducted by M. Scheltema for Greening Australia Inc. 1990). Of these, approximately half will not have survived, due mainly to water stress and insect attack. To achieve the target of 500 million surviving trees within ten years would require planting 100 million trees each year. To allow for replacements (assuming a 30-year life span in narrow belts) would require 33 million trees to be planted *every* year.

The direct and indirect value of these vegetation belts would determine the extent to which this strategy is economically driven. The direct value requires a harvestable commodity such as wood or essential oils. Possibly more important is the indirect value they have in ameliorating climatic extremes and thereby improving the productivity of crops, pastures, and animals (Brandle et al. 1988; Kort 1988; Kubicki et al., Chapter 9).

Between the basic network of vegetation belts described above, there is the potential for the inclusion of other forms of perennial vegetation. These include herbaceous plants, grasses, and shrubs which are of direct commercial value to agriculture as well as increasing water use, and these are described in the following section on nutrient cycling.

The other aspect of the water cycle that concerns us is water quality. The conditions for life demand that water moving through a system meets certain requirements. Indicators include pH, electrical conductivity (as an indicator of salinity), total nitrogen content, total phosphorus content, suspended solids, and turbidity (Mitchell and Stapp 1986). These are being monitored for ground and surface water in the wheatbelt by schools and government agencies (Brown 1990) and could serve as useful indicators of the impact of current farming practice and the effectiveness of a strategy such as described above. Table 8.1 lists some ecological parameters for hydrologic balance and water quality.

8.4.2 Nutrient Cycling

A.R. Main (Chapter 7) detailed the marked differences in the nutrient cycle between natural and agricultural systems. A characteristic of

DESIGN AND LOCATION OF REVEGETATION

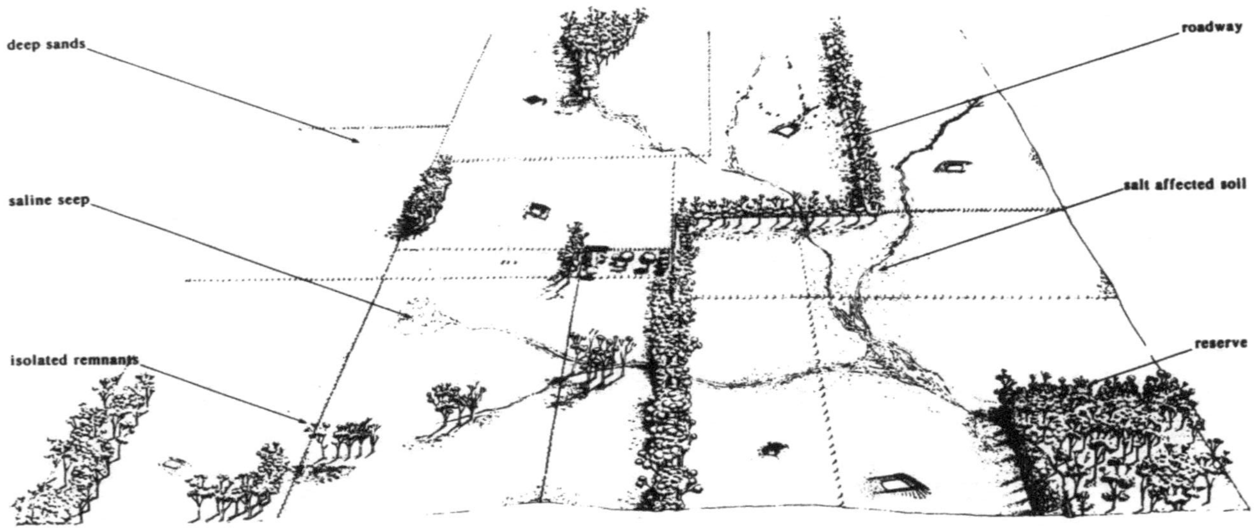

(a) **The Base Plan**
showing drainage, structures, vegetation, major problem areas and aspect.

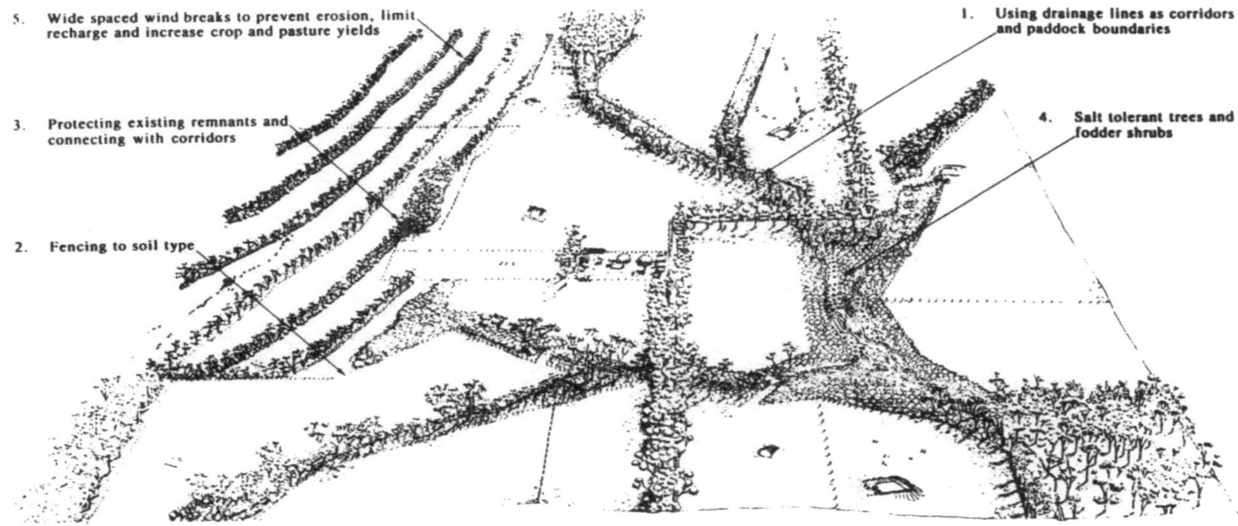

(b) **A Long Term Revegetation Plan**

Figure 8.3. A stylised representation of a central wheatbelt farm illustrating the transition from the present landscape (a) to a mosaic of agricultural, natural, and seminatural systems (b). Key elements of the transition are: 1. Reinstated drainage lines used as conservation corridors and field boundaries; 2. Coincidence between soil types and management unit; 3. Protection of existing remnants in vegetation corridors; 4. Productive revegetation of land affected by secondary salinity; 5. Alley-farming; a network of wide-spaced shelterbelts (10 to 25 × their height apart) to prevent wind erosion, limit recharge of ground water, increase crop and pasture yield, and act as conservation corridors. (From Lefroy et al. 1991.)

natural systems is that the nutrient cycle is essentially closed. From the fixation of carbon and nitrogen and the extraction of nutrients in soil through to the main redistribution pathways of herbivores, predators, and decomposers, the time that nutrients are retained within the system is maximised while losses are minimised. In agricultural systems, the continuing harvest of biological material results in a net export of nutrients which have to be replaced if agriculture is to continue. There is a considerably body of information on nutrient cycling in agricultural and natural, partly forested, systems (e.g., Frissel 1977; Likens et al. 1977; Borman and Likens 1979; Waring and Schlesinger 1985). There is little such information for the wheatbelt of Western Australia, with data on nutrient cycling being restricted mainly to the more mesic jarrah (*Eucalyptus marginata*) forest closer to the coast (Hingston et al. 1988). The desired outcome in an agricultural system is to close the nutrient cycle as much as possible by replacing exported nutrients, fostering biological recycling, and restricting losses due to leaching and erosion. While it is true that agriculture cannot be considered ecologically sustainable at the global level since it is dependent on inputs of nutrients that are being used at a faster rate than they are being deposited (as is the case with phosphorous), we are concerned initially with effects that can be measured at the landscape level.

8.4.2.1 Minimising Inputs

One approach to reducing inputs would be to change the agricultural system so that the plants grown are matched more closely to the ability of the environment to support them. This could be achieved by selecting more nutrient-efficient crop and pasture varieties and by making greater use of indigenous plants in agriculture. Most of the native vegetation in the wheatbelt is adapted to soils with low levels of phosphorous and nitrogen in particular (Bowen 1981; Bettenay 1984; Lamont 1984), whereas the cereal and grain legume crops and annual pastures presently grown are heavily dependent on additions of macro and micronutrients (Burvill 1979). Native plants are generally poor sources of forage for domestic animals since, with the exception of a very few endemic grasses (Gardner 1942), they are characteristically low in the digestible cellulose and protein compounds typical of agricultural plants. Instead, they are typically high in indigestible cellulose and lignin and often contain unpalatable and toxic organic compounds such as tannins and oxalates. Exploitation of native plants would, therefore, rely on farming indigenous animals such as the kangaroo or emu or be based on nonedible plant products such as essential oils, flowers, or wood. While this would still require the replacement of exported nutrients, it could lower the input requirements. In the meantime, however, we are concerned with grain, wool, and domestic livestock as the main products and, consequently, on the continued dependence on the input of nutrients, phosphorus in particular. The current average level of phosphorus applied to wheatbelt cropping soils is approximately 4.8 kg ha^{-1} each season while, for nitrogen, the level is 4.1 kg ha^{-1} (Australian Bureau of Statistics 1990).

8.4.2.2 Fostering Biological Recycling

The need for applied nitrogen in this region has been reduced by the use of legumes in pastures and, more recently, grain legume crops in rotation with cereal crops (Puckridge and French 1983). However, apart from acknowledging the importance of the rhizobium bacteria, little attention is paid to the status of other beneficial soil microorganisms in the Western Australian wheatbelt, other than recognising that much of conventional agricultural practice is at odds with a high level of biological activity in the soil (Abbott et al. 1979). This is despite recognition that the recycling of nutrients is essentially a biological process (Power 1990), that microbial activity is important for the improvement in the stability of soil aggregates and hence soil structure (Smith and Elliot 1990), that free-living bacteria can be responsible for fixing significant levels of nitrogen (Alexander 1977), and that mycorrhizal fungi, which have been shown to improve phosphate uptake by some species, are associated with most agricultural plants (Abbott and Robson 1982). From an ecological perspective, and particularly with respect to nutrient cycling and improving soil structure, the husbandry of soil microorganisms becomes as important to agriculture as the husbandry of plants and animals above the ground.

Parker (1989) detailed the steps that have led to a decline in biological activity in wheatbelt soils since the development of agriculture. Briefly stated, they involve removal of surface cover, frequent disturbance of the top 10 cm through ploughing, compaction of subsoil by heavy machinery, and replacement of deep-rooted perennial trees and shrubs with shallow-rooted annual plants. A reversal of this process requires the following three changes in practice:

(i) Retention of residue cover on the soil surface. This performs the same role as leaf litter in a natural system in reducing evaporation, reducing surface temperatures in summer, maintaining sufficient moisture in the surface for soil macrofauna, and acting as a source of organic matter to feed the biotic system.

(ii) A reduction in cultivation. The consequences of regular disturbance of the top 10 cm of soil have been the elimination of the soil macrofauna (such as ants, termites, and earthworms; see Hobbs et al., Chapter 4; A.R. Main, Chapter 7), a reduction in the size of water stable soil aggregates (Smith and Elliot 1990), and reduced infiltration of water and oxygen through disruption of root channels leading to the subsoil (Potter and Meyer 1990). Compaction of the subsoil at a depth of between 20 cm and 40 cm by heavy machinery has also reduced the infiltration of water and hence the growth of plant roots.

(iii) The use of deeper rooted plants. These will recycle nutrients from depth, particularly those leached below the root zone of annuals, and act as a focus for microbial activity in the rhizosphere throughout the profile.

The first two of these objectives, cover retention and reduced cultivation, could be achieved through the more widespread use of minimum tillage systems (Douglas 1987; Smith and Elliot 1990), providing that the benefits would not be negated by the reliance on herbicides harmful to soil flora and fauna, particularly nitrogen-fixing bacteria and mycorrhizal fungi (Trappe et al. 1984). The third objective, that of reintroducing perennial plants, would appear to be in conflict with a system based on rotations of annual crops and pastures. However, this could be achieved through the use of herbaceous perennial plants, both legumes and nonlegumes, either during the pasture phase (providing that it lasted for three to four years in order to obtain benefit from deeper root penetration) or as permanent cover with occasional annual crops over-sown. Used as temporary cover, the beneficial effects of the old root channels on biological activity through infiltration and aeration of the subsoil may last into several subsequent crop and pasture phases (Parker 1989). Potential also exists for the production of grain from well-adapted perennial grasses, using the same approach as the development of intermediate wheat grass (*Thinopyrum intermedium*) as a perennial grain crop in the United States (Wagoner 1990). Their particular value would be on land where annual cropping cannot be justified due to low production or erodability. Such perennial crops would increase nutrient cycling by achieving all three of the above objectives; however, the risk of harbouring root diseases of annual cereal crops would be a major consideration.

In addition to perennial pasture plants and perennial grain crops, shrubs as a source of browse fodder also have a role in such a system and are already being used in the wheatbelt. Several thousand hectares of *Atriplex* species are planted on salt-affected land each year (Malcolm 1986) and a similar area of the leguminous evergreen tree, *Chamaecytisus palmensis* (Tagasaste), is planted on sandy soils each year (Oldham and Moore 1989). The more widespread use of these and other perennial plants suited to a different range of soil types and used in conjunction with the tree belts would further improve the efficiency of nutrient cycling and increase water use.

8.4.2.3 Limiting Losses from the System

In addition to minimising inputs and fostering recycling processes, the other means of closing the nutrient cycle as much as possible is reducing the loss of nutrients from the system by leaching and soil erosion. Loss of applied fertiliser through leaching can be reduced by judicious timing and placement of fertiliser to maximise uptake and minimise losses (Power 1990), by using less soluble formulations that release more slowly (Hillman et al. 1990), or, more simply, by matching fertiliser additions to plant uptake on particular soils. The latter can be achieved by farming to soil type, which allows fertiliser to be applied at the most appropriate level and minimises potential losses through over-application, where field boundaries and soil type boundaries vary markedly. Loss of soil and hence organic matter and nutrients through water erosion could be reduced by a combination of factors already mentioned: minimum tillage carried out on the contour in combination with water-harvesting techniques and tree belts to decrease runoff, plus occasional rotations of deep-rooted herbaceous perennials to increase infiltration and water use; and the use of browse shrubs or perennial grain crops on soils with high erodability.

Loss of soil and hence organic matter and nutrients through wind erosion can similarly be reduced by a combination of the practices listed above, with the permanent vegetation belts being especially important. Significant increases in crop growth through shelter effects have been measured downwind of shelter belts to a distance of approximately ten times the height of the shelterbelt, with some effects continuing to 25 times the height (Kort 1988). Increased crop and pasture growth would result in increased nutrient uptake. A network of vegetation belts would also give permanent protection against soil loss, particularly important in autumn when dry annual pasture cover is at its lowest level. A demonstrated increase in the profitability of grain and wool production in the wheatbelt environment under such a system would reinforce, economically and ecologically, the broadacre alley-farming approach.

Table 8.2 lists some examples of ecological parameters for assessing the nutrient cycle. As the processes involved occur mainly in the soil and are difficult to measure directly, the parameters used are indirect, with the exception of number 3 which would rely on the availability of a simple assay technique for soil biological activity (Anderson 1978; Paoletti et al. 1991).

8.4.3 Energy

We are dealing here with the flow of energy that drives the cycling of water, nutrients, and ecosystems in general. However, it is well-recognized that agricultural systems differ from natural systems in terms of energy flow. The natural energy available from the sun is the same for both at any given location, but considerable additional inputs go into the agricultural system. These enter directly in the form of liquid fossil fuels and indirectly as petroleum-based chemicals and the energy used in manufacturing and transporting all other farm inputs, notably fertilisers and machinery.

The energy efficiency of agriculture can be improved in three main areas: firstly, by reducing the level of supplementary inputs; secondly, by improving the conversion of available solar energy; and, thirdly, by farms producing energy from the sun, wind, or biomass. Agricultural enterprises cannot be considered sustainable in a global sense while they remain heavily dependent on large inputs of energy derived from nonrenewable sources.

8.4.3.1 Reducing Energy Inputs

While Mollison (1988) has developed small-scale farming systems which are net energy producers, the challenge remains to adapt these to grain production in semiarid climates. In the meantime, the greatest potential for lowering inputs comes from reducing the consumption of liquid fuels and petroleum-based chemicals.

By world standards, Australian dryland cereal production is very energy efficient. The ratio of calories produced per calorie of energy input consumed is 4:1 compared to a ratio of 1:10 for dryland corn production in North America (Watt 1982). Ways of further reducing energy inputs include reducing fuel consumption through minimum tillage, contour farming, and the use of perennial crops. Perennial crops, in particular, though their application may be limited in area, also offer opportunities to reduce the use of herbicides. A small number of Australian farmers have already eliminated the use of herbicides in the production of annual cereals in order to meet the standard of the National Association for Sustainable Agriculture Australia for the sale of certified organic produce (National Association for Sustainable Agriculture Australia Limited 1990). There is little published material comparing these farmers with conventional cereal growers in terms of farming practice and profitability; however, it is apparent that weed control is achieved by a combination of practices which were commonly used in ley farming prior to the widespread use of herbicides. These include longer rotations, more intensive management of grazing for weed control during the pasture phase, and occasional mechanical control of grass weeds by "topping" (i.e., high mowing) prior to seed set. In addition, pasture legumes are sometimes resown more frequently, and there is the advantage today of access to newer varieties of crop and pasture plants and more efficient machinery. In one study, it was apparent that, although grain yields were lower using this approach, these were compensated for by lower input costs, which included a halving in the cost of fuel and the cost of machinery depreciation (Wynen 1988). This approach may also increase soil biological activity, where it replaces chemicals that normally suppress such activity. However, it is more likely to be economic rather than ecological factors that act as incentives for a wider interest in this approach to dryland farming.

A reduction in the use of phosphate fertilisers is, however, less likely. The soils of the Western Australian wheatbelt are characteristically low in phosphate and supplementation from more concentrated sources seems inevitable. Opportunities for lowering levels of imported mineral phosphate are through plant breeding and better management of the mycorrhizal fungi that improve the efficiency of uptake in many plants (Abbott and Robson 1982). The likelihood of lessening dependence on supplemental nitrogen, however, is much greater. The recently reported, artificially encouraged association of the nitrogen-fixing organism *Azospirillum* with the roots of wheat plants (Beardsley 1991) holds some hope for agriculture if it can be successfully transferred to the field. Combined with more widespread use of perennial and annual legumes, it is conceivable that nitrogen supplementation could be eliminated in the future.

8.4.3.2 Increasing the Conversion of Solar Energy

Agriculture in semiarid climates uses only a fraction of incoming solar flux because crops and pastures are composed of annual species which are actively growing for only five or six months of the year. The growing season in the wheatbelt extends from the break of season in April/May with the onset of winter rains until the end of spring in October (McFarlane et al., Chapter 6). Crop and pasture species do not attain full ground cover until some time after the start of the season and, if pastures are grazed during germination and early growth, they may never reach full cover (Fels 1990). Available energy is, therefore, underutilized, but this could be overcome in several ways. Improving the agronomy of annual crops and pastures could ensure full cover is reached early in the season and is maintained thereafter. In the case of crops, this means making the best use of opening seasonal rains by sowing early and using longer season varieties and minimum tillage techniques. In the case of pastures, this can be achieved by encouraging seed set in the previous season and by deferring the grazing of pastures until there has been time for germination and establishment without competition from grazing animals. Deferred grazing is dependent on alternative sources of feed at a critical time in the year, which can be achieved by increasing the proportion of perennial crops and pastures.

The use of *Atriplex* species and Tagasaste (*Chamaecytisus palmensis*) have already been mentioned in the context of water use and nutrient cycling. Used as a source of browse feed in autumn, they fill the gap between late summer crop residues and established winter pastures and, in this way, also improve the interception and use of energy when annual plants are inactive. The selection of herbaceous perennial forage plants and perennial grain species for a wide range of soil types would further increase the potential for improving energy conversion. Table 8.3 gives possible parameters which could serve as indicators of energy efficiencies.

8.4.3.3 Power from the Sun, Wind, and Biomass

The third possibility is the production of energy as a source of income. Commoner (1990) argued that, as the earth intercepts each day 1000 times as much energy as it uses, there can be no such thing as an energy crisis, simply an energy conversion crisis.

Ethanol and methane are two possible substitutes for liquid fossil fuels in powering farming operations or road transport, which make up 12 percent of the total energy consumption in Australian agriculture (Watt 1982). We could also see wheatbelt farmers generating income from sun and wind as they do now from soil by selling electrical power from these sources into the regional grid (Newman 1990). In the long term, increases in the price of fossil fuels as supplies inevitably run down will increase the pressure to reduce their consumption and to use alternatives such as methane and ethanol. The generation of electricity by alternative means is likely to occur sooner.

8.4.4 Diversity

The function of diversity in natural systems is still poorly understood. A.R. Main (Chapter 7) argued that diversity can be seen as a mechanism that ensures that the functions in an ecosystem will be operating under all the environmental conditions so far experienced. In this sense, the variety of organisms present represents a form of memory of all past cyclic and stochastic events that have shaped the biota we see today. Consequently, future response to change will be determined to some extent by the adaptations to past changes. Diversity in a genetic and structural sense is often seen as imparting stability to an ecosystem (Watt 1973); however, as A.R. Main (Chapter 7) pointed out, as stability is a function of time, observed stability in an ecosystem may be an artifact of the human time scale. Resilience, or the ability to withstand stress and persist, may therefore be a more useful way of describing the consequence of diversity. If we accept that diversity imparts resilience by increasing the probability that ecosystem functions will be maintained, the question becomes: "what are our objectives in a predominantly agricultural landscape?" One is simply the long-term retention of the native biota. Another is to increase the diversity within the agricultural component of the landscape. A third is to create an intermediary or seminatural system that helps to achieve the first objective by including endemic species. This system also contributes to the cycling of water and nutrients and increases the efficiency of energy use, which increase agricultural productivity through shelter effects. In this way, the real interaction between conservation and agricultural land use in terms of ecosystem services would be realised (Ehrlich and Mooney 1983; Norse 1987).

8.4.4.1 Retention of the Native Biota

In the small remaining areas of natural vegetation, grazing by domestic livestock and rabbits (*Oryctolagus cuneatus*) and predation by introduced species, notably the fox (*Vulpes vulpes*) and the feral cat (*Felis catus*), have degraded these areas and been responsible for extinctions (Hobbs et al., Chapter 4). The first stage in the retention of the regional biota is to fence these areas to exclude domestic animals and to eradicate rabbits, foxes, cats, and other feral animals. The second stage is the appreciation of the role of disturbance in the maintenance of diversity in natural systems, as pointed out by A.R. Main (Chapter 7). Consequently, in addition to protection, these areas will require some intervention in order to restore the necessary conditions for physical and biological disturbance. Fire (a form of physical disturbance) will require management and predation (a form of biological disturbance) will require the reintroduction of native plant and animal species.

8.4.4.2 Increased Diversity in Agriculture

The agricultural system at present is characterised by a low level of diversity in species, in form, and in longevity. The plant population of the vast proportion of the wheatbelt landscape can be accounted for by probably less than 25 species. The majority are exotic annual weeds with the balance being the introduced pasture species (subterranean clover, [*Trifolium subteraneum*], annual grasses, and annual species of *Medicago*) and crops (wheat, barley, lupins, and field peas). In contrast, the species richness of some areas of the original sandplain vegetation is amongst the highest in southern Australia (Lamont et al. 1984). Increasing the diversity within the agricultural system could be achieved by increasing the range of plant species, grazing animals, and soil organisms. The potential role that perennial crops and pastures can play has already been mentioned in discussing the water and nutrient cycle (8.4.1 and 8.4.2), and their inclusion would increase not only species number but also plant form and longevity, which all contribute to the resilience of the system. The use of endemic herbivores, such as emus and kangaroos, would further increase the range of plant species suitable as sources of forage on farms. Increasing the activity of the soil biota by the reintroduction of species and by the methods mentioned in discussing nutrient cycling offers advantages in improving the efficiency of nutrient uptake and is recognized as being a vital process in regenerating partially degraded landscapes (Allen 1988; Whitford 1988) and in biodynamic farming (Podalinsky 1985).

8.4.4.3 Creating Seminatural Systems

Perhaps the greatest potential for increasing diversity comes from the creation of the seminatural system that was described initially in the discussion on restoring balance to the water cycle (Fig. 8.3). This network of permanent vegetation, primarily trees for water use and wind protection, should include, as a condition of diversity, as many endemic species as possible and be aligned, wherever practical, to add to existing remnants and act as a corridor between them. As well as tree species, a variety of plant forms (shrubs, annual and perennial grasses, and herbaceous perennials) would need to be included to encourage the colonisation of these areas by endemic fauna. The goal should therefore be to produce a partially synthetic ecosystem which contains representatives of all functional groups, has conservation value, is self-sufficient, and enhances the required ecosystem services (Naveh 1988; Hobbs and Hopkins 1990). An effective measure of the degree of diversity achieved, and hence the resilience of the system, could be the presence of key species of birds and other animals that would reflect the extent to which food chains have been reestablished. Suggested indicators for diversity are shown in Table 8.4.

8.5 Economic Parameters

In economic terms, the decision-making framework in Figure 8.2 is concerned with criteria as they apply at the farm level. However, as many of the ecologically determined land use criteria will not be economically viable at the farm level, we must also consider the criteria that apply at regional and national levels. If society chooses to see the basic ecological conditions of land use met, then either greater structural decline in the rural society must be accepted or society as a whole must pay in some form, such as through government intervention.

8.5.1 Economics at the Farm Level

The farmer is concerned with the impact of land use on three criteria common to business in general. These are profit and loss, cash flow (the sources and uses of funds), and equity (the balance sheet). Minimum levels in each of these must be met for the business to remain commercially viable. Attempting also to meet ecological objectives raises the issue of how these might affect the economic viability of the farm over time.

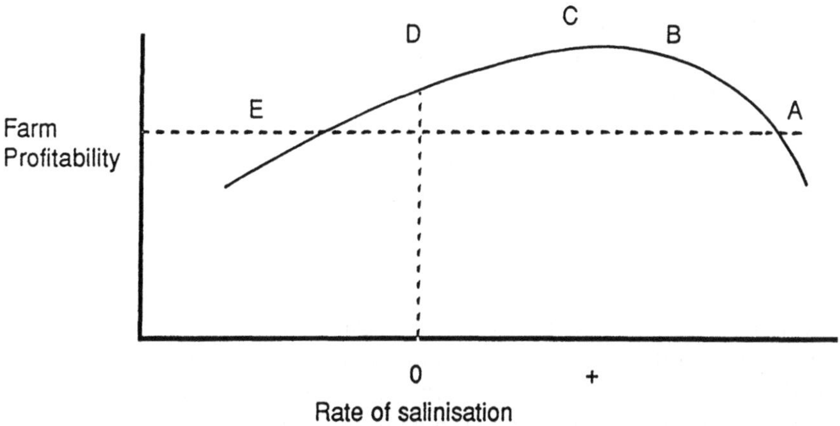

FIGURE 8.4. A schematic representation of the relationship between farm profitability and the rate of soil salinisation. The broken horizontal line represents the break-even point in farm profitability. A change in practice from point A to point B involves an increase in profitability and a decrease in the spread of salinity. In moving from point B to C, profit remains unchanged and the spread of salinity further decreases. Moving from point C to points D and E involves further reduction in salinity at the expense of declining profitability. (Based on data from Salerian [1990]).

Figure 8.4 uses the theoretical example of soil salinisation to illustrate the different types of change in attitude and behaviour that may be necessary to bring about a reduction in the rate of ecological degradation. A change from point A to point B on the curve is relatively simple, as it involves both an increase in profitability and a decrease in the rate of salinisation. An example of this would be planting salt-tolerant forage plants that both slow the rate of soil salinisation and provide a more cost effective source of autumn stock feed than conventional sources such as grain. This alternative feed source is only profitable up to the point that the seasonal feed shortage is satisfied. Planting beyond that point (from point B to point C) would further reduce salinisation but not increase profitability. In this case, overall farm profitability is maintained, ecological objectives are partly met, and, as is currently the case, there are income tax deduction benefits provided by the government for adopting recognised conservation practices. Moving from point C to point D is more difficult, as this results in a reduction in profit. Motivation for such a move would need to come from the belief that future production will be stabilised or increased or from the knowledge that the rate of soil salinisation has been reduced to zero. An example of a practice that might follow this pattern is a tree-planting programme higher in the landscape to slow the rate of groundwater recharge. A further move from point D to E requires changes in farm practice that either raise the curve or lower the breakeven point. An example of this might be the adoption of the organic farming methods mentioned earlier (8.4.3.1) that lowered the cost of inputs and raised the value of farm produce, thereby increasing profit despite a slight decline in yield (Wynen 1988).

The direct and indirect benefits of the trees as they mature may effectively shift the curve to the left and so bring about their own rewards. In the meantime, some profit must be forgone, and it is at this point that temporary intervention by society may be desirable (see Fig. 8.5).

Changes in land use practice will impact differently on profit, cash flow, and equity. Consequently, annual projections for each of these need to be made for ten to twenty years into the future in order to compare alternative practices. A means of computing such projections for several simultaneous changes in land use and determining the preferred mix of options is discussed by Kubicki et al. (Chapter 9).

8.5.2 Economics and the Social Optimum

The failure of market forces to protect the landscape is evident in the current level of degradation. The reasons for this have been widely discussed by economists (McConnell 1983; Blyth and McCallum 1987; Chisholm 1987; Kirby and Blyth 1987a,b; Quiggin 1987). They include the short-term planning horizons within agriculture, the lack of understanding of the complexities of the landscape, underinvestment in research, intervention by governments in commodity and input prices, and a long-term decline in the terms of trade. The last of these has been of particular importance. As the prices received by farmers for their produce have been increasing at a slower rate than the cost of inputs, the land has been put under increasing pressure in the search for increased productivity, and the impact of land degradation often accelerated.

The fundamental cause, however, is that the dominant system we have for valuing land (and nature in general) is based on its value as a resource for human beings and as a commodity of exchange, not on its intrinsic value. Recognition of its intrinsic value involves recognising that the essential processes that determine the integrity of nature often operate at vastly different scales of time and space to human activity and that the processes themselves are the unifying principle between all living and nonliving things in the world (Fox 1990). This acknowledgement of intrinsic value suggests that the condition of these processes, rather than the status of things, becomes the basis for a system of values.

That values other than a purely resource value exist in society is apparent in the difference between the observed levels of land degradation and those increasingly being demanded by society. Society's expectation in this context is referred to by economists as the socially optimal land

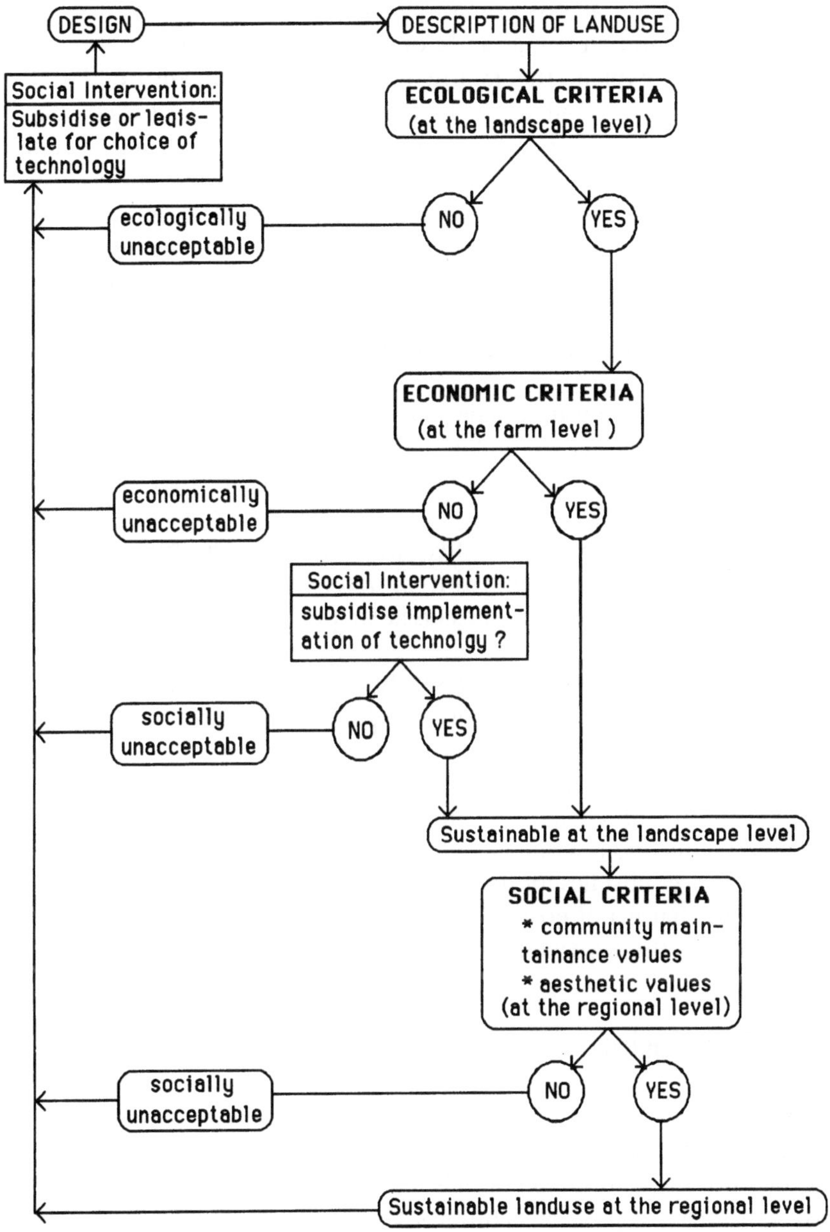

FIGURE 8.5. A model for the integration of ecological, economic, and social values in a landscape dominated by agriculture.

use. Some economists acknowledge that the socially optimal land use cannot be determined by economic analysis alone since it is a mix of social, economic, ecological, and aesthetic values (Bishop 1978; Anderson 1982; Izac 1986; Quiggin 1987). The market system has failed to protect the land as it is primarily concerned with the resource value of the land to human beings and not in its intrinsic value.

In order to explain the failure of the market to ensure that the actual standards of land use match those demanded by society, some economists have introduced the concepts of the option value, bequest value, and existence value of land. These reflect the value society places on preventing varying degrees of land degradation. The option value is the value placed on preventing irreversible damage, the bequest value is that placed on preventing any reduction in productive capacity, and the existence value is that placed on preventing any degradation whatsoever and so goes beyond the pure resource value (Pearce 1983; Kirby and Blyth 1987b).

Collectively, society may have higher option, bequest, and existence values than individual landholders. In that case, society may be willing to make the direct economic sacrifices necessary in order to see these values maintained. An example of this would be where landholders who use recognised land protection measures are exempted from land tax or rates on that land, are exempted from other fixed costs unrelated to the level of production, or are subsidised in the implementation of those measures.

Those forms of land use that satisfy ecological (i.e., the intrinsic or process) criteria but are not economically viable become candidates for direct intervention by society. Two examples of such intervention are shown in Figure 8.5. In the first case, there is legislation against the use of ecologically unacceptable technology or subsidies for the development of desirable technology such as renewable energy sources. In the second case, there are subsidies for the implementation of ecologically acceptable technology that is not economically viable at the farm level in the short term, such as the revegetation of recharge areas.

8.6 Conclusion

Subjecting agriculture to a simple ecological analysis in terms of the cycling of water and nutrients, the efficiency of energy use, and the role of diversity helps to highlight deficiencies in the present system of land use and arrive at possible solutions. Using such an approach, it may be possible to develop a system of values that emphasises the integrity of these processes rather than the value of land as a resource for any particular hierarchy of organisms.

In examining ways to restore the balance to the water cycle, close the nutrient cycle as much as possible, improve energy use efficiency, and increase diversity, it becomes apparent that some solutions impact on several processes simultaneously. Central to these solutions in this particular landscape is a pattern of revegetation that blurs the distinction between agricultural land and nature reserves by having, at the same time, nature conservation values, functional ecosystem values of benefit to agriculture, and a direct positive impact on agricultural production. In this way, it is possible to envisage an integrated landscape that would replace what Bradby (1989) described as "ecological apartheid" where, in the extreme case, on one side of the fence human activity is unrestrained and the limits of natural processes exceeded while, on the other side, human activity is prohibited and ecosystem functions remain largely unappreciated.

The extent to which this integration occurs will depend on two factors, the economic viability within the ecological parameters and society's willingness to support measures that are not economically viable at present. Some of the practices mentioned will become viable as input prices rise. Others dependent on technology as yet unavailable are a function of research efforts. In particular, research that establishes whether the described revegetation strategy can increase crop and pasture productivity in the wheatbelt, as it does in similar environments elsewhere, is a key factor in determining the extent to which this strategy is economically driven.

The rate and direction of change in society's attitudes to optimal land use will determine the degree to which it is socially reinforced.

Acknowledgement. Some of this work (E.C. Lefroy and J. Salerian) was undertaken as part of projects funded by the National Soil Conservation Program. We also wish to thank the following for their comments on the manuscript: David Bicknell, Graeme Arnold, Andrew Campbell, John Ives, Don McFarlane, Ann Hamblin, Graeme Robertson, Lex Parker, Kathy Velterop, and Martine Scheltema.

References

Abbott, L.K. and Robson, A.D. (1982) The role of vesicular arbuscular mycorrhizal fungi in agriculture and the selection of fungi for inoculation. *Australian Journal of Agricultural Research* 33:389–408.

Abbott, L.K., Parker C.A., and Sills, I.D. (1979) Changes in the abundance of large soil animals and the physical properties of soils following cultivation. *Australian Journal of Soil Research* 17:343–53.

Alexander, M. (1977) *Introduction to Soil Microbiology*. John Wiley and Son, New York.

Allen, M.F. (1988) Below ground structure: A key to reconstructing a productive arid ecosystem. In E.B. Allen (ed.) *The Reconstruction of Disturbed Arid Lands*. Westview Boulder, Colorado. pp. 113–115.

Allen, T.F.H. and Starr, T.B. (1982) *Heirarchy: Perspectives for Ecological Complexity*. University of Chicago Press, Chicago.

Anderson, J.R. (1978) Some methods for assessing pesticide effects on nontarget soil microorganisms and their activities. In I.R. Hill and S.J.L. Wright (eds.) *Pesticide Microbiology*. Academic Press, New York. pp. 313–501.

Anderson, T.L. (1982) The new resource economics: Old ideas and new applications. *American Journal of Agricultural Economics* 64:928–934.

Australian Bureau of Statistics (1990) *Agricultural Land Use and Selected Inputs. Western Australian Season 1988–1989*. Australian Bureau of Statistics, Perth.

Beardsley, T. (1991) A nitrogen fix for wheat. *Scientific American* 264:8.

Berry, W. (1977) *The Unsettling of America: Culture and Agriculture*. Sierra Club Books, San Francisco, California.

Bettenay, E. (1984) Origin and nature of the sandplains. In J.S. Pate and J.S. Beard (eds.) *Kwongan: Plant Life of the Sandplain*. University of Western Australia Press, Nedlands. pp. 57–68.

Bishop, R.C. (1978) Endangered species and uncertainty: The economics of a safe minimum standard. *American Journal of Agricultural Economics* 60(1): 10–18.

Blyth, M.J. and McCallum, A.J. (1987) Onsite costs of land degradation in agriculture and forestry. In A. Chisholm, and R. Dumsday (eds.) *Land Degradation Problems and Policies*. Cambridge University Press, Cambridge. pp. 79–98.

Borman, F.H. and Likens, G.E. (1979) *Pattern and Process in a Forested Ecosystem*. Springer-Verlag, New York.

Bowen, G.D. (1981) Coping with low nutrients. In J.S. Pate and A.J. McComb (eds.) *The Biology of Australian Plants*. University of Western Australia Press, Perth. pp. 33–64.

Bradby, K. (1989) New challenges in the bush. In J. Arnold (ed.) *The Bush Comes to the City*. Fitzgerald Biosphere Project, Perth. pp. 61–71.

Brandle, J.R., Hintz, D.L., and Sturrock, J.W. (1988) *Windbreak Technology*. Elsevier, New York.

Brown, B., Hanson, M.E., Liverman, D.M., and Meredith, R.W. Jr. (1987) Global sustainability: Towards definition. *Environmental Management* 11:713–719.

Brown, K. (1990) Ribbons of blue. *Land Management Society Newsletter, Winter 1990*, Perth.

Burvill, G.H. (ed.) (1979) *Agriculture in Western Australia 1929–1979*. University of Western Australia Press, Nedlands.

Chisholm, A.H. (1987). Abatement of land degradation: Regulation vs economic incentives. In A. Chisholm, and R. Dumsday (eds.) *Land Degradation Problems and Policies*. Cambridge University Press, Cambridge. pp. 223–247.

Commoner, B. (1990) *Making Peace with the Planet*. Victor Gollancz, London.

Douglas, L.A. (1987) Effect of cultivation and pesticide use on soil biology. In P.S. Cornish and J.S. Pratley (ed.) *Tillage – New Directions in Australian Agriculture*. Australian Society of Agronomy. Inkata Press, Melbourne. pp. 308–317.

Ehrlich, P.R. and Mooney, H.A. (1983) Extinction, substitution, and ecosystem services. *Bioscience* 33:248–254.

Fels, H. (1990) Grazing animals in mixed farming systems. *Land Management Society Newsletter, Spring 1990*, Perth.

Fox, W. (1990) *Towards a Transpersonal Ecology*. Shambala, Boston.

Frissel, M.J. (1977) Cycling of mineral nutrients in agricultural ecosystems. *Agroecosystems* 4:1–345.

Gardner, C.A. (1942) The vegetation of Western Australia with particular reference to climate and soils. *Journal of The Royal Society of Western Australia* 28:11–87.

Gleissman, S.R. (1990a) Quantifying the agroecological component of sustainable agriculture. In S.R. Gleissman (ed.) *Agroecology: Researching the Ecological Basis for Sustainable Agriculture.* Springer-Verlag, New York. pp. 366–370.

Gleissman, S.R. (1990b) Agroecology: Researching the basis for sustainable agriculture. In S.R. Gleissman (ed.) *Agroecology: Researching the Basis for Sustainable Agriculture.* Springer-Verlag, New York. pp. 3–10.

Good, R.B. (1985) Preplan and the management of national parks. In J.R. Ford (ed.) *Fire Ecology and Management in Western Australian Ecosystems.* Western Australian Institute of Technology, Bentley. pp. 147–151.

Greenwood, E.A.N. (1988) The hydrologic role of vegetation in the development and reclamation of dryland salinity. In E.B. Allen (ed.) *The Reconstruction of Disturbed Arid Lands.* Westview, Boulder, Colorado. pp. 205–233.

Greenwood, E.A.N. and Beresford, J.D. (1979) Evaporation from vegetation in landscapes developing secondary salinity using the ventilated chamber technique. 1. Comparative transpiration from juvenile *Eucalyptus* above saline groundwater seeps. *Journal of Hydrology* 42:369–382.

Hillman, K., Lukatelich, R.J., and McComb, A.J. (1990) The impact of nutrient enrichment on nearshore and estuarine ecosystems in Western Australia. *Proceedings of the Ecological Society of Australia* 16:39–53.

Hingston, F.J., Connell, A.M., and Groves, T.S. (1988) Nutrient cycling in the jarrah forest. In *The Jarrah Forest. A Complex Mediterranean Ecosystem.* Kluwer, Dordrecht. Netherlands. pp. 155–177.

Hobbs, R.J. and Hopkins A.J.M. (1990). From frontier to fragments. European impact on Australia's vegetation. *Proceedings of the Ecological Society of Australia* 16:93–114.

Hynes, H.B.N. (1960) *The Biology of Polluted Waters.* Unversity of Liverpool Press, Liverpool.

Inhaler, H. (1976) *Environmental Indices.* John Wiley and Son, New York.

Izac, A.M.N., (1986) Resource policies, property rights, and conflicts of interest. *Australian Journal of Agricultural Economics* 30:23–31.

Kessel, S.R. (1979) *Gradient Modeling: Resource and Fire Management.* Springer-Verlag, New York.

Kirby, M.G. and Blyth, M.J. (1987a) An economic perspective on government intervention in land degradation. In A. Chisholm and R. Dumsday (eds.) *Land Degradation Problems and Policies.* Cambridge University Press, Cambridge. pp. 213–222.

Kirby, M.G. and Blyth, M.J. (1987b) Economic aspects of land degradation in Australia. *Australian Journal of Agricultural Economics* 31:154–174.

Kort, J. (1988) Benefits of windbreaks to field and forage crops. *Agriculture Ecosystems and Environment* 22/23:165–190.

Lamont, B.B. (1984) Specialised modes of nutrition. In J.S. Pate, and J.S. Beard (eds.) *Kwongan: Plant Life of the Sandplain.* University of Western Australia Press, Perth. pp. 126–145.

Lamont, B.B., Hopkins, A.J.M., and Hnatiuk, R.J. (1984) The flora – Composition, diversity and origins. In J.S. Pate and J.S. Beard (eds.) *Kwongan: Plant Life of the Sandplain*. University of Western Australia Press, Perth. pp. 27–50.

Lefroy, E.C., Hobbs, R.J., and Atkins, L.J. (1991) *A Revegetation Guide to the Central Wheatbelt*. Western Australia Department of Agriculture. Perth.

Likens, G.E., Borman, F.H., Pierce, R.S., Easton, J.S., and Johnson, N.M. (1977) *Biogeochemistry of a Forested Ecosystem*. Springer-Verlag, New York.

Lockeretz, W. (1988) Open questions on sustainable agriculture. *American Journal of Alternative Agriculture* 3:174–181.

Lowrance, R., Hendrix, P.F., and Odum, E.P. (1986) A hierarchical approach to sustainable agriculture. *American Journal of Alternative Agriculture* 1:169–173.

Malcolm, C.V. (1986) Production from salt affected soils. *Reclamation and Revegetation Research* 5:343–361.

Mauldon, R.G. and Schapper, H.P. (1974) *Australian Farmers under Stress, in Prosperity and Recession*. University of Western Australia Press, Nedlands.

McConnell, K.E. (1983) An economic model of soil conservation. *American Journal of Agricultural Economics* 65:83–89.

McHarg, I.L. (1969) *Design with Nature*. Doubleday, New York.

Mills, D.H. (1972) *An Introduction to Freshwater Ecology*. Oliver and Boyd, Edinburgh.

Mitchell, M.K. and Stapp, W.B. (1986) *Field Manual for Water Quality Monitoring: An Environmental Education Program for Schools*. Thomson Shore, Dexter, Michigan.

Mollison, B. (1988) *Permaculture: A Designer's Manual*. Tagari, Tyalgum, New South Wales.

National Association for Sustainable Agriculture Australia (1990) *Standards*. National Association for Sustainable Agriculture Australia, Sydney.

National Research Council (1989) *Alternative Agriculture*. National Academy Press, Washington, D.C.

Naveh, Z. (1988) Multifactorial reconstruction of semiarid Mediterranean landscapes for multipurpose land uses. In E.B. Allen (ed.) *The Reconstruction of Disturbed Arid Lands*. Westview, Boulder, Colorado. pp. 234–256.

Newman, P. (1990) Future energy supplies. In *Sustainable Agriculture for Western Australia – Fact or Fantasy?* Symposium Proceedings, Western Australian Branch of Australian Institute of Agricultural Science, Perth.

Norse, E.A. (1987) Habitat diversity and genetic variability: Are they necessary ecosystem properties. In R.S. Draggon, J.J. Cohrssen and R.E. Morrison (eds.) *Preserving Ecological Systems: The Agenda for Long Term Research and Development*. Preager, New York. pp. 93–111.

Oldham, C.M. and Moore, P. (1989) Tagasaste (*Chamaecytisus palmensis*), an evergreen fodder tree, in grazing systems of Mediterranean-type climates. *Proceedings of the 16th International Grasslands Conference*, Nice.

Paoletti, M.G., Fauretto, M.R., Stinner, B.R., Pursington, F.F., and Bates, J.E. (1991) Invertebrates as bioindicators of soil use. *Agriculture, Ecosystems, and Environment* 34:341–362.

Parker, C.A. (1989) Soil biota and plants in the rehabilitation of degraded agricultural soils. In J.D. Majer (ed.) *Animals in Primary Succession: The Role of Fauna in Reclaimed Lands*. Cambridge University Press, Cambridge. pp. 423–438.

Parr, J.F., Stewart, B.A., Hornick, S.B., and Singh, R.P. (1990) Improving the sustainability of dryland farming systems. *Advances in Soil Science* 13:1–8.

Pearce, D.W. (1983) *Cost-Benefit Analysis*. Macmillan, New York.

Peck, A.J. and Hurle, D.H. (1973) Chloride balance of some farmed and forested catchments in southwestern Australia. *Water Resources Research* 9:643–657.

Pickett, S.T.A., Collins, S.L., and Armesto, J.J. (1987) Models,mechanisms, and pathways of succession. *Botanical Review* 53:335–371.

Podalinsky, A. (1985) *Biodynamic Agriculture: Introductory Lecture Series. Volume 1.* Gavemer, Melbourne.

Potter, C.S. and Meyer, R.E. (1990) The role of soil biodiversity in sustainable dryland farming systems. *Advances in Soil Science* 13:241–252.

Power, J.F. (1990) Fertility management and nutrient cycling. *Advances in Soil Science* 13:131–149.

Puckridge, D.W. and French, R.J. (1983) The annual legume pasture in cereal ley farming systems of southern Australia: A review. *Agriculture, Ecosystems, and Environment* 9:229–267.

Quiggin, J. (1987) Land degradation: Behavioural causes. In A. Chisholm and R. Dumsday (ed.) *Land Degradation Problems and Policies*. Cambridge University Press, Cambridge. pp. 203–212.

Robertson, G.A. (1988) Soil conservation districts – Land users solving mutual problems. In P. Newman, S. Nevile, and L. Duxbury (eds.) *Case studies in Environmental Hope*. Environmental Protection Authority, Perth. pp. 89–95.

Salerian, J. (1990) *Economic Evaluation of Soil Salinity in a Subcatchment in Western Australia*. Unpublished Ph.D. Thesis, University of Western Australia, Perth.

Savory, A. (1988) *Holistic Resource Management*. Island Press, Washington, D.C.

Schofield, N.J., Stoneman, G.L., and Low, I.C. (1988) Hydrology of the jarrah forest. In B. Dell, J.J. Havel, and N. Malajczuk (eds.) *The Jarrah Forest. A Complex Mediterranean Ecosystem*. Kluwer, Dordrecht Netherlands. pp. 179–201.

Smith, J.L. and Elliot, L.F. (1990) Tillage and residue management effects on soil organic matter dynamics in semiarid regions. *Advances in Soil Science* 13: 69–88.

Tilman, D. (1988) *Plant Strategies and the Dynamics and Structure of Plant Communities*. Princeton University Press, Princeton, N.J.

Tilman, D. (1990) Constraints and tradeoffs: Toward a predictive theory of competition and succession. *Oikos* 58:3–15.

Tisdell, C. (1988) Sustainable development: Differing perspectives of ecologists and economists and relevance to LDC's. *World Development* 16:373–384.

Trappe, J.M., Molina, R., and Castellano, M. (1984) Reactions of mycorrhizal fungi and mycorrhiza formation to pesticides. *Annual Review of Phytopathology* 22:331–359.

Uhl, C. and Jordan, C.F. (1984) Succession and nutrient dynamics following forest cutting and burning in Amazonia. *Ecology* 65:1476–1490.

Unger, P.W. (1990) Conservation tillage systems. *Advances in Soil Science* 13: 27–68.

Wagoner, P. (1990). Perennial grain: New use for intermediate wheatgrass. *Journal of Soil and Water Conservation* 45:81–82.

Walker, B.H. (1989) Diversity and stability in ecosystem conservation. In D. Western and M.C. Pearl (eds.) *Conservation for the Twenty-first Century*. Oxford University Press, Oxford, pp. 121–130.

Walker, L.R. and Chapin, F.S. (1987) Interactions among processes controlling successional change. *Oikos* 50:131–135.

Walters, C.J. and Hilborn, R. (1978) Ecological optimization and adaptive management. *Annual Review of Ecology and Systematics* 9:157–188.

Waring, R.H. and Schlesinger, W.H. (1985) *Forest Ecosystems: Concepts and Management*. Academic Press, Orlando, Florida.

Watt, K.E.F. (1973) *Principles of Environmental Science*. McGraw Hill, New York.

Watt, M. (1982) *An Energy Analysis of the Australian Food System*. Ph.D. Thesis, Murdoch University, Perth.

Whitford, W.G. (1988) Decomposition and nutrient cycling in disturbed arid ecosystems. In E.B. Allen (ed.) *The Reconstruction of Disturbed Arid Lands*. Westview, Boulder, Colorado. pp. 126–141.

Williams, R. (1990) An open letter to the agricultural community on defining sustainability. *Journal of Soil and Water Conservation* 45:91–93.

Wynen, E. (1988) An economic comparison of sustainable and conventional farms in southeastern Australia. In *Sustainable Agriculture: A Viable Alternative*. National Association for Sustainable Agriculture, Sydney. pp. 25–35.

Yeomans, P.A. (1958) *The Challenge of Landscape*. Keyline, Sydney.

Yeomans, P.A. (1964) *Water for Every Farm*. Murray, Sydney.

Chapter 9

Determining the Long-term Costs and Benefits of Alternative Farm Plans

A. Kubicki, C. Denby, M. Stevens, A. Haagensen, J. Chatfield*

9.1 Introduction

The emphasis of this chapter is on the integration of physical, production, financial, and social issues into farm management plans. The process needs to include the comparison of long-term costs and benefits of the various alternatives available, and the determination of the economic viability of the farm.

No attempt has been made to define desirable agriculture. That is left to the community and to those with individual farm interests (Lefroy et al., Chapter 8). To do so would be to focus attention away from the integration process. Instead, these considerations are seen as alternatives which can be presented, analyzed, and decided upon.

9.2 Definition of Farm Planning

The term "Farm Planning" is used to refer to the arrangement of the farm's physical features, its land use, and the financial resources into operational management plans. The options available depend on business, family, and personal objectives, their strengths and weaknesses, their disposition toward these alternative practices, and the current and future financial positions.

To achieve planned changes, it is necessary to induce managers to change their attitudes and behaviour to particular practices by showing that financial independence is still possible. Because of this, plans need to be developed on the farm. Consultants can be used for technical assistance to provide additional information, to indicate the implications

*We acknowledge the assistance of Christine Easton in the digitising of farm maps

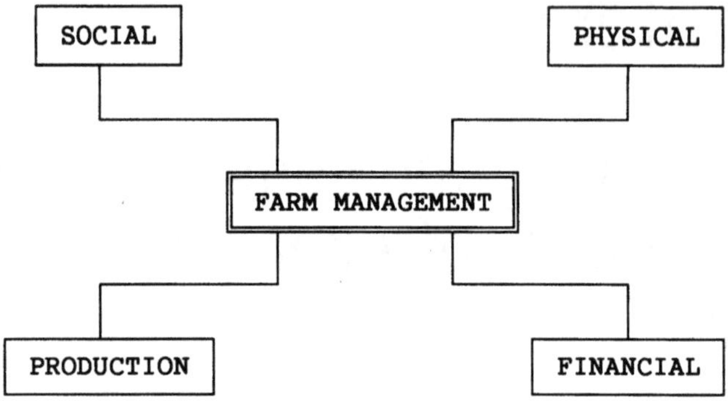

FIGURE 9.1. Operational environment of farm planning.

of actions, and to incorporate broader issues such as impacts on catchment hydrology.

For the property to sustain agriculture in the future, systems must be developed that can cope with current and future environmental pressures in a way that allows farmers to remain viable. It is this system that the farm plans address. They need to be dynamic and require review depending on changing internal and external conditions. The operational environment is summarised in Figure 9.1.

9.3 Factors to be Integrated

9.3.1 Social Factors

The social environment is the most important as it relates to objectives and expectations. Unfortunately, social objectives are rarely recorded and tend to be generalised. Community, family, individual, and business interests need to be included. Community considerations include long-term productivity, aesthetics, soil erosion and siltation, a reduction of salinisation, and a balance with the ecosystem in terms of reafforestation, provision of nature corridors, and use of remnant vegetation. Family considerations include life style, independence and dignity, employment, security, and superannuation. Individual goals include improving the farm's productivity and being a capable provider and an efficient operator. Business considerations, be they corporate or partnership, include flexibility in management and operation, efficiency, provision for lean times, and financial independence.

Obviously, a number of interests are being represented, each with their own hidden agendas. If external parties become involved, they generally do so without a full understanding of management objectives or budget implications. Their interests are specific and "sold" to man-

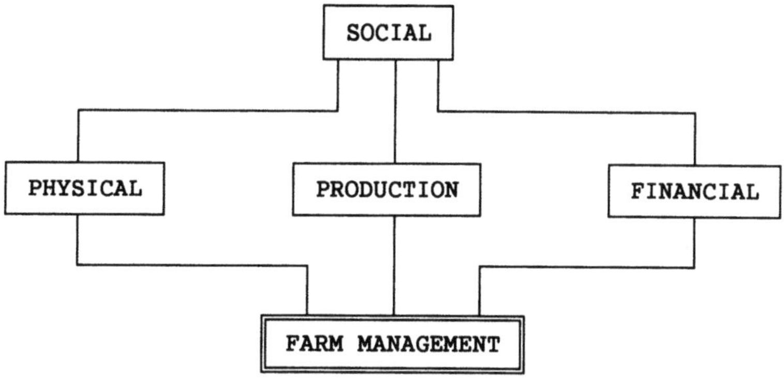

FIGURE 9.2. Management objectives as part of farm resources.

agement as "the best option" but only take a partial account of the operating environment. This can lead to distortions and redefinition of management objectives.

This lack of precise definition, common understanding, and agreement often leads to disagreement and dissatisfaction during implementation. For example, there is community consensus that conservation and agricultural sustainability are desirable goals. However, there does not appear to be agreement as to exactly what these are or what farmers should do to achieve them (Morrison 1990; Lefroy et al., Chapter 8). It is, therefore, best to define or redefine social objectives in as clear terms as possible. This includes translating options into physical plans, estimating their effect on production, and estimating their costs and benefits. Such definition will lead to a clearer understanding and greater agreement and will provide a means by which targets, measurements, and judgments as to achievements can be made. Using that argument, Figure 9.1 is rearranged as in Figure 9.2. An example would be the goal of arresting the spread of salinity by proposing actions that act on the contributing factors of surface and subsurface drainage and subsoil permeability (Nulsen, Chapter 5; McFarlane et al., Chapter 6). Possible solutions include drainage and water usage by perennials. The choice, density, and positioning of the perennials are a function of the rate of recharge and soil permeability. The solutions require consideration of land and financial resources, machinery type and width, and management practices to take advantage of the new production system (Kubicki et al. 1991).

9.3.2 Physical Factors

The physical environment is usually the first and most common focus in farm planning, as it is the most obvious and graphically illustrates

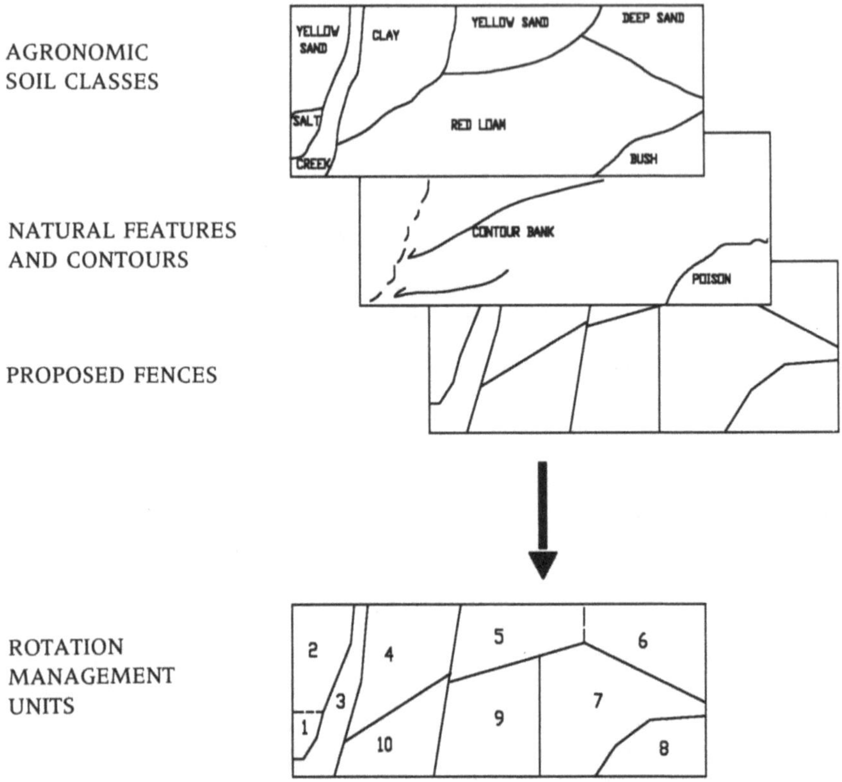

Figure 9.3. Development of Rotation Management Units.

the symptoms of previous actions. Soil types, topographical features, groundwater intake and discharge areas, rock outcrops, and remnant vegetation are the building blocks of developing layouts that are physically practical (Gunn et al. 1991; Wells and King 1989).

The procedures involved are ubiquitous, well-understood, and practiced in the form of land conservation farm planning and soil surveying (Gunn et al. 1991). For that reason they will not be elaborated on here, but their relationships are summarised in Figure 9.3 which shows how layers of information are used to develop convenient management units. For the convenience of further analysis, the smallest so-derived area of land that can be conveniently managed is referred to as a Rotation Management Unit (RMU). There can be more than one RMU per paddock, but there cannot be more than one soil class per RMU.

9.3.3 Production Factors

Traditionally, production data tend to be collected after mapping the farm's physical resources. "Appropriate" practices are recommended, and considerable time is often spent on the technicalities. Appropriate practices, however, relate to management objectives and expertise, as well as to the physical and financial environment from the investment. Seasonal conditions also affect the yield levels, and these conditions need to be included in a long term analysis (Bathgate et al. 1991; Kingwell et al. 1992). These issues are covered more fully in Section 9.4.1.

9.3.4 Financial Factors

The traditional conservation approach has centred mainly on the rearrangement of physical resources to fit specific goals such as fence planning for water erosion, salinity, or flooding control. What has generally been missing is some form of financial analysis to justify the adoption of the plan. Justification has usually been on "sound agricultural practices," and the plan design assumes that it is of benefit in the long term; rarely does it test for it. The analyses should indicate a range of strategies and their outcomes in terms of their physical and financial context. Which option is adopted is then a matter for the individual farmer and family unit to decide.

The economics involved in farm planning, however, have a lot more to offer. They are the relationships between farm investments and productivity. They tie the production methods (rotations and practices), resources (physical, machinery, and labour), costs and returns (cash reserves), and risks (management abilities and assessments) into a whole farm analysis.

Direct costs such as additional fencing and construction of earthworks are included and, if an economic analysis is performed, it is normally on a year-in-year-out budget basis. Often, indirect benefits such as efficiency of operations, crop yields, pasture-carrying capacity, and the effects of different rotations on water usage are discussed but only superficially treated in the analysis.

Unfortunately, to a large extent, the implications of the strategies are unknown or are specified in nonproduction terms, and therefore difficult to evaluate. Also, the options considered "best" are generally more difficult and demanding on management. Many organisations have avoided these issues, often leaving considerable problems with implementation. From a business viewpoint, though, more profitable outcomes are preferred, as they produce the greatest amount of financial flexibility which, in turn, provides greater management and implementation flexibility.

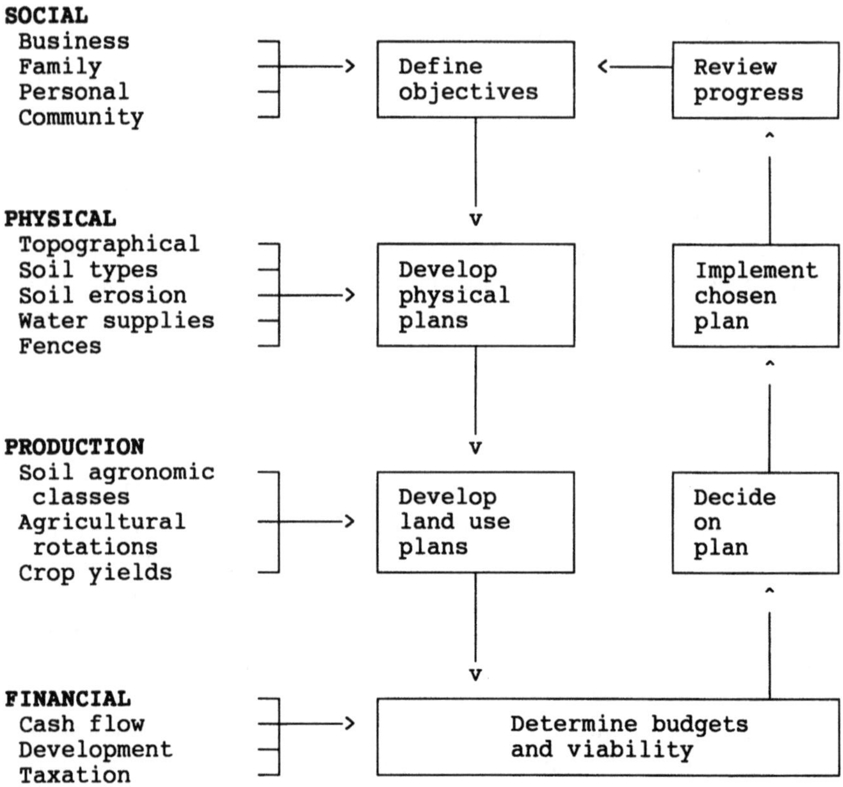

FIGURE 9.4. Processes of farm planning.

With good planning and advice, it should be possible to develop achievable management plans that address environmental issues while maintaining farm viability. The main processes involved in this whole farm planning approach are summarised in Figure 9.4. These steps can be passed through the ecological and economic filters depicted in Figure 8.2 (Lefroy et al., Chapter 8).

9.4 FARMULA: An Integration Model

The FARMULA Simulation Model has been used to process the analysis for this case study. FARMULA, a long-term, whole farm cost-benefit model was developed by the Western Australia Department of Agriculture to address the issues of "improving farm profitability through increasing productivity and reducing degradation" (Kubicki et al. 1991; see also Bathgate 1990; Kubicki and Denby 1990; Kingwell and Pannell 1987; Kubicki 1991). It is well-suited to comparing plans that have different physical layouts, cost, and income patterns.

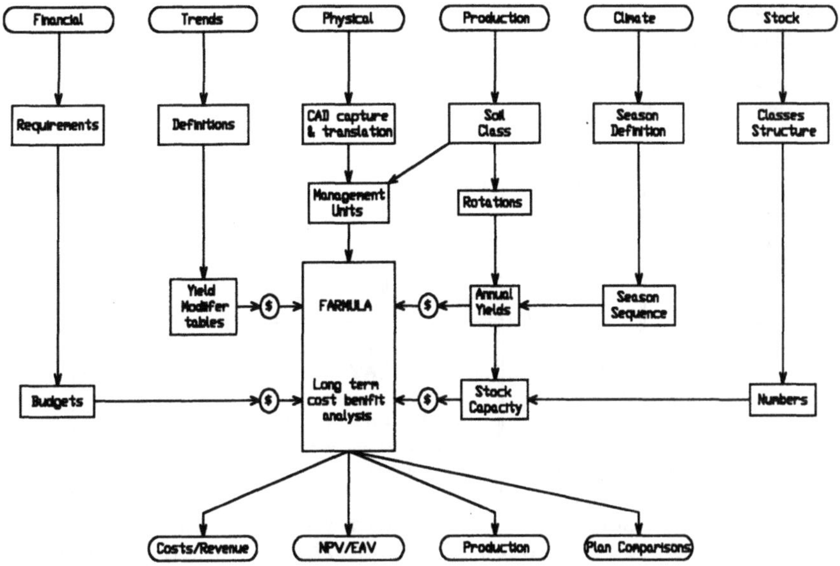

FIGURE 9.5. Overview of the FARMULA decision aid model.

Factors limiting productivity and profitability need to be identified, and strategies developed to overcome them. These strategies become the alternative plans, and their implications, are estimated and compared. An overview of FARMULA in given in Figure 9.5.

9.4.1 Development of Production Data

For FARMULA, production relationships are presented in a hierachial manner. The initial mapping is into soil types which are then grouped into soil classes representing areas that are managed and respond in a similar manner. Each soil class is assigned rotations which should include existing practices as well as proposals addressing long-term sustainability and conservation issues. The data set is completed by adding the crop yield and stock-carrying capacities to the practices.

As the analysis is long-term, some allowance for seasonality needs to be made. This is best via a set of seasonal descriptions, along with the production variation. Figure 9.6 summarises these relationships. Costs shown in the example are in Australian dollars. DSE refers to a Dry Sheep Equivalent (an adult male or nonlactating adult female).

The production data so collected can be attached to the Rotation Management Units (RMU) as defined in Section 9.3.2. For each RMU, a proposed plan exists, be it real or imaginary. It is possible to use these proposed plans to expand management intentions for the long term and evaluate outcomes.

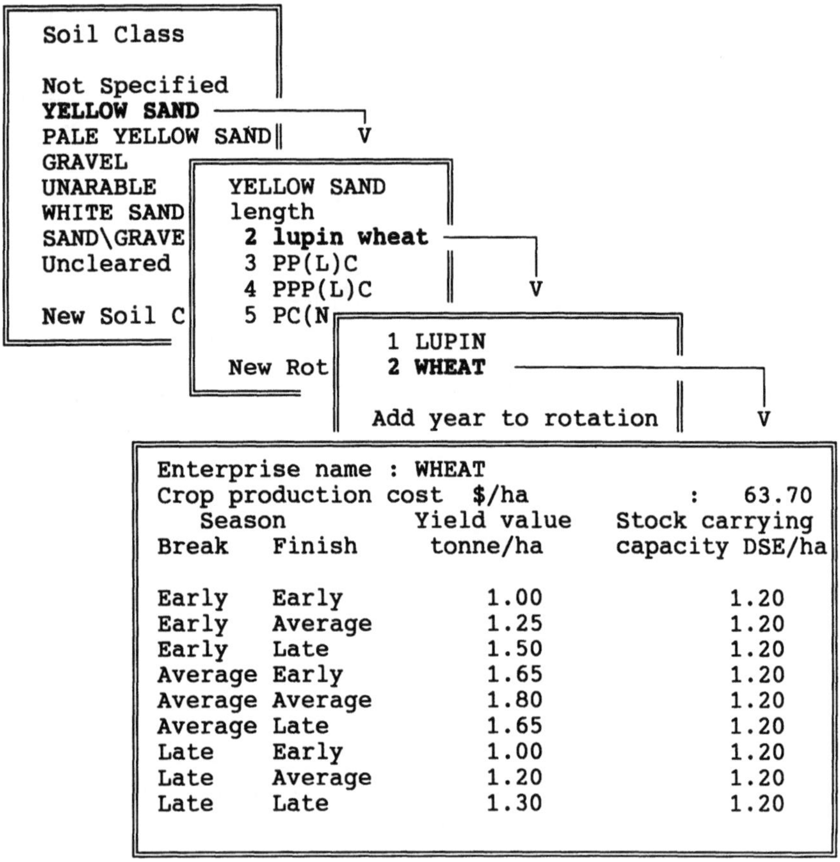

FIGURE 9.6. Production data relationships for FARMULA.

Table 9.1 provides an example of an area which is progressively going saline. The planned treatments are continuous barley, followed by salt-tolerant grasses and the planting of salt-tolerant trees. A deep-ripping treatment is planned for years 5 and 15 to assist in grass and tree establishment.

In Table 9.1, the seasonal sequence is a continuous run of average "start to season" and "finish to season." This is shown by the rows of "A" against Break and Finish. Different season-sequences such as Poor (P), Good (G), Early (E), or Late (L) can be included in the analysis. The seasonal describers and sequence can be changed using the model's database editors. The "C" for the first row against Enterprise indicates that crops are planned for that year. The second row expands this to indicate the type of crop, where "B" is barley, "S" is salt-tolerant species, and "T" is trees.

TABLE 9.1. Proposed sequence for a Rotation Management Unit.

Management Unit: Clay soil near saline seep	
Management Unit area (ha)	103.56
Year 15 Season Average break Average finish	
Enterprise Name: Wheat	
Estimated Crop Production Cost	$6,000
Estimated Price Received	$12,893
Estimated Yield (tonne)	68.00
Stock Carrying Capacity (DSE)	1300.00
Yield Modifier: Deep Ripping	
Total Area Affected (ha)	67.00

Year	1	2	3	4	**5**	6	...	14	**15**	16	17	18	19	20
Break	A	A	A	A	**A**	A	...	A	**A**	A	A	A	A	A
Finish	A	A	A	A	**A**	A	...	A	**A**	A	A	A	A	A
Enterprise	C	C	C	C	**C**	C	...	C	**C**	C	C	C	C	C
	B	B	B	B	**S**	S	...	S	**T**	T	T	T	T	T
Modifier					**D**		...		**D**					

9.4.2 Yield Modifiers

Soil treatments may be applied that affect production in a detrimental or beneficial way. These, like erosion, may be superimposed on the system or, like stubble mulching, may be adopted by management. These have been termed yield modifiers. Figure 9.7 shows how a wind erosion event can be treated as a yield modifier. It shows a large yield decrease in the year of the event, followed by recovery as organic matter, fertiliser levels, and soil structure are rebuilt. Different levels of recovery such as never, partial, or whole may be possible. In Table 9.1, the "D" in the row against modifier indicates that deep ripping has been applied, which affects yield. Further examples of modifiers are given in Tables 9.2 and 9.3.

Yield modifier data are shown in tabular form in Table 9.2, which provide an example of how yields and carrying capacity are slowly decreasing because of saline creep. Similarly, soil treatments may have beneficial short- and long-term effects on yield. Examples include treatments such as drainage, contour working, deep ripping, and stubble mulching. Table 9.3 displays an example.

9.4.3 Comparison of Farm Plans

Three types of information are needed to present an adequate picture of the long-term profitability:

- an adequate time for the benefits to be realised;
- a measure of short- and long-term profitability; and
- an indication of the cash flow patterns.

TABLE 9.2. Effects of gradual salinisation on yield.

Yield Modifier: Long-term degradation
Modifier identification character: L
Initial cost in first year $/ha: 0.00

Year	Yield % mod.	Stock carrying %	Year	Yield % mod.	Stock carrying %
1	100	100	11	92	96
2	100	100	12	92	95
3	99	100	13	91	95
4	98	100	14	90	95
5	97	99	15	89	92
6	96	98	16	88	92
7	96	98	17	87	90
8	95	97	18	86	90
9	94	97	19	85	90
10	93	96	20	84	90

TABLE 9.3. Beneficial affects of a soil treatment on yield.

Yield Modifier: Deep ripping
Modifier identification character: D
Initial cost in first year $/ha: 50.00

Year	Yield % modification	Stock carrying % modification
1	130	120
2	125	120
3	120	120
4	120	110
5	110	110

Plans may need to be chosen for further examination, comparison, presentation to management and/or funding authorities and for implementation. The most profitable options are likely to be implemented first, but that decision needs to consider management objectives and available resources.

Obviously, different plans will have different income and asset patterns. To be meaningful, the results of plan comparisons will need to be expressed in understandable, everyday terms. This is done by comparing the profitability of alternative farm plans with the existing one and determining their Net Present Value (NPV) and their Equivalent Annual Values (EAV). The Net Present Value (NPV) is the total value of a plan expressed in today's terms. The Equivalent Annual Values (EAV) of a plan are a reworking of the NPV to express it in annual values, using today's terms. The EAV is not just the NPV divided by the number

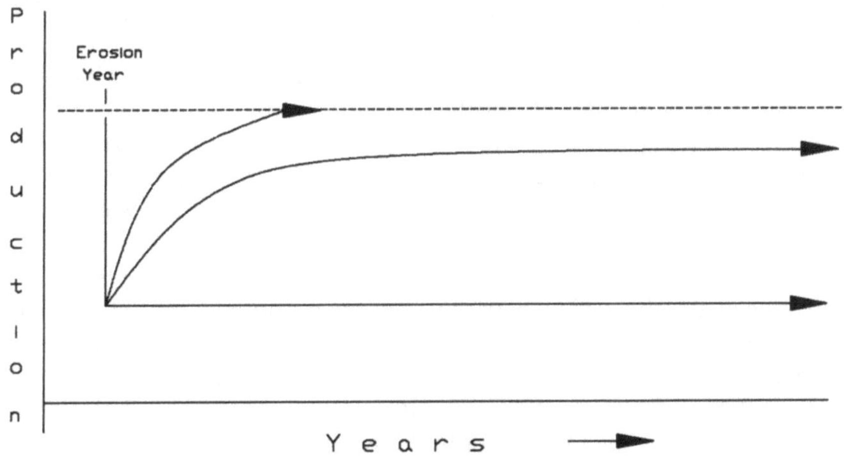

FIGURE 9.7. Effects of a wind erosion event on production (see text for discussion).

of years involved, as the calculations must include the changing value of money over time.

The analysis will generally be based on long-term averages or on estimates of future prices. Obviously, uncertainty about future prices, seasons, interest rates, and community pressures will influence decisions. However, while future changes will often make the absolute NPV values inaccurate, the relative differences between the existing and compared plans is less likely to change.

The procedures described are a decision support process that has been used on the following case study using the FARMULA computer program.

9.5 Case Study

9.5.1 Background

"Liverseys" is located seven kilometres west of Tammin on the Great Eastern Highway in the central wheatbelt area of Western Australia. It is 2700 hectares is size and is described as a dryland, wheat and sheep farming property. "Liverseys" is farmed by the third generation of the Chatfields who come from a long line of pioneering stock and who originally began clearing bush at Tammin in the early 1890s. The property is managed as a partnership, with the partners having diverse speciality roles.

"Liverseys" is predominantly sandplain (1077 ha) with clays on the hilltops (229 ha) and shallow sands over grey clays on the valley floors (695 ha). The climate is Mediterranean, and the farm receives an average annual rainfall of 360 mm, predominantly during the winter months. Growing season is late May/early June to late November.

Development was initiated in 1906 when a flat was cleared of its York gum (*Eucalyptus loxophleba*) and Tea tree (*Melaleuca* sp.) vegetation. Excellent hay crops were reported to have been produced. Salt first appeared about 1917 and these areas were not cropped after 1926. The salt spread up the valley floor but has been confined in the past 10 years. Affected areas are now fenced from livestock, and samphire (*Halosarcia* sp.) has spread completely, covering the bare sand.

A salt lake to the west of the property was reported to have been fresh before the 1900s and a haven for bird life. That lake now has a thick salt crust in summer and the surrounding vegetation has disappeared.

There are seepages around the base of the yellow sand hills. These areas have been successfully treated through a combination of drains and strategic tree planting.

Clearing of the sandplain hills commenced the 1950s. Little remnant vegetation now remains on the sandplain, as the norm in those initial years was wholesale clearing which was also encouraged by government legislation and incentives. A severe rabbit infestation reinforced the need to remove vegetation on a whole-scale basis. (Myxomatosis and trace elements came too late to save the sandplain.)

A significant fragment of banksia (*B. prionotes*) and woody pear (*Xylomelum angustifolium*) survived and is protected and provides a feeding ground for a prolific bird population. A fenced remnant of salmon gum (*Eucalyptus salmonophoia*) remains in isolation. Underneath their canopy are charcoal pits – the remains of an era when charcoal was produced to provide fuel during the war years. Protection of remnant *Melaleuca uncinala* thickets has resulted in natural regeneration of the species.

In the 1970s, it was recognised that productivity was decreasing due to land deterioration. This was mostly in the form of the valley floors going saline, hillside seepages, and reduced and variable paddock production. Seeing their resource decline spurred the partners on to "do something" rather than just "talk about it." A major tree planting scheme was inaugurated in 1980 with the objective of annually propagating and planting in excess of 10,000 trees, shrubs, and saltbush on strategic "problem" areas (George 1991). Some 150,000 trees/shrubs have now been successfully established on "Liverseys." Over the decade, this enterprise has developed into a number of off-farm businesses including a farm tree nursery producing some 300,000 trees annually, a conservation machinery business producing the Chatfield planter, which has included the design and construction of 73 tree planters, and tourism.

The initial objective, however, has not been lost. Though originally focussing on problem areas, it has broadened with on-ground experience to the development of conservation farm plans and a realisation that farming to soil type and a balance of production and conservation are vital. This approach required critical consideration of paddock performance and a change in management practices to match land capability more closely. The case study examines some of these ideas and their economic implications.

9.5.2 Management Objectives

The broad management objectives are to:

- provide a basis for the families' physical and economic survival;
- have sufficient income to educate children;
- protect the farm's remnant vegetation;
- link tree planting to provide corridors for conservation and ultimately to link these with neighbour's vegetation corridors throughout the catchment and Shire;
- halt the occurrence of wind blasting;
- revegetate degraded areas such as the saline valley floors;
- maintain flexibility in physical and management operations; and
- maintain the core sheep flock structure for the stud and main flock.

The factors limiting the attainment of these objectives have been recognised, in particular:

- very high personal drawings, owing to the number of families involved;
- high education costs associated with the upbringing of seven children;
- a gradual but consistent decline in production as parts of the farm suffered degradation due to wind blasting and salinisation;
- rising saline water table and increased salinisation in the valley floors;
- variable lupin yields;
- increasing cost of operations;
- decreasing commodity prices; and
- variable seasons

A solution adopted to reduce the high personal drawings was the specialisation of labour and the use of acquired expertise to provide off-farm income. This included diversification into the:

- establishment and operations of a tree nursery;
- design and manufacture of tree planting equipment;
- provision of a tree establishment and planting service;
- provision of a conservation planning and survey service;
- breeding and supply of stud rams; and into
- a concrete contracting service.

These activities are consistent with the overall objectives of conservation and are effective in reducing the dependency on the farm as the sole supplier of income. Although related to the total farm business, most of these activities are seen as separate businesses and will not be included in this analysis. Instead, the analysis focusses on the objective to increase farm profitability through increasing productivity and reducing degradation.

The analysis presented here compares four plans:

- The existing property layout and agronomy;
- A proposed layout and practices for the reduction of wind blasting;
- A proposed layout and practices for the reduction of salinisation; and
- A proposed layout and practices for the reduction of salinisation and wind blasting.

The proposed layouts need to include a physical rearrangement of the paddocks. However, to keep the options comparable, paddock refencing has been kept to a minimum, while allowing changes consistent with the principles of efficient operations and farming to soil types. New fencing is mainly associated with the adoption of tree planting and the provision of grazing control in establishment years.

9.5.3 Soil Agronomic Classes

The bases of the farm plans are the soil agronomic classes. Twelve soil types were identified for the farm, then combined and remapped as eight soil agronomic classes. This simplified the production data requirements and eased the analysis. The soil legend for the property (Table 9.4) has been coded to fit the descriptions presented by Lantzke (1990) and the landforms defined by McArthur (Chapter 2).

9.5.4 Physical Farm Plans

Two of the physical plans developed for the property are shown in Figures 9.8 and 9.9. They give the current fence layout and the long-term proposal. The long term includes additional establishment of trees, more intense farming to soil types, and remnant vegetation and tree corridors. Considerable use has been made of contour working and tree planting to lower the water tables in salt-affected areas.

9.5.5 Flock Management and Wool Prices

The stock plan consists of two merino sheep enterprises, a general purpose flock and a stud flock. The arrangement has proved successful and no changes are planned. For this analysis, flock compositions and

TABLE 9.4. Soil and vegetation of the Tammin case study.

Description	Landform	Vegetation
White sand Low production from cereals, lupins and subclover. Suitable for Tagasaste.	Ulva	Banksia, *Banksia prionotes.* Christmas tree, *Nuytsia florifunda.* Low shrubs.
Yellow sand Suitable for cereals and lupins. Subclover persists but is sparse. Suitable for Tagasaste.	Ulva	Banksia, *Banksia prionotes.* Woody pear, *Xylomelum angustifolium.*
Good cereal, lupin, and subclover production.	Ulva	Black tammar, *Allocasurina campestris.*
Shallow cemented laterite Very poor cereal, lupin, and subclover production.	Ulva	Black tammar, *Allocasurina campestris.* Acacia, *Acacia* sp.
Seepage Poor cereal, lupin, and subclover production.	Collgar	Tea tree, *Melaleuca* sp. Acacia, *Acacia* sp.
Grey loamy sand over clay Poor to good cereal and medic production. (Possible gypsum reponses)	Booraan	White gum, *Eucalyptus wandoo.* Salmon gum, *E. salmonophloia.*
Red brown loam over clay Good cereal, medic, and field pea production.	Merredin	Gimlet, *Eucalyptus salubris.* Salmon gum, *E. salmonophloia.*
Salt	Baandee	Samphire, *Halosarcia* sp. Saltbush, *Atriplex* sp.

numbers are kept constant, with the exception of minor changes caused by developments while implementing the planting of shelterbelts.

The current stock-carrying capacity of the farm averages about 4800 DSE (Dry Sheep Equivalents) on an annual basis. Because of concerns about the effects of overgrazing and wind erosion, a more conservative grazing pressure has been adopted. Actual stock carried average 4000 DSE, or about 80% of the farm's carrying capacity (Fig. 9.10). This ratio is representative of many farms in the district and has the added advantage of providing flexibility to adjust numbers quickly in the event of poor as well as favourable seasons.

Replacement rams for the general flock are obtained from the stud at no cost and this is different from the norm. However, as the property is

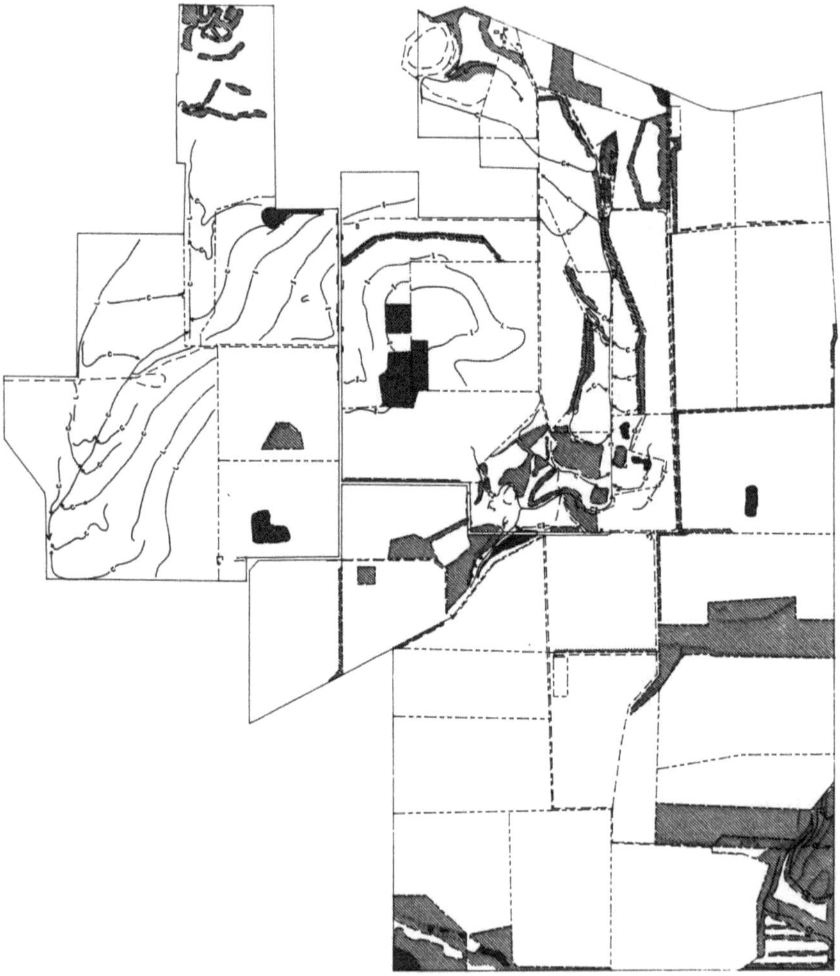

FIGURE 9.8. Plan of existing layout of example property, "Liverseys."

supporting a stud operation, hand-feeding costs at $4.00 per head for all classes of stock are considerably higher than normal, which would be more in the order of about $2.00 per head. All other costs and sheep parameters are considered typical and are summarised in Table 9.5. A summary of the flock values are given in Figure 9.11, which is a screen representation from the FARMULA program.

9.5.6 Tree Shelterbelts

The long-term plan is to provide wind protection to cropped paddocks through tree plantings as shelterbelts. Plantings are proposed to be as

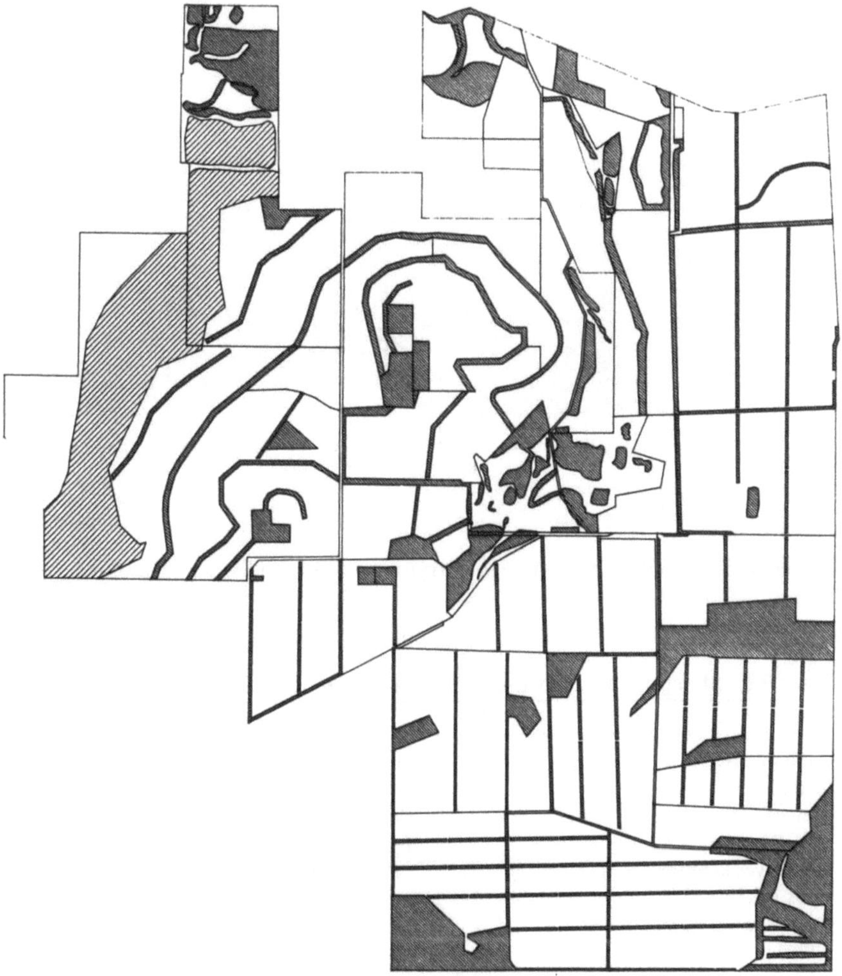

Figure 9.9. Plan of proposed layout of example property "Liverseys."

perpendicular to prevailing winds as possible and are composed of three rows of trees, a row each of:

- Tagasaste (*Chamaesystus palmensis*) for coverage from ground level;
- *Eucalyptus* species for height; and
- Tagasaste to complete the triangular cross-sectional shape.

The design provides a wind barrier starting from ground level, thus reducing the wind-tunnelling effect experienced with many single-, two-, and some three-row systems. As the trees are planted into old paddocks, soil surface scalping is required to reduce competition by weeds.

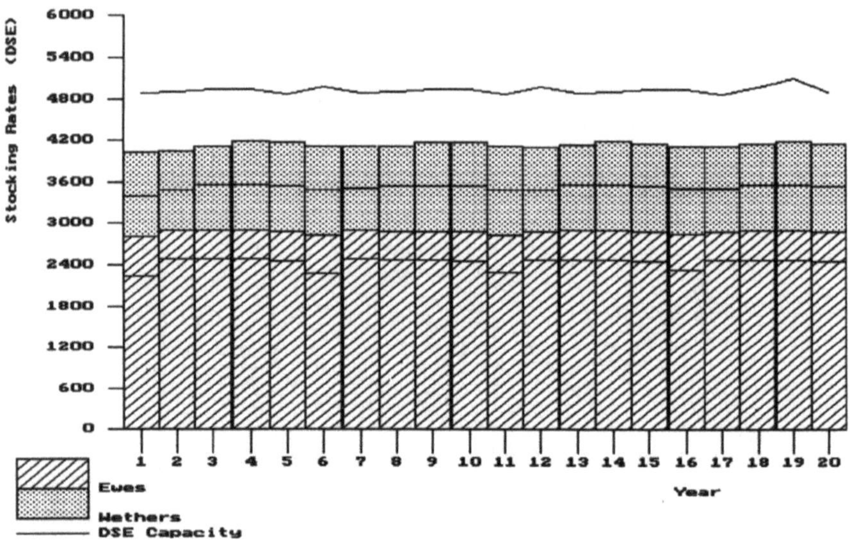

FIGURE 9.10. Relationship between carrying capacity and stock numbers.

TABLE 9.5. Sheep parameters used in the analysis.

Parameter	Rate	Parameter	Rate
Lambing rate %	80	Livestock tax $/head	4.0
Death rate		Opening values $/head	10.0
Ewes	4	Wool price $/kg greasy	3.0
Wethers and weaners	3	Husbandry costs $/head	
Years ewes kept for	5	Ewes and wethers	0.6
Years wethers kept for	2	Weaners	1.2
Wool cuts kg/animal		Hand feeding $/head	4.0
Ewes	5.5	Sale prices $/head	
Wethers	5.5	Cull ewes	5.0
Weaners	4.0	Hoggets and weaners	12.0
Lambs	1.5	Wethers	22.0
DSE rating		Wether purchases	20.0
Ewes	1.5	Ram purchases (from stud)	0.0
Wethers and weaners	1.0	Years rams kept for	5

Weed scalping, fertiliser application, and tree planting are achieved in a one-pass operation using the Chatfield Tree Planter which is designed and manufactured on the property. A number of the Chatfield Tree Planters have now been manufactured and are being utilised by other farmers and contractors, as well as being available for export. Using the machinery and establishment techniques developed, tree growth following plantings has been rapid, with low mortality rates.

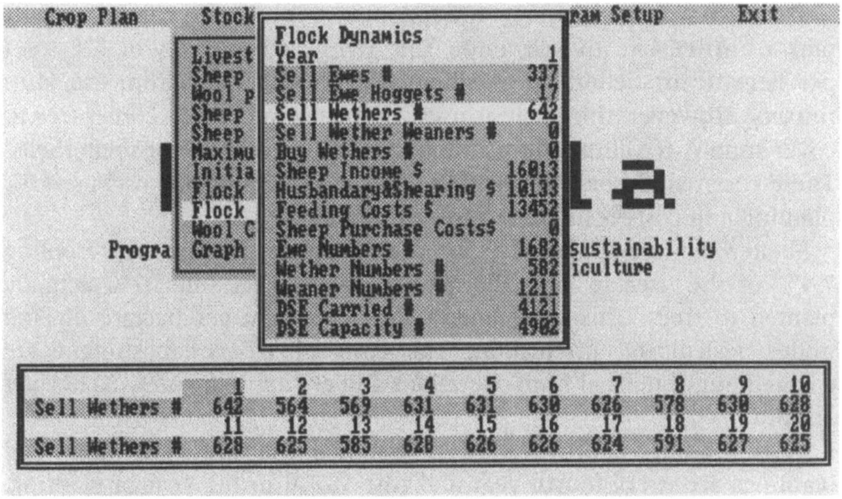

FIGURE 9.11. Flock values for year 1 of the existing plan.

Shelterbelt plantings are proposed to protect 1850 hectares of the property. The main enterprises are wheat (804 to 860 ha per annum), lupins (116 to 153 ha per annum), grass-based pastures (630 to 650 ha per annum), and legume-based pastures (416 to 427 ha per annum). The figures in brackets indicate the range of each enterprise over the years. New plantings are on an annual basis and are protected from grazing stock by the use of double electric fencing consisting of four plain wires powered by batteries and solar boosters. In total, 92.6 km of shelterbelts are proposed.

Tree pruning is not planned unless emergency conditions such as drought occur. Tagasaste has proved to be a most suitable species for inclusion in the windbreaks. It has the added benefit of being a suitable tree fodder that can be managed through grazing, cutting, baling, and feeding in feedlots. To increase the potential conservation value of the shelterbelts, tagasaste can be replaced in some areas by native shrubs such as *Acacia* sp.

9.5.6.1 Costs

A number of costs are associated with the adoption of shelterbelts:

- The supply of tree seedlings or seeds, fertiliser, and the actual tree planting operation;
- Supply, erection, and maintenance of fencing to provide grazing protection during the establishment years;
- The agricultural production foregone on the land planted to trees and their associated buffers; and
- The opportunity cost of capital.

Trees are spaced about 5 m apart along the rows, with rows 4 m apart plus a further 4 m to each fence. This gives a tree density of 378 trees per hectare of shelterbelt. Seedlings are provided free from the farm nursery. However, this analysis uses contractor rates of $1.20 per tree to cover supply, fertiliser, and planting ($453 per kilometre of shelterbelt). These costs would be considerably reduced if farm labour was used for planting and costs restricted to tree purchases.

Shelterbelts are proposed to be at 200 metre spacings, representing 7.4% of the useable land allocated to shelterbelts and 3.7% actually planted to trees. This represents a cost of $22.70 per hectare of land under agricultural production. The costs of tree establishment are covered out of normal farm operations and are approximately $2185 per annum.

In this analysis, however, it is assumed that tree purchases and plantings are every fourth year and cost $8740 in the year of planting. This is different than the current method used on "Liverseys" but is used here as it allows the protective fencing to be purchased once, then reused as trees mature and new plantings are initiated.

Electric fencing is used to provide grazing protection and is given a total life of 20 years. A purchase cost of $330 per kilometre for materials and freight is used. Erection and relocation costs are taken as nil as these are performed by on-farm labour. Solar panels are included in the estimates at the rate of one energiser unit ($300) per 50 kilometres of fencing. In total, 185 kilometres of electric fencing will need to be erected. As the fencing is reusable, 37 kilometres is purchased initially for $10,854 for materials, solar panels, and freight. In this analysis, that amount is taken out as a loan at the commencement of the plan at 18% interest and amortised over 17 years.

All costs associated with conservation are tax deductible in the year of expenditure. Taxation adjustments have been made using the partnership scale and assuming that two partners operate the business, which will be more typical of farms in the Tammin area. The opportunity cost of capital has been taken as nil as farm improvement is a stated management goal, as is the desirability of retaining the farm as a productive unit for the long term.

9.5.6.2 Benefits

A number of benefits from the use shelterbelts were identified:

- Increased crop and pasture production;
- Increased efficiency of grazing;
- Increased animal production;
- Reduction of animal losses;
- Reduction of hand feeding costs;

- Decreased frequency and extent of crop losses due to wind blasting or wind erosion;
- Decreased operating costs due to decreased cropped area;
- Provision of a timber asset and the possibility of other tree products;
- Increased real estate value;
- Acquiring of new skills; and
- Provision of a tourist focus.

The effects of the shelterbelts on agricultural production are expected to vary with the age of the trees by reducing the wind effects and desiccation of crops, pastures, and stock and by providing protection from soil erosion. These effects are estimated on a whole-paddock basis to range from nil in the first three years, increasing to 6% by years 8 to 10, to about 12% by year 16, and remaining at about 12% from there onwards. The analysis includes these influences on production as a yield modifier.

Associated with shelterbelt adoption is a reduction in the paddock's arable area and a corresponding reduction in crop operating costs. This is the method adopted in the analysis. The alternative is to maintain the cropping programme by including another paddock and forgoing some pasture production. It is expected that many farmers would select that option, especially considering current sheep and wool prices. That option is not examined here.

Shelterbelts would reduce the exposure of crops, pasture, and stock to the elements, resulting in a reduction in the rate of pasture and stubble deterioration. This rate is unknown and is not included in the analysis. However, it is most likely that grazing efficiency will improve, along with increased animal body weights and wool production. That benefit is estimated at 2% and has been added to the carrying capacity of the treated areas.

Managers have often stated that the real benefit of the improved pasture utilisation is a reduction in hand feeding costs. This reduction is most opportune, as it coincides with times of high animal husbandry demands and reduces demands on farm labour. This analysis includes a reduction in hand feeding costs from $4.00 to $3.50 per head for all classes of stock but does not include a reduction in labour costs.

9.5.6.3 Financial Implications of Shelterbelts

The financial outcomes of adopting shelterbelts as a long-term strategy to increase profitability were estimated using the techniques outlined and the results are summarised in Table 9.6. The results are shown corresponding to the years trees are proposed for planting. Each of the plans are summarised on its own merits, as well as providing a comparison against the previous plan as to relative profit or as an estimate of what is foregone. Obviously, the NPV and EAV absolute

Table 9.6. Financial implications of adopting shelterbelt plantings.

Item	Year tree shelterbelts planted					
	Previous	1	5	9	13	17
Hectares protected	Nil	382	742	1,107	1,473	1,850
Value of plans						
Net income	1,742,996	1,760,454	1,741,641	1,717,801	1,694,111	1,674,470
Net Present Value	986,018	982,667	967,173	951,787	940,748	932,925
Equivalent Annual Value	82,890	82,608	81,306	80,012	79,084	78,427
Change to plans						
Net income	0	17,000	(1,000)	(25,000)	(49,000)	(69,000)
Net present Value	0	(3,000)	(19,000)	(34,000)	(45,000)	(53,000)
Equivalent Annual Value	0	0	(2,000)	(3,000)	(4,000)	(4,000)

Values in backets are relative losses

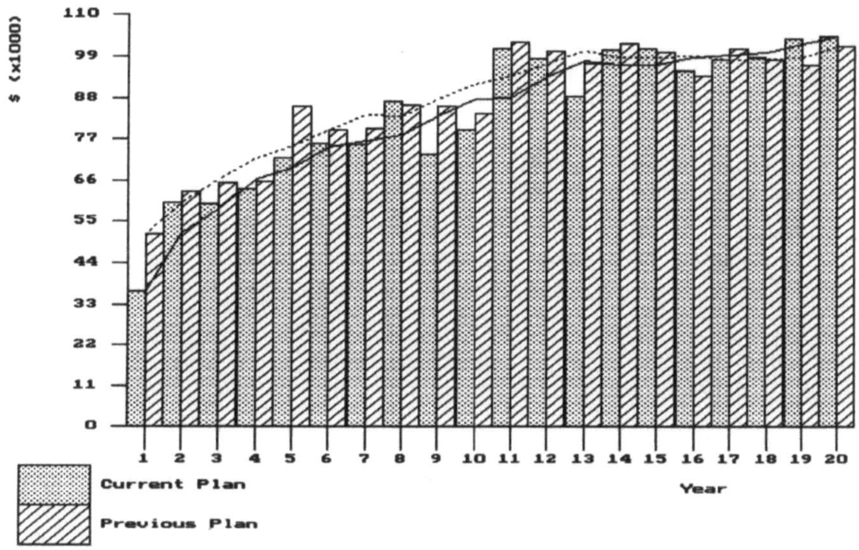

FIGURE 9.12. Comparison of the net incomes of the Previous (...) and Current plan (—).

values will change with different prices and productivity estimates. However, all plans will be subject to the same fluctuations and it is likely that the relativities will remain.

All plans are profitable, although adopting shelterbelts results in a reduction in profit. The adoption of some shelterbelts has been beneficial to net income, but not to the plan's relative NPV. The adoption of more plantings has resulted in a relative decrease in profitability. This is partly illustrated in Figure 9.12 which compares the net income patterns of the two programmes. The costs of the plantings can clearly be seen in years 1, 5, 9, 13, and 17. There is an apparent merging of the net incomes, partly due to discounting procedures and to the increased production resulting from maturing trees. In this and subsequent figures, "previous plan" refers to the no change option, and "current plan" refers to the proposals.

The general increase in net income shown in Figure 9.12 is due to the progressive payments of a \$300,000 machinery loan, and the stepped change is due to projected termination of education expenses. The proposed plan in Figure 9.12 also shows small but consistant reductions in net income up to year 18. From there onwards, it appears that the pattern is reversed. These small reductions would accummulate to a substantial relative loss over the assessment period, and Figure 9.13, which compares the cummulative balances, shows that more clearly.

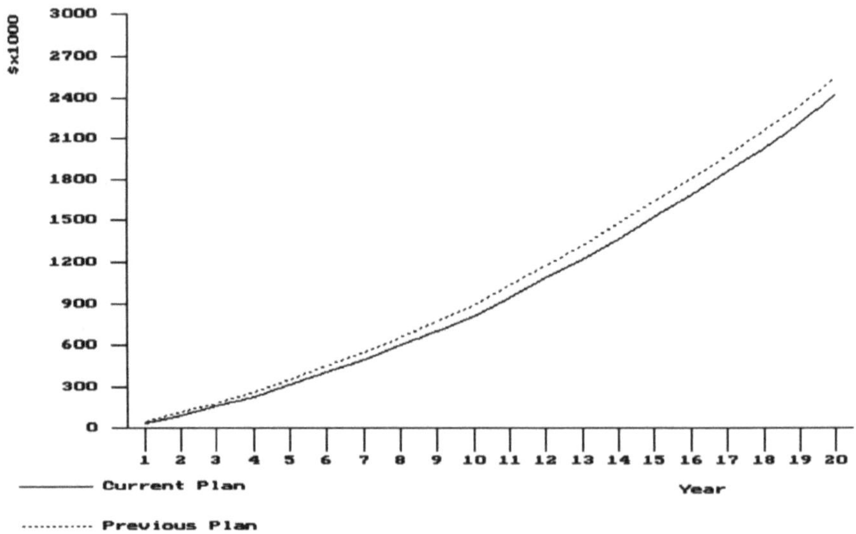

FIGURE 9.13. Comparison of cumulative balances, with (current) and without (previous) revegetation.

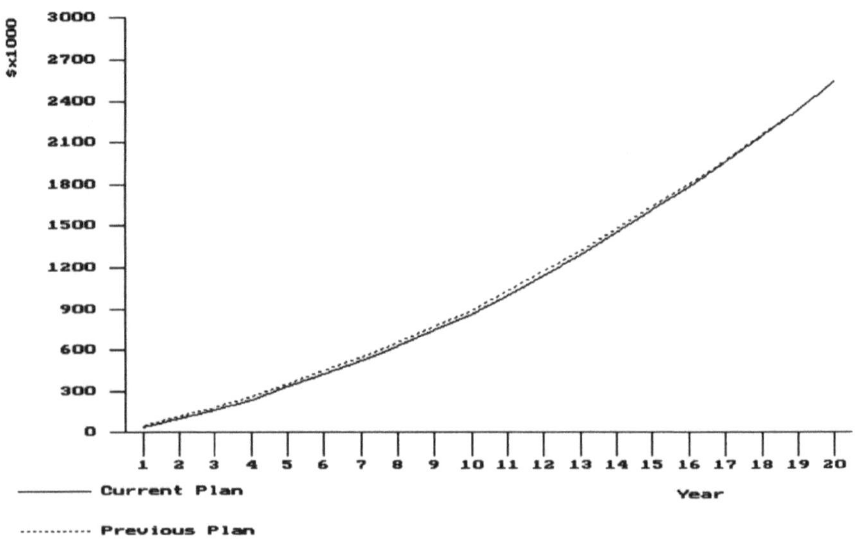

FIGURE 9.14. Comparison of cumulative balance for first planting.

On the data provided, the initial reaction is to reject the proposals. However, that does not address the objective of sustainable agriculture, and the issue needs to investigated further, taking into account the effects of wind events.

TABLE 9.7. Sensitivity of plans to shelterbelt establishment costs.

Item	Cost per protected ha		
	Nil	5.00	22.70
Change to plan			
Net income	(41,000)	(48,000)	(69,000)
Net Present Value	(34,000)	(39,000)	(53,000)
Equiv. Annual Value	(3,000)	(3,000)	(4,000)

Values are relative to the current plan

Referring again to Table 9.6, the first group of plantings achieved an increase in net income over the assessment period and produced a similar level of profit, as shown in Fig. 9.14. This indicates that there may have been insufficient time for the accumulation of benefits, or there is some sensitivity as to the selection of areas that require protection.

9.5.6.4 Sensitivity of Plans to Shelterbelt Costs

As the desirability and need for utilising tree-based shelterbelts were clearly stated, a sensitivity analysis to their costs was provided. Two options were examined, a zero cost, and a highly reduced cost.

The zero option is considered as conservation issues are of world interest and, to propagate those activities, political pressure could eventually allow a rebate system to operate, rather than current system of taxation deduction of expenses. The highly reduced cost option is included to allow for changing technology such as direct seeding using farm labour and equipment. In this case, the cost per protected hectare is reduced from $22.70 to $5.00. Table 9.7 summarises these results which include $10,854 for the purchase of electric fencing.

Reducing costs has naturally resulted in increased profit, but it is still lower that the original plan. The decrease under the Nil option relates to the loss of agricultural production from the land area allocated to the trees and their associated buffers, i.e., the 7.4% mentioned in Section 9.5.6.1. This decrease occurs despite the production boosts mentioned in Section 9.5.6.2.

9.5.6.5 Frequency and Effect of Wind Years

The reason for establishing shelterbelts was to reduce the productivity losses associated with wind years. These detrimental affects are estimated to occur at three levels of intensity:

- Low: sand blasting of crops and pastures, no soil removal;
- Medium: sand blasting, removal of soil fines which include fertiliser and organic matter; and

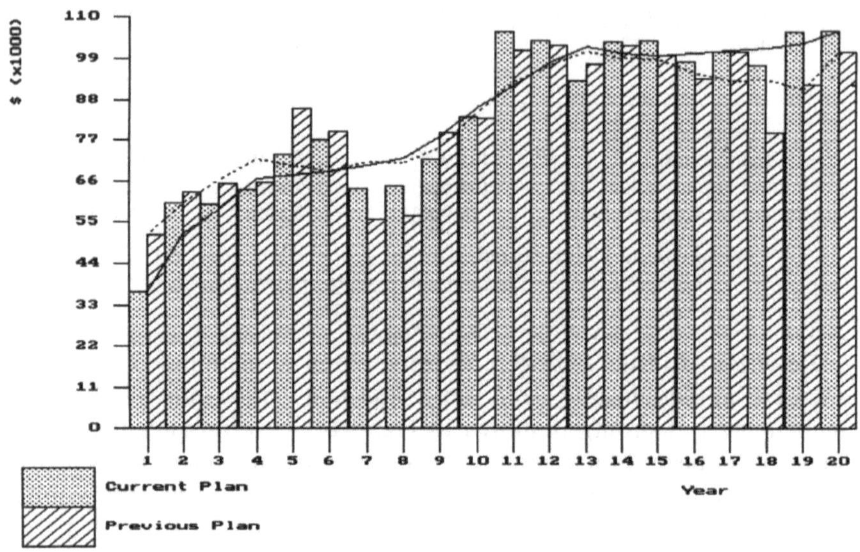

FIGURE 9.15. Comparison of net income of the Previous (. . .) and the proposed planting (Current —) plans subject to wind.

- High: sand basting, removal of soil fines and some removal of soil surface.

Sand blasting is assumed to reduce the yields of crops and pastures only in the year the blasting occurs. That reduction is estimated to be 20%. There are no additional costs such as nitrogen top-dressing or reseeding of affected crops.

Medium wind erosion is assumed to reduce crop yields to 60%, 85%, and 95% of normal, for the wind erosion year and the two years following it. Similarly, pasture yields are reduced to 85% and 95% compared to normal. Additional costs associated with partial reseeding and fertiliser application are estimated to be $17.00 per ha.

High wind erosion is assumed to reduce crop yields to 40%, 75%, 85%, 90%, 95%, and 97% of normal, for the wind erosion year and the six years following it. Similarly, pasture yields are reduced to 75%, 85%, 90%, and 95% compared to normal. Additional costs associated with reseeding and fertiliser application are estimated to be $35.00 per ha.

A review of the meteorological records, and discussion with farmers, indicates that wind years are expected to occur about three to five times within a 20-year period. A dice was thrown to select those years and their severity.

In this analysis, years 7, 8, 11, and 18 drew medium, medium, low, and medium wind damage. A low-wind year is taken to affect 10%, whereas a medium year is taken to affect 20%, of the farm. These effects

Table 9.8. Wind effects on productivity.

Item	Current	Proposed
Change to plan		
Net Income	(82,000)	(66,000)
Net Present Value	(48,000)	(57,000)
Equiv. Annual Value	(4,000)	(5,000)

were included and the plans reanalyzed. Figure 9.15 compares the net incomes of the existing and proposed plans, both subject to the wind years. Table 9.8 summarises those changes compared with no wind effects on the original plan as a reference.

The cost of shelterbelt plantings is almost offset by the production and income losses experienced with the wind years. An alteration in:

- the productivity boost due to the presence of trees;
- an increase in the occurrence of sand blasting or wind erosion;
- an increase in the severity of sand blasting or wind erosion; or
- an increase in real estate values

would alter the relativity of the plans and favour the establishment of shelterbelts.

9.5.7 Salinisation of Valley Floors

Salinisation of the valleys and lower slopes was first observed in 1917, some 11 years after first clearing, and has progressed since. These areas are subject to high hydraulic pressures, receive groundwater inflow, and will become waterlogged and salt-affected if left unattended (Nulsen, Chapter 5; McFarlane et al., Chapter 6). George (1991; personal communication) estimates that if the valleys are left untreated, land use would be restricted to grazing, and productivity would decline to 50 percent of the nonsaline pasture yield on 30 percent of the area and to 10 percent on the remaining 70 percent. Cropping would not be an option.

Obviously, the adoption of practices that have high-water usage must be the primary management objective if the degradation is to be arrested or reversed. Trees have been the strategy selected. It is estimated that about 30 to 40 percent of the valley areas need reforestation, with 80 to 160 trees per hectare.

The saline areas no longer form a major component of the productive capacity of the farm and, owing to the agronomic problems with the revegetation of saline areas, the high risk of failure, and the difficulties of subsequent management, it is unlikely that these areas will be treated as anything other than reclamation and experimental sites. Costs on the saline areas have been kept to a minimum by treating them as large paddocks with a minimum of internal fencing. Tree plantings have used seedlings and direct sowing to develop experimental techniques and experience with reclamation.

Under these circumstances, productivity and economic considerations are secondary. The main interest in the financial assessment is whether the costs associated with the activities threaten farm viability or whether alternative practices would result in increased profits, while maintaining the primary water usage objective.

Revegetation of the areas has been commenced and is similar to that described by Schofield and Scott (1991) and George (1991). Agricultural use is expected to be restricted to grazing, and yield levels could return to about 30 to 40 percent of nonsaline conditions within four years of reforestation and to 100 percent within 10 years. The remaining 60 percent receives benefits from the reforestation programme, and productivity could be returned to full nonsaline levels for both crop and pasture within 10 years. However, the crops selected and the frequency of cropping would need to be carefully monitored as the area could easily be mismanaged, undoing revegetation efforts.

The following financial analysis is, therefore, of general interest, rather than for management consideration and recommendation. Two plans are evaluated:

(a) not treating the valley areas and allowing deterioration to continue; and
(b) revegetating the areas with trees and salt bushes.

9.5.7.1 Perennial Pastures

Perennial pastures such as *Phalaris* sp. and Lucerne have a high establishment cost, high prospects of establishment failure, and poor water usage in the first two years and would fail to persist in the Tammin area due to insufficient summer rainfall. They were not considered.

Saltbush (*Atriplex* sp.) is a desirable inclusion as it is salt tolerant and a useful summer fodder crop. It has poor water usage in the first few years but, once established, uses considerably more water than bare saline soil. For Tammin, saltbush is estimated to produce about one tonne of 60 percent digestible dry matter per annum and can support 18 DSE per hectare for two months in autumn. This provides the ability to carry more sheep, to keep them in better condition, to reduce nutrient-related wool fibre weaknesses, and to reduce supplementary feeding costs.

The availability of autumn grazing from salt land pastures is particularly valuable as it provides sheep feed during a time of normal pasture shortage, provides additional management flexibility, and allows the carrying capacity of normal annual pastures to be increased as a result of pasture deferment (Salerian et al. 1987; Howard 1988; Barrett-Lennard et al. 1990). In this example, the carrying capacity of the farm

TABLE 9.9. Financial implications of revegetating saltland areas.

Value of plans	Previous	Proposed	Difference
Net income	1,742,996	1,791,123	48,000
Net Present Value	986,018	923,234	6,000
Equivalent Annual Value	82,890	83,420	0

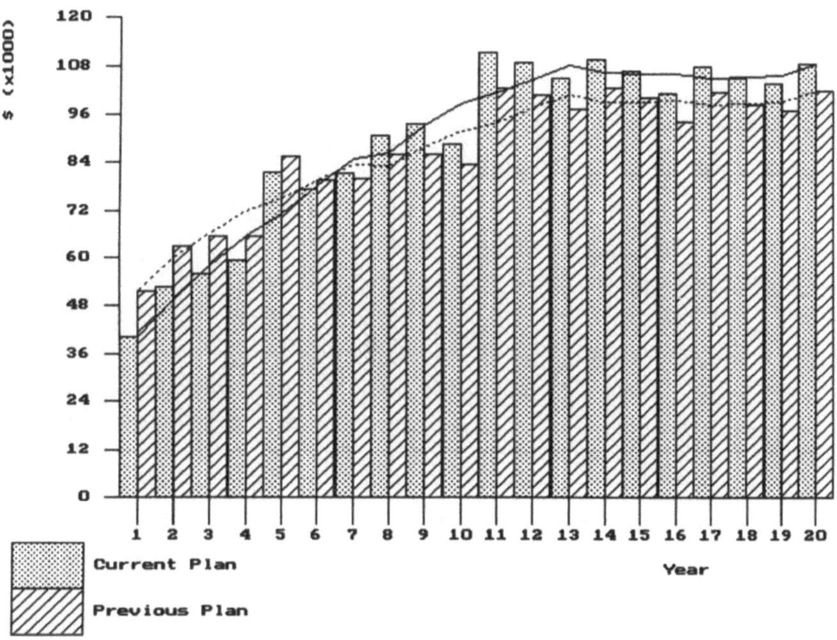

FIGURE 9.16. Comparison of net income of plans with (Current —) and without (Previous . . .) salt land revegetation.

is not increased due to management policies. Instead, the additional feed will be used to provide supplementary feed, thus reducing animal feeding costs.

The cost reductions used are $32 per hectare for saltbush, and $10.80 per hectare for trees, once they are in full production. It is estimated that 20 percent of that grazing rate is achieved by year 3, 50 percent by year 5, 100 percent by year 10, and that rate maintained from there onwards. These values are considered conservative. Only direct costs of seed, fertiliser, fuel, and repairs are included. This is estimated at $64.00 per planted hectare. Saltbush establishment costs are taken to be $133 per treated hectare.

Trees are proposed to be planted in strips rather than as randomly scattered single trees. This provides more flexibility in future utilisation

TABLE 9.10. Financial implications of including all revegetaton programs.

Value of plans	Previous	Proposed	Difference
Net income	1,742,996	1,715,068	(28,000)
Net Present Value	986,018	934,779	(52,000)
Equivalent Annual Value	82,890	78,549	(4,000)

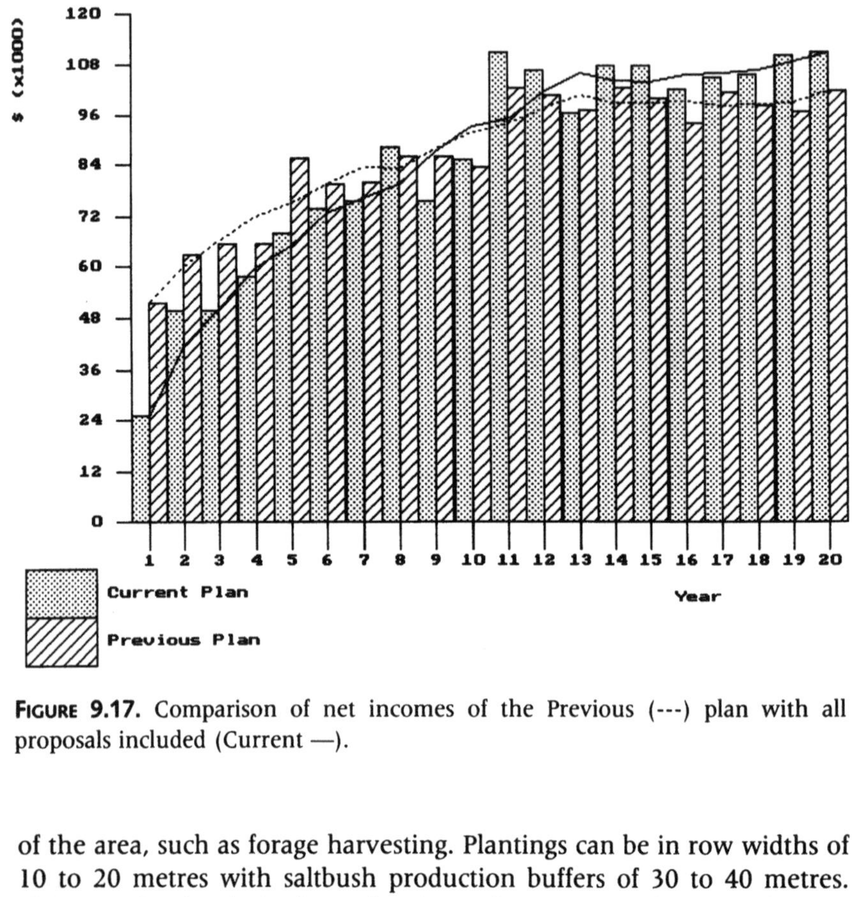

FIGURE 9.17. Comparison of net incomes of the Previous (---) plan with all proposals included (Current —).

of the area, such as forage harvesting. Plantings can be in row widths of 10 to 20 metres with saltbush production buffers of 30 to 40 metres. This provides the desired tree density and tree spacing for ground water control.

The analysis used a total establishment cost of $50,477, treated as a loan over 10 years at 18 percent interest. The production benefits are progressive as previously outlined and included for the twenty-year analysis period. As the land revegetated to trees and saltbush is already saline or expected to go saline, it is regarded as unsuitable for normal agricultural production, and is given an opportunity value of nil. Table 9.9 summarises the results of the plan comparisons.

The proposal to revegetate the salt-affected areas is profitable, even using conservative production estimates. The change in NPV is significant, as is the effect on net income, which is shown in Fig. 9.16. The net income of the revegetation proposal (current plan) is slightly lower for the first six years. From there onwards, as grazing becomes available, the trend is reversed.

9.5.8 Combining Wind and Salinisation Proposals

The final assessment includes the conservation proposals of shelterbelts and the revegetion of the saline valleys. The results of that analysis are given in Table 9.10. Combining the proposals results in a farm plan that is profitable but where the profits are reduced. There are significant improvements in NPV and net income from this combined approach, particularly when compared to the use of shelterbelts only. Figure 9.17 shows the net income patterns of the existing (previous) and combined (current) plans.

The comments mentioned in section 9.6 are relevant, especially those relating to the time period for the analysis and the benefits of the later planted shelterbelts not being fully expressed. The indications are very favourable, and an improvement in any of the additional benefits previously mentioned would favour the adoption of the conservation programmes.

9.6 Underestimation of Productivity Benefits

It appears that, although a 20-year time frame has been used, the cumulative productivity benefits are still being underestimated, at times by up to fifty percent. Additional analysis indicates that the time frame should be extended a further 20 to 80 years. This needs to be resolved and incorporated into FARMULA. In the interim, the additional productivity benefits can be treated as improvements to the farm's asset value and discounted to present day terms. When those factors are taken into consideration, the proposal of utilising shelterbelts will be enhanced.

The degree to which the productivity benefits are underestimated is dependent on the interest rates as well as on the time period involved. The interest rates used were 14% for bank interests rate and 8% for inflation. The real interest rate was calculated to be 5.56%.

9.7 Farm Asset Value

The current asset value of the property is estimated at $1,100,000. The proposed improvements would increase that value and make it more attractive to potential buyers. Discussions with real estate agents indicate that a premium of up to 20% could be possible. However, this is unsubstantiated by land prices, sales, or current demand. These increases are not included as sale of the property is not intended. Nevertheless, they should be remembered, as they are asset items that could be borrowed against.

9.8 Emergence of New Opportunities

Other benefits have emerged from implementing the plans, such as the acquiring of specialised skills and the development of tourism interest. The additional skills have enabled the development of a tree nursery, the design, manufacture, and sale of Chatfield Tree Planters, and the provision of a tree planting consultancy and contracting service. This diversification into three additional businesses has gone a long way towards the attainment of the survival objective for the three families. The skills gained have attracted interest outside of the district, resulting in frequent visits on a consulting basis to other parts of the state or within Australia. In addition, the practices utilised on the farm have attracted visitors to observe and discuss the techniques used and their effectiveness. These visits are now extending to the general public and a tourist trade is developing. An opportunity exists to increase that trade and to diversify the consultancy business further.

Although the consultancy and tourism aspects have considerable financial potential, they are not included in the analysis as they are treated as a separate activities and, if included, could appear as intangible goals for other farmers. Nevertheless, they are real and represent options previously unknown and unavailable to management.

9.9 Conclusions

The inclusion of conservation or other long-term development considerations within a farm plan is not easy due to social, physical, and financial considerations. Assessments need to be made, using different time scales, costs, commodity prices, and interest rates that require processing on an annual basis to reflect cash flow realistically. The issues are compounded as costs are often short-term and relatively easy to determine, whereas benefits are long-term and difficult to estimate.

Existing tools such as year-in-year-out and development budgets were not suitable for this type of whole farm analysis. When used, they often give wrong answers which appear to be mainly related to the identification of benefits, their changing value over time, and the averaging of cost and benefit data. To help overcome this, the FARMULA decision aid was developed and was used to evaluate the case study presented. The analysis procedure is valuable in that it focusses on the productivity and financial assumptions of the physical and agricultural practices involved in the farm plans. It documents those assumptions and allows review and reference. It highlights the strengths and weaknesses of plans and knowledge and forces users to qualify and quantify the issues involved.

Not surprisingly, the results indicate that highly productive areas have a high-opportunity cost. Allocating them to conservation works requires considerable yield boosts from the surrounding land and long time periods to compensate for their removal from normal production. Practices that help to make nonproductive land more productive have very high benefits, and the financial returns are substantial and rapid. In this regard, the results of analyses made confirmed commonly held beliefs and "gut feelings" about the inclusion of conservation practices and planning. However, the added focus on productivity and the qualification and quantification of assumptions are valuable in allowing further discussion and rational assessments. Detailed analysis of the short-and long-term costs and benefits of conservation measures provide land managers with a basis for rational decision making.

Long-term cost-benefit analysis, such as undertaken with FARMULA, is of value to planners and managers as they provide time and monetary values for decision making. Where physical data are not available, the sensitivity analysis has been most useful to provide guidelines for further research and extension.

References

Barrett-Lennard, E.G., Bathgate, A.D., and Malcom, C.V. (1990) Salt land agronomy in Western Australia – The present scene and directions for future research. *Salinity Workshop, 5th Australian Soil Conservation Conference*, Perth.

Bathgate, A.D. (1990) A whole farm model for determining the profitability of strategies which reduce the risk of wind erosion in the south coast of Western Australia *Western Australian Department of Agriculture Division of Resource Management Technical Report* 110.

Bathgate, A.D., Robinson, S., and Kingwell, R.S. (1991) Adjusting farm strategy to allow for seasonal conditions. *Journal of Agriculture – Western Australia* 32:31–35.

Frost, F. and Howell, M. (1990) *Land Management Manual: Merredin Advisory District*. Western Australian Department of Agriculture, Perth.

George, R. (1991) Reclaiming sandplain seeps by planting trees. *Journal of Agriculture – Western Australia* 32:18–23.

Gunn, R.H., Beattie, J.A., Reid, R.E., and van de Graaff, R.H.M. (eds.) (1988) *Australian Soil and Land Survey Handbook. Guidelines for Undergoing Surveys.*

Howard, B. (1988) The value of saltbush. *Management Matters. Miscellaneous Publication 16/88, No. 4.* Western Australian Department of Agriculture, Perth.

Kingwell, R.S. and Pannell, D.J. (eds.) (1987) *MIDAS, A Bioeconomic Model of a Dryland Farm System*. Pudoc, Wageningen.

Kingwell, R.S., Morrison, D.A., and Bathgate, A.D. (1992) Effect of climatic risk on dryland farm management. *Agricultural Systems* (in press).

Kubicki, A. (1991) *FARMCARE*. Occasional paper, Western Australian Department of Agriculture, Perth.

Kubicki, A. and Denby, C.G. (1990) Farmcare. *Proceedings 4th Australian Joint Conference on Artificial Intelligence.*

Kubicki, A., Denby, C.G., Haagensen, A., and Stevens, M. (1991) *FARMULA: User and Reference Manual.* Western Australian Department of Agriculture, Perth.

Lantzke, N. (1990) *Soils of the Northam Advisory District. Volume 1. The Zone of Ancient Drainage.* Western Australian Department of Agriculture, Perth.

Morrison, D. (1990) Sustainable agriculture. Fact or fantasy. *Australian Institute of Agricultural Science,* Perth.

Salerian, J., Malcolm, C., and Pol, E. (1987) The economics of salt land agronomy. *Western Australian Department of Agricultural Division of Resource Management Technical Report* 56.

Scofield, N. and Scott, P. (1991) Planting trees to control salinity. *Journal of Agriculture – Western Australia* 32:3–10.

Wells, M. and King, P. (1989) Land capability assessment methodology. *Western Australian Department of Agriculture Land Resource Series No.* 1.

Chapter 10

Conservation Management in Fragmented Systems

R.J. Hobbs, D.A. Saunders and A.R. Main

10.1 Introduction

Previous chapters in this book have discussed the physical environment of the central wheatbelt of Western Australia within which human settlement took place, the governmental, financial, and social conditions which initiated settlement and conditioned human behaviour in the region, and the environmental consequences of human activities. The later chapters also set out the principles by which boundary conditions might be met for the natural systems of the wheatbelt to be self-regulating and self-replacing and for farming to become economically and ecologically sustainable.

For resources to be devoted to conservation management *per se*, both at a community and individual farm level this has to be justifiable in economic or social terms. The minimum requirement is that such areas do not make individual farms nonviable, and the optimum would be that expenditure on conservation also enhances the profitability of the agricultural enterprise. The analyses by Kubicki et al. (Chapter 9) indicate that at least the minimum requirement can be met at the individual farm level and that conservation measures can increase profitability in the long term. At a broader community level, expenditure on conservation will depend on public and political perceptions of its importance. A community which is aware of conservation issues and of the importance of its natural heritage will be more willing to expend resources on conservation management.

However, these are not the only constraints. The inconsistencies of government policy and activities of financial institutions have, in the past, had a greater influence on land management practices than environmental constraints. There are also likely to be episodes in the future when conditions imposed by financial institutions and government will lead farmers to take actions whose adverse consequences can be predicted in the context of the social and environmental changes already seen. Thus, the management of farmland and conservation areas is under two sets of constraints; firstly, the ecological and economic limitations discussed by Lefroy et al. (Chapter 8), and, secondly, the constraints imposed by institutions divorced from, or uninformed about, the consequences of their actions, both on the natural environment and on rural society. While Lefroy et al. (Chapter 8) argued that ecological constraints must be met as a priority, this will only be possible when the community as a whole, and government and financial institutions in particular, have a reasonable understanding of the rationale for conservation management. Hence, education must be a continuing and large component of conservation management.

This is, therefore, the context within which conservation management must operate. Ecological considerations determine what actions ought to be taken, but, frequently, political and economic decisions determine what actions are actually carried out. In this chapter, we examine the ecological principles of conservation management (i.e., what decisions *ought* to be taken) from the viewpoint of the individual remnants and of the remnants combined into conservation networks within the agricultural landscape.

10.2 Management of Individual Remnants

Main (1987) reviewed current management problems associated with remnants in the Western Australian wheatbelt. Passive management or benign neglect is the primary style of management of remnant biota at present. Most of the remnants present in the Kellerberrin study area, and in much of the wheatbelt of Western Australia, have received little or no active management since they were isolated by clearing. Wallace and Moore (1987) discussed the difficulties encountered by the Department of Conservation and Land Management, the agency responsible for the management of designated nature reserves. Lack of adequate staffing means that management is largely reactive, responding to crises and complaints and dealing with people management and matters other than the primary biological aims of conservation management. Few of the designated nature reserves have management plans as yet, and few are adequately surveyed. The dynamics of most of the vegetation types and faunal assemblages are poorly understood. It is against this background of a general lack of knowledge, inadequate resources, and degrading conservation values that management must be carried out.

The situation becomes worse when remnants on private land are considered. As about 80% of the remnant vegetation in the central wheatbelt is on private property, these remnants must be taken into account because of their importance to the overall conservation network. The lack of knowledge of how to manage remnant vegetation extends to the private landholders but has often been accompanied by a lack of appreciation of the value of native vegetation and an understanding of basic ecological processes. Fortunately, this situation is changing, and many private landholders are keenly interested in learning how best to retain and manage their remnant areas. Breckwoldt (1983) has provided a good introduction to wildlife conservation on farms, which points out, as do other chapters in this book, that conservation measures need not be divorced from the agricultural activities on farms.

Simple retention of a remnant will not necessarily ensure the continued existence of the remnant as a functional ecosystem. As detailed by Hobbs et al. (Chapter 4), the fragmentation process has resulted in large changes in the landscape surrounding the remnant. Management must thus consist of two components; firstly, the internal processes within the remnant must be managed (i.e., vegetation dynamics, fire regime, etc.), and, secondly, the influences impinging on the remnant from the surrounding landscape must also be managed (i.e., nutrient inputs, weed invasion, feral animals, etc).

Management of remnant vegetation is complex because of the mosaic nature of the landscape. Except in the case of very small remnants, an individual remnant is rarely made up of just one vegetation type, and thus there cannot be a simple management prescription for the whole remnant. Nevertheless, there are a few straightforward guidelines which apply to remnant vegetation generally and arise from a consideration of the effects of fragmentation, as discussed by Hobbs et al. (Chapter 4).

10.2.1 Management of External Factors

Because ecological processes in remnants are now mainly driven by external forces, external influences pose the biggest threat to small remnant areas and thus should be accorded higher priority in the first instance. Some of the external influences are relatively easy to deal with at the level of the individual remnant, but others, such as hydrological processes, can only be tackled in the context of the overall landscape, as discussed by McFarlane et al. (Chapter 6) and Lefroy et al. (Chapter 8). These factors can be of overriding importance and, without some attempt to deal with landscape-level processes, any actions at the individual remnant level could prove futile. Management of landscape-level processes requires cooperation and communication between different managers and owners and a recognition that conservation areas and agricultural lands are inextricably interlinked. Tackling conservation problems without, at the same time, tackling the problems of degradation on agricultural land is basically a waste of time; remnant vegetation, no matter how well it is managed, will not persist within a rapidly degrading landscape (Hobbs and Saunders 1991; Hobbs et al. 1992).

Given that attempts are made to rectify the larger-scale problems, there are still many local effects to be considered. Firstly, it is clear that successful conservation of remnant vegetation and its associated fauna depends heavily on the exclusion of domestic stock. Fencing is thus a prerequisite, but this is one of the most expensive aspects of remnant protection. In Western Australia, a protection scheme for remnant vegetation has been in operation, which offered subsidised fencing of remnant areas. New forms of electric fencing which offer cheaper alternatives to traditional fences are now available to protect remnant vegetation. In addition to control of domestic livestock, rabbits should also be controlled, and this is achieved relatively easily using poison baits. The poison 1080 (sodium monoflouroacetate) is used extensively in Western Australia because it is largely specific to nonnative species. Native fauna are resistant to the poison due to their long coevolution with the suite of *Gastrolobium* species (Papilionaceae), collectively known as poison bushes, which are present throughout the southwest of the state (King et al. 1978, 1981; Mead et al. 1985).

Poison baiting is also used to control other nonnative species, particularly foxes, which have been shown to have a severe effect on the populations of several marsupial species. Experimental removal of foxes by baiting has resulted in dramatic increases in the number of animals such as the black-footed rock wallaby (Kinnear et al. 1988). The major problem with this approach is that the poisoning has to be continuous to deal with reinvasion by foxes of cleared areas.

Management of weed invasion poses greater problems, due to the continuous exposure of remnant areas to propagules from surrounding agricultural areas. In general, it seems that prevention is easier than the cure, and the maintenance of intact native vegetation canopies and their judicious management can successfully prevent weed invasion (Cooke 1983; Breckwoldt 1986). In low-nutrient systems, continued resistance to weed invasion depends on there being no large increases in soil nutrient levels, and this, in turn, depends on the successful prevention of nutrient transfer from adjacent agricultural land through the adoption of soil conservation strategies.

Where weeds have already invaded, successful weed control depends on the treatment of the cause rather than the symptoms (Cale and Hobbs 1991). Simple spraying of weed species can be effective in the short term, but this must be accompanied either by some method of removing accumulated nutrients or by a rapid regrowth of a native plant canopy. Without this, weed species will rapidly reestablish. Removal of nutrients could be achieved by sequestering off in repeated harvests of weed biomass, as is done in parts of Europe (Marrs 1985; Bakker 1987), by chemical means (e.g., precipitating out excess phosphorous by adding calcium) or by removing the upper layer of soil (mainly for use in revegetation). This latter method gives rapid results and has been used successfully in the higher rainfall areas of southwestern Australia and has the added advantage of removing most of the weed seed stored in the soil (Loney 1990).

Bradley (1988) has advocated the use of weed removal techniques rather than chemical spraying; this approach is useful where there is considerable voluntary support, as around urban areas, but will be less useful where large areas and sparse populations are involved. The disturbance associated with weed extraction may also have unwanted detrimental effects.

Where weed control is carried out in conjunction with attempts to rehabilitate degraded remnant areas, there may also be a need for reseeding or replanting of native species, depending on the availability of seed either on adult plants or in the soil for natural regeneration. Methods used will be similar to those used for revegetation, as discussed in Section 10.5.

10.2.2 Management of Internal Processes

The management of internal processes depends on the types of vegetation present, and, as yet, there is little detailed information available for effective management strategies to be formulated. Minimising interventive management is one option available, but A.R. Main (Chapter 7) has argued that lack of disturbance (either biotic or abiotic) will result in reduced species richness. The extinction of predators and herbivores documented by Hobbs et al. (Chapter 4) is likely to have weakened biotic interactions within remnants, which may also lead to communities dominated by fast-growing, competitively superior species, unless interventive management is carried out. Disruption of biotic interactions and natural disturbance regimes, coupled with the small size of many remnants, means that some form of periodic interventive management may be essential to maintain the diversity and function of native communities.

Fire is the major tool available to managers because of its relative ease of application and cheapness. It is possible that other options such as rolling or mowing may be viable alternatives in some cases, but there has been no research into these. While fire is simple to apply, the optimal fire regime is difficult to determine since information on the response of wheatbelt vegetation types is only slowly being accumulated. It is not possible to extrapolate from the higher rainfall areas where more information is available, since the dynamics of the wheatbelt systems are liable to be quite different. For instance, Kessel et al. (1984) and Burrows et al. (1987) have shown that fuel accumulation curves in wheatbelt woodlands dominated by *Eucalyptus wandoo* level off at very low fuel loadings, possibly due to high levels of termite activity. Fire-planning models used elsewhere thus cannot be successfully applied to wheatbelt vegetation. Despite their early promise (Kessel et al. 1984; Hopkins 1985), such models have not been successfully developed for wheatbelt vegetation, partially due to the large amounts of data on fuel loadings that are required to run the models successfully.

Early data from our studies in the Kellerberrin area indicate that fire can stimulate the regeneration of the major tree species and that plant species diversity is increased postfire by the establishment of fire-following species, such as *Dampiera eriocephala, Goodenia affinis, Actinotus superbus*, and *Keraudrenia integrifolia*. The limited results to date show few adverse effects of fire on the vegetation, although some components of the fauna may be adversely affected by more intense fires (A.R. Main, unpublished data) and data are, as yet, only short-term. The results to date are from individual fires, however, and do not provide a complete picture of the effects of variation in the different components of the fire regime or the possible interactions between fire and grazing animals in determining postfire vegetation development. As discussed by Hobbs et al. (Chapter 4), in the face of uncertainty concerning fire effects, it is desirable to have a range of approaches, including periodic burning of some areas and retention of unburnt vegetation in others.

Information from similar types of vegetation elsewhere may be useful in predicting general vegetation response. For instance, the heath or kwongan vegetation of the northern sandplains, north of Perth, is very similar in structure to the heath vegetation found in the wheatbelt, and detailed information is available on various components of the vegetation response to fire (Bell et al. 1984; Delfs et al. 1987; van der Moezel et al. 1987). While floristics differ markedly between the two areas, predictions of vegetation response based on a functional grouping of species should be possible (e.g., Noble and Slatyer 1980; Moore and Noble 1990), although this has not yet been attempted. However, the northern sandplain areas remain relatively unfragmented, and fire effects can be significantly modified in the fragmented wheatbelt system, due to changes in fire characteristics and through interactions with adjacent agricultural areas.

As an example, the interaction between fire and weed invasion is important. Fire is frequently considered to be a factor in increasing the likelihood of weed invasion of native vegetation in southwestern Australia (e.g., Bridgewater and Backshall 1981; Christensen and Burrows 1986), and Hobbs and Atkins (1990) have shown that fire in woodland vegetation can, in some cases, lead to an increased abundance of weedy species, albeit temporarily. Fire has the effect of providing a temporary increase in nutrient availability (Hester and Hobbs 1992), but this does not necessarily lead to increased weed invasion. Studies at the remnant edges indicate that the weed abundances after fire roughly approximate those found prior to fire (Hester and Hobbs 1992). Hobbs and Atkins (1991) found little evidence for increases in weed abundance after fire in heath vegetation, except where the fire reached a roadverge. Where the roadside vegetation had been disturbed by road maintenance, weeds became very prolific. Thus, fire combined with an other disturbance will exacerbate the invasion process.

This raises questions concerning the manufacture of firebreaks and roads within reserves; these may actually increase management problems by facilitating weed invasion. In other parts of Western Australia, the manufacture of management roads has been implicated in the spread of dieback disease, *Phytophthora cinnamomi* (Weste and Law 1976; Hill 1990; Wills 1992). In addition to these considerations, studies elsewhere have indicated that even relatively narrow roads can constitute effective barriers to components of the fauna (Mader 1984; Mansergh and Scotts 1989; Andrews 1990). Careful consideration must therefore be given to the further subdivision of remnant areas by firebreaks and roads.

A further consideration in terms of fire regimes in fragmented habitats is the interrelationship between the remnant areas and the surrounding landscape. A frequently held belief has been that fires usually start in remnant vegetation and spread out into agricultural land, thus causing damage and economic loss. In fact, the opposite is usually true, and the majority of fires in remnant vegetation start outside the remnant, frequently due to deliberate or accidental human activities (Burrows et al. 1987). There is the danger that small remnant areas will be completely burnt out by accidental fires, and, in that case, the fauna may have difficulty in surviving, due to the lack of possibility to disperse away from the fire area.

10.2.3 Rehabilitation

Another aspect of management within remnant areas that is increasing in importance is the rehabilitation of degraded areas. Restoration ecology is growing as a discipline (Cairns 1986; Allen 1988; Jordan et al. 1988), but there are, as yet, few general principles available. The ease with which rehabilitation can be carried out depends on the level of disturbance the vegetation has experienced (Hobbs and Hopkins 1990). The desired result is also important; it is rarely possible to reproduce a functioning system identical to that present prior to disturbance, and hence some level of success has to be established. For instance, a useful goal may be the development of a cover of self-sustaining vegetation, not necessarily containing all of, or the same, species present prior to disturbance. Bridgewater (1990) has pointed out that synthetic or nonnatural vegetation has an important role to play in today's human-dominated landscapes.

In the case of remnant areas in the Western Australian wheatbelt, areas to be rehabilitated include those that have been grazed by stock, invaded by weeds, cleared for agriculture and subsequently abandoned, or used for gravel extraction. These problems and others occur across Australia (Buchanan 1989). Stock exclusion and weed control seem to be prerequisites for rehabilitation, but there may be a requirement for reseeding or replanting of native species if these are no longer present in the area or no longer produce or have stores of viable seed. Where soil structure is significantly modified, for instance, by stock trampling or by gravel extraction, a ground treatment such as ripping or scarification is required. Knowledge of the germination and establishment requirements is needed for the rehabilitation process to be successful; such knowledge is slowly being accumulated for a small proportion of the wheatbelt's rich flora (e.g., Lefroy et al. 1991).

10.3 Management of Conservation Networks

Management of individual remnants in isolation is liable to fail to meet conservation objectives for a number of reasons. The conservation resources of the wheatbelt region are, in common with most other regions, scattered over many different remnants with different ownerships and primary uses. Many different managers are involved in remnant management, including conservation agencies, private owners, road authorities, and local authorities. There is no guarantee that all managers will act in a coordinated fashion (Hobbs et al. 1992). However, maintaining representative samples of native communities and populations of rare and endangered species demands that more than just the remnants specified as nature reserves be conserved. For instance, in the central wheatbelt, much of the remaining woodland is on private land and roadverges (Arnold and Weeldenberg 1991), and roadverges constitute an important component of habitat for sections of the fauna (Cale 1990). There is also evidence emerging that a collection of remnants may constitute a conservation network for species such as the kangaroos and some birds (Arnold et al. 1991; Saunders and de Rebeira 1991), while other species (e.g. small lizards, invertebrates) are likely to constitute of a series of populations isolated on remnants. Opdam (1991) has recently pointed out that the conservation problems posed by these different sets of species are quite distinct.

As mentioned by Hobbs et al. (Chapter 4), many populations of animals are too small and too isolated to be viable in the short term, let alone the long term. These are species which are totally dependent on remnant vegetation for their survival. They do not use the agricultural matrix; in fact, it may be a barrier to movement. These species are the species of highest conservation priority in the wheatbelt because they will not persist without active management. Until these populations are linked by vegetation to enable them to move between remnants, we need to establish the distribution and abundance of species on remnants and then monitor the remnants with high conservation value to record the changes in the remnant-dependent species. The control of predators, particularly foxes, will be necessary within the remnants of high conservation priority, as has been done for the black-footed rock wallaby (Kinnear et al. 1988). Fox control will also have to be included in reintroduction attempts, as has been done with the numbat, *Myrmecobius fasciatus* (Friend 1990). In fact, the control of exotic predators and reintroduction may become a normal management activity until we can reintegrate heavily cleared and fragmented rural landcapes.

An important aspect of network management is the management of corridor vegetation. Most corridors in the wheatbelt are present along roadsides and fence lines. Corridors are essentially similar to remnant areas, except that the management problems are exacerbated due to their long, thin nature. The influence of edge effects and external processes is particularly acute in corridors, and hence management has to be modified accordingly (Loney and Hobbs 1991). Management practices such as periodic burning, which are acceptable in larger remnant areas, may not be acceptable in corridors because of their interactions with other disturbances such as road maintenance. Ecological processes may also be significantly altered in narrow corridors (e.g., van Schagen et al. 1992) and, hence, corridors require special care. If, as we have suggested, the corridor network is important in conservation terms, corridor maintenance and reestablishment has to be an integral component of conservation management. Corridor management is still in its infancy, but interest in corridors in the wheatbelt landscape is increasing, and the development of corridor networks (Hussey et al. 1991) is one concrete way that the local community can aid in conservation while also helping to solve the agricultural problems discussed by Nulsen, Chapter 5 and McFarlane et al., Chapter 6.

The current conservation network is probably inadequate to maintain the biotic diversity of the region (Hobbs et al., Chapter 4) and further extinctions are likely, both locally and regionally. The conservation network is also currently degrading due to the impingement of external factors; this is likely to speed up the loss of biotic diversity. However, the current network should not be viewed as static. Indeed, there is a tremendous potential for increasing the conservation resources of the region by taking advantage of the surge of community interest in revegetation, as discussed in Section 10.4. Incorporation of design features which render revegetation programs useful for nature conservation could add to the area and connectivity of the existing network.

10.4 Revegetation

Revegetation, or the reintroduction of native vegetation into areas previously cleared and used for agricultural production, is an activity which can greatly add to the existing conservation resources of the wheatbelt region. Revegetation purely for conservation purposes is rarely considered by the farming community but is regarded as a viable option to deal with the major agricultural problems as discussed by Nulsen, Chapter 5 and McFarlane et al., Chapter 6. If revegetation for these purposes can also incorporate features valuable for conservation, then we will be able to make progress in the direction of integrating conservation and agriculture. This seems possible, although development of the correct procedures is very much in its early stages.

A major consideration is the identification of priority areas for revegetation. McFarlane et al. (Chapter 6) have already discussed the location of revegetation for salinity and waterlogging control. For conservation purposes, anything which leads to increased areas of natural or seminatural vegetation will be of benefit. However, it is clear that some areas require more attention than others. Thus, for instance, vegetation which is currently underrepresented in the remnant system, particularly the woodland types, should be targeted. There are good arguments for siting revegetation so that it extends the area of existing remnants, i.e., around existing remnant boundaries. This not only increases overall area but also moves the remnant edge out a bit and provides a buffer strip. This function can be carried out by vegetation other than that already in the remnant; for instance, a productive timber belt could be established at the edge of a woodland reserve. The importance of establishing corridor networks has already been discussed; again, these can be incorporated into the design of shelterbelt systems.

As with rehabilitation, the goal of revegetation should be clear. It is very unlikely that we will ever be able to reproduce a completely natural system, but the goal should at least be a self-sustaining vegetation which provides some conservation benefit. This makes sense not only ecologically, but also economically, since we do not want to have to repeat the repair process in another 20 to 30 years.

Methods of revegetation include natural regeneration, direct seeding, and replanting. These follow a gradient of intervention and cost. Natural regeneration is possibly the preferred method in conservation terms since it would allow the reestablishment of native species and provenances, but it is debatable if it is a viable option for revegetating land extensively modified for agriculture for long periods. This is mainly because of the large changes in soil characteristics and microclimate that have been noted earlier by Hobbs et al., Chapter 4 and Nulsen, Chapter 5. Our studies have indicated that native vegetation can reestablish in areas used for agriculture for short periods, but that, even then, only a subset of the original biota can return (Hobbs and Atkins, unpublished data). Long periods of cultivation and fertiliser addition make it difficult for native species to reestablish.

Direct seeding is an attractive option for revegetation because of its relatively low cost and the possibility of reintroducing complex mixes of species. It has achieved considerable success in some areas of south-western Australia (Loney 1990), predominantly in zones of higher and more predictable rainfall than the central wheatbelt. Direct seeding in these lower rainfall areas has achieved mixed success, largely depending on adequate timing and amount of rainfall. Improved seeding techniques and increased knowledge of species' requirements should greatly enhance the value of direct seeding in this area.

Planting of established stock has been the preferred method of revegetation in the wheatbelt, and good techniques are available to assure its success. A number of mechanical planters allow rapid planting and good establishment. Planting and seeding success depends on adequate weed and animal control and a good enough water supply to ensure that the young plant can grow sufficiently to endure its first summer. Cultivation and water-harvesting techniques are useful here. Care of seedlings after establishment may also be required, particularly in terms of insect attack. It takes many years for a tree to grow but only a short time for it to be defoliated. Insect outbreaks can severely affect revegetation programs, and a neglected part of species selection is an assessment of resistence to herbivory. It may be that, in the long term, herbivory resistance is a more important characteristic than growth rate.

Whether revegetation is by seeding or planting, the choice of species is important. For success, the species clearly have to be matched with the soil type into which they are being sown. The reestablishment of woodland species from heavy valley soils will not be easy in the sandier soils at the top of the slopes. Local species should be used wherever possible, and local seed should be used to maintain genetic integrity. This involves extensive seed collection and a knowledge of plant phenology. Linking species for revegetation with their correct location in the landscape is vital (Lefroy et al. 1991).

This does not, however, imply that revegetation should be tied only to the species thought to be present in the area before agriculture. Other species not native to the area may perform better than native species in terms of growth, evapotranspiration, or salinity tolerance. Some species not native to the area may be very easy to establish and give good results quickly. Use of these species will maintain the enthusiasm of farmers just starting out in revegetation, whereas poor results or slow growth of native species may dampen that enthusiasm. Also, conditions may have deteriorated to such an extent that it is now impossible to reestablish the preexisting vegetation. Common examples of this are found on salmon gum soils (*Eucalyptus salmonophloia*) that are now too saline to support salmon gums. The choice of species thus has to be pragmatic; the rule should be to use the species native to the area wherever possible but not to rule out the use of other species where necessary. A cover of nonnative species is, in general, better than a rapidly degrading landscape, unless the nonnatives themselves start to pose a threat in terms of the invasion of native vegetation. To date there has been little evidence of any introduced tree species invading natural vegation in the Western Australian wheatbelt, although this has occurred elsewhere (Groves 1986; MacDonald et al. 1989), and careful consideration of this problem is required before any species can be used in revegetation.

Planting a mixture of tree and understorey species will greatly enhance the conservation value of the revegetation. Emphasis has been predominantly on tree species for revegetation and many guides are available on planting trees (Department of Arts, Heritage and Environment 1985; Venning 1988; Cremer 1990). However, tree cover without any understorey really only provides half an answer, since shrubs provide essential food supplies and habitats for fauna, add to the shelter effect in windbreaks, especially when the trees grow taller, and can be effective for increasing evapotranspiration.

10.5 Priority Setting

A primary concern of conservation management in fragmented systems is the development of priorities for remnant retention and management. Priority setting will be determined by the goals set for conservation management, and these may include the retention of representative examples of all native ecosystems in an area, the maintenance of species diversity, and, in particular, the preservation of rare and endangered species. In order to do this, we need to know the distributions of species and ecosystems and be able to select areas which represent them. Detailed survey information is available from only a few remnant areas in the wheatbelt (Kitchener et al. 1980a,b, 1982; Kitchener 1982), and a full inventory is next to impossible to achieve within a sensible time frame. Alternative approaches to providing at least some information on remnant characteristics include the use of remote sensing (Hobbs et al. 1989) and a modelling technique for determining the minimum subset of existing remnants required to represent the diversity of the region (Margules et al. 1988; Margules and Stein 1989).

Both of these approaches are being applied to the central wheatbelt and will yield useful information on which remnants are the most important in the regional context and which should therefore receive priority for retention and management. These approaches provide only part of the answer, however, because they give data on individual remnants in isolation. If, as we are suggesting, remnant networks are important for the retention of the biota, greater consideration will also have to be given to the spatial patterning and connectivity of remnants in relation to one another. In addition, remnants which do not score highly in terms of representativeness or other measures may, in fact, be very important when viewed in a landscape context (e.g., for soil conservation purposes). Most corridors would probably fall into this category.

The need for a landscape approach to remnant retention and management is clear (Saunders et al. 1991) but is only slowly being tackled by research and management. However, the need for this approach makes the setting of priorities even more difficult. Should resources be directed mostly at the largest remnants which are in the best condition, to degraded remnants in need of rehabilitation, to corridors which are difficult to maintain but may (or may not) have an important linkage role, or to revegetation programs which may enhance the overall conservation value of the landscape? Clearly, the answer is that all these components should be considered simultaneously, but limited resources rarely allow this to happen.

However, priority setting can become easier when all interested managers and owners participate and conservation management is shared by the whole community. Each party will have particular priorities; for instance, the conservation authority has, as its priority, the retention and management of conservation reserves, whereas the farmer has, as a priority, revegetation to combat land degradation and the shire's priority is to maintain roadside (= corridor) vegetation. If the different parties can be guided to work towards a common goal, all aspects of conservation management can be tackled.

This, therefore, brings in another activity which has to have a very high priority, as mentioned in Section 10.1, i.e., education. Integrated conservation management will be achieved only if managers and the community as a whole are aware of the problems, have possible solutions available to them, and have the will to implement these solutions. While this type of awareness is increasing, there is still a long way to go, and everyone involved in conservation research and management should be involved in communication to all sections of the community. At the same time, the community as a whole should be involved in conservation management. Involvement of the local community in management and planning means that it is more likely that the conservation objectives are met, since they have a vested interest in the area and are "on the ground" most of the time. Local community and voluntary organisation involvement also greatly increases the manpower and resources available for conservation.

10.5 Conclusions

Conservation management world-wide must increasingly deal with the consequences of fragmentation and must recognise the importance of placing conservation reserves firmly in the context of the surrounding landscape. It is no longer sufficient to manage what is going on inside remnant areas. Most of the management problems arise outside the remnants in the surrounding altered matrix. Some of these problems can be tackled locally, but often they must be tackled in the landscape as a whole. This implies that conservation management must move away from the management of individual patches and more towards an integrated landscape management. The importance of remnant networks for the conservation of representative assemblages and to allow the movement of mobile faunal elements also necessitates an integrated approach to the management of multiple remnants, irrespective of their ownership. The setting of priorities for management of conservation networks is a difficult task, but management can be tackled in an integrated way so that various components are dealt with simultaneously. The existing conservation network can be enhanced by taking advantage of revegetation programs, but this involves an increased awareness of conservation problems and solutions by both the managers and the community as a whole. Increasing community awareness must be a priority activity because, without community understanding and support, conservation objectives will not be reached.

Acknowledgments. We thank G. Arnold, C. Margules, J. Short, G. Smith, and K. Wallace for constructive comments on the manuscript.

References

Allen, E.B. (ed.) (1988) *The Reconstruction of Disturbed Arid Ecosystems – An Ecological Approach*. Westview, Boulder, Colorado.

Andrews, A. (1990) Fragmentation of habitat by roads and utility corridors: A review. *Australian Zoologist* 26:130–141.

Arnold, G.W. and Weeldenberg, J.R. (1991) The distributions and characteristics of remnant native vegetation in parts of the Kellerberrin-Tammin-Trayning and Wyalkatchem shires of Western Australia. *CSIRO Division of Wildlife and Ecology Technical Memo* 33.

Arnold, G.W., Weeldenberg, J.R., and Steven, D.E. (1991) Distribution and abundance of two species of kangaroo in remnants of native vegetation in the central wheatbelt of Western Australia and the role of native vegetation along roadverges and fence lines as linkages. In D.A. Saunders and R.J. Hobbs (eds.) *Nature Conservation 2: The Role of Corridors*. Surrey Beatty and Sons, Chipping Norton, NSW. pp. 273–280.

Bakker, J.P. (1987) Restoration of species-rich grassland after a period of fertiliser application. In J. van Andel, J.P. Bakker, and R.W. Snaydon (eds.) *Disturbances in Grasslands. Causes, Effects and Processes*. Junk, Dordrecht. pp. 185–200.

Bell, D.T., Hopkins, A.J.M., and Pate, J.S. (1984) Fire in the kwongan. In J.S. Pate and J.S. Beard (eds.) *Kwongan. Plant Life of the Sandplain*. University of Western Australia Press, Nedlands. pp. 178–204.

Bradley, J. (1988) *Bringing Back the Bush. The Bradley Method of Bush Regeneration*. Landsdowne, Sydney.

Breckwoldt, R. (1983) *Wildlife in the Home Paddock*. Angus and Robertson, North Ryde, NSW.

Breckwoldt, R. (1986) *The Last Stand – Managing Australian Remnant Forest and Woodland*. Australian Government Publishing Service, Canberra.

Bridgewater, P.B. (1990) The role of synthetic vegetation in present and future landscapes of Australia. *Proceedings of the Ecological Society of Australia* 16:129–134.

Bridgewater, P.B. and Backshall, D.J. (1981) Dynamics of some Western Australian ligneous formations with special reference to the invasion of exotic species. *Vegetatio* 46:141–148.

Buchanan, R.A. (1989) *Bush Regeneration: Recovering Australian Landscapes*. TAFE, NSW, Sydney.

Burrows, N.D., McCaw, W.L., and Maisey, K.G. (1987) Planning for fire management in Dryandra forest. In D.A. Saunders, G.W. Arnold, A.A. Burbidge, and A.J.M. Hopkins (eds.) *Nature Conservation: The Role of Remnants of Native Vegetation*. Surrey Beatty and Sons, Chipping Norton, NSW. pp. 305–312.

Cairns, J. (1986) Restoration, reclamation, and regeneration of degraded or destroyed ecosystems. In M.E. Soulé (ed.) *Conservation Biology. The Science of Scarcity and Diversity*. Sinauer, Sunderland, Mass. pp. 465–484.

Cale, P. (1990) The value of road reserves to the avifauna of the central wheatbelt of Western Australia. *Proceedings of the Ecological Society of Australia* 16:359–367.

Cale, P. and Hobbs, R.J. (1991) Condition of roadside vegetation in relation to nutrient status. In D.A. Saunders and R.J. Hobbs (eds.) *Nature Conservation 2: The Role of Corridors*. Surrey Beatty and Sons, Chipping Norton, NSW. pp. 353–362.

Christensen, P.E. and Burrows, N.D. (1986) Fire: An old tool with a new use. In R.H. Groves and J.J. Burdon (eds.) *Ecology of Biological Invasions: An Australian Perspective*. Australian Academy of Science, Canberra. pp. 57–66.

Cooke, B.D. (1983) Rabbit control and the conservation of native vegetation on roadsides. In F.J. Van der Sommen (ed.) *Trees in the Rural Environment: Towards a Greenprint for South Australia*. Faculty of Natural Resources, Roseworthy Agricultural College, Roseworthy, SA. pp. 4.6–4.7.

Cremer, K.W. (ed.) (1990) *Trees for Rural Australia*. Inkata, Melbourne.

Delfs, J.C., Pate, J.S., and Bell, D.T. (1987) Northern sandplain kwongan: Community biomass and selected species response to fire. *Journal of the Royal Society of Western Australia* 69:133–138.

Department of Arts, Heritage and Environment (1985) *Think Trees Grow Trees* Australian Government Publishing Service, Canberra.

Friend, J.A. (1990) The numbat *Myrmecobius fasciatus* (Myrmecobiidae): History of decline and potential for recovery. *Proceedings of the Ecological Society of Australia* 16:369–377.

Groves, R.H. (1986) Invasion of Mediterranean ecosystems by weeds. In B. Dell, A.J.M. Hopkins, and B.B. Lamont (eds.) *Resilience in Mediterranean-type Ecosystems.* Junk, Dordrecht. pp. 129–145.

Hester, A. and Hobbs, R.J. (1992) Influence of fire and soil nutrients on native and non-native annuals at remnant vegetation edges in the Western Australian wheatbelt. *Journal of Vegetation Science* 3:101–108.

Hill, T.C.J. (1990) Dieback disease and other *Phytophthora* species in the northern kwongan. In A.A. Burbidge, S.D. Hopper, and S. van Leeuwen (eds.) *Nature Conservation, Landscape and Recreation Values of the Lesueur Area.* Environmental Protection Authority, Perth. pp. 89–97.

Hobbs, R.J. and Atkins, L. (1990) Fire-related dynamics of a *Banksia* woodland in southwest Western Australia. *Australian Journal of Botany* 38:97–100.

Hobbs, R.J. and Atkins L. (1991) Interactions between annual and perennial vegetation components in a Western Australian wheatbelt reserve. *Journal of Vegetation Science* 2:643–654.

Hobbs, R.J. and Hopkins, A.J.M. (1990) From frontier to fragments: European impact on Australia's vegetation. *Proceedings of the Ecological Society of Australia* 16:94–114.

Hobbs, R.J. and Saunders, D.A. (1991) Reintegrating fragmented landscapes: A preliminary framework for the Western Australian wheatbelt. *Journal of Environmental Management* 33:161–167.

Hobbs, R.J., Saunders, D.A., and Arnold, G.W. (1992) Integrated landscape management: A Western Australian perspective. *Biological Conservation* in press.

Hobbs, R.J., Wallace, J.F., and Campbell, N.A. (1989) Classification of vegetation in the Western Australian wheatbelt using Landsat MSS data. *Vegetatio* 80:91–105.

Hopkins, A.J.M. (1985) Planning the use of fire on conservation lands in southwestern Australia. In J. Ford (ed.) *Fire Ecology and Management in Western Australian Ecosystems.* Western Australian Institute of Technology, Perth. pp. 203–208.

Hussey, B.M.J., Hobbs, R.J., and Saunders, D.A. (1991) *Guidelines for Bush Corridors.* CSIRO Division of Wildlife and Ecology, CALM, MRD, and RCC, Western Australia.

Jordan, W.R. III, Peters, R.L. II, and Allen, E.B. (1988) Ecological restoration as a strategy for conserving biological diversity. *Environmental Management* 12:55–72.

Kessel, S.R., Good, R.B., and Hopkins, A.J.M. (1984) Implementation of two new resource management information systems in Australia. *Environmental Management* 8:251–270.

King, D.R., Oliver, A.J., and Mead, R.J. (1978) The adaption of some Western Australian mammals to food plants containing fluoroacetate. *Australian Journal of Zoology* 26:699–712.

King, D.R., Oliver, A.J., and Mead, R.J. (1981) *Bettongia* and fluoroacetate: A role for 1080 in fauna management. *Australian Wildlife Research* 8:529–536.

Kinnear, J.E., Onus, M.L., and Bromilow, R.N. (1988) Fox control and rock wallaby population dynamics. *Australian Wildlife Research* 15:435–450.

Kitchener, D.J. (1982) Predictors of vertebrate species richness in nature reserves in the Western Australian wheatbelt. *Australian Wildlife Research* 9:1–7.

Kitchener, D.J., Chapman, A., Dell, J., Muir, B.G., and Palmer, M. (1980a) Conservation value for mammals of reserves in the Western Australian wheatbelt. *Biological Conservation* 18:179–207.

Kitchener, D.J., Dell, J., Muir, B.G., and Palmer, M. (1982) Birds in Western Australian wheatbelt reserves – Implications for conservation. *Biological Conservation* 22:127–163.

Kitchener, D.J., Chapman, A., Dell, J., Muir, B.G., and Palmer, M. (1980b) Lizard assemblage and reserve size and structure in the Western Australian wheatbelt – Some implications for conservation. *Biological Conservation* 17:25–62.

Lefroy, E.C., Hobbs, R.J., and Atkins, L.J. (1991) *A Revegetation Guide for the Central Wheatbelt.* Western Australian Department of Agriculture, Perth.

Loney, B. (1990) Direct seeding in Western Australia – Its status and future. In *Sowing the Seeds. Direct Seeding and Natural Regeneration Conference Proceedings.* Greening Australia, Deakin, Australian Capital Territory. pp. 103–112.

Loney, B. and Hobbs, R.J. (1991) Management of vegetation corridors: Establishment, maintenance, and rehabilitation. In D.A. Saunders and R.J. Hobbs (eds.) *Nature Conservation 2: The Role of Corridors.* Surrey Beatty and Sons, Chipping Norton, NSW. pp. 299–311.

MacDonald, I.A.W., Loope, L.L., Usher, M.B., and Harmann, O. (1989) Wildlife conservation and the invasion of nature reserves by introduced species. In J. Drake, F. di Castri, R. Groves, F. Kruger, H. Mooney, M. Rejmánek, and M. Williamson (eds.) *Biological Invasions: A Global Perspective* John Wiley and Son, Chichester. pp. 215–255.

Mader, H.-J. (1984) Animal habitat isolation by roads and agricultural fields. *Biological Conservation* 29:81–96.

Main, A.R. (1987) Management of remnants of native vegetation – A review of problems and the development of an approach with reference to the wheatbelt of Western Australia. In D.A. Saunders, G.W. Arnold, A.A. Burbidge, and A.J.M. Hopkins (eds.) *Nature Conservation: The Role of Remnants of Native Vegetation.* Surrey Beatty and Sons, Chipping Norton, NSW. pp. 1–13.

Mansergh, I.M. and Scotts, D.J. (1989) Habitat continuity and social organisation of the mountain pygmy possum restored by tunnel. *Journal of Wildlife Management* 53:701–707.

Margules, C.R. and Stein, J.L. (1989) Patterns and distributions of species and the selection of nature reserves: An example from eucalypt forests in southeastern New South Wales. *Biological Conservation* 50:219–238.

Margules, C.R., Nicholls, A.O., and Pressey, R.L. (1988) Selecting networks of reserves to maximise biological diversity. *Biological Conservation* 43:63–76.

Marrs, R.H. (1985) Techniques for reducing soil fertility for nature conservation purposes: A review in relation to research at Roper's Heath, Suffolk, England. *Biological Conservation* 34:307–332.

Mead, R.J., Oliver, A.J., King, D.R., and Hubach, P.H. (1985) The coevolutionary role of fluoroacetate in plant-animal interactions in Australia. *Oikos* 44:55–60.

Moore, A.D. and Noble, I.R. (1990) An individualistic model of vegetation stand dynamics. *Journal of Environmental Management* 31:61–81.

Noble, I.R. and Slatyer, R.O. (1980) The use of vital attributes to predict successional changes in plant communities subject to recurrent disturbances. *Vegetatio* 43:5–21.

Opdam, P. (1991) Metapopulation theory and habitat fragmentation: A review of holarctic breeding bird studies. *Landscape Ecology* 5:93–106.

Saunders, D.A. and de Rebeira, C.P. (1991) Values of corridors to avian populations in a fragmented landscape. In D.A. Saunders and R.J. Hobbs (eds.) *Nature Conservation 2: The Role of Corridors*. Surrey Beatty and Sons, Chipping Norton, NSW. pp. 221–240.

Saunders, D.A., Hobbs, R.J., and Margules, C.R. (1991) Biological consequences of ecosystem fragmentation: A review. *Conservation Biology* 5:18–32.

van der Moezel, P.G., Loneragan, W.A., and Bell, D.T. (1987) Northern sandplain kwongan: Regeneration following fire, juvenile period, and flowering phenology. *Journal of the Royal Society of Western Australia* 69:123–132.

Van Schagen, J.J., Hobbs, R.J., and Majer, J.D. (1992) Defoliation of trees in roadside corridors and remnant vegetation in the Western Australian wheatbelt. *Journal of the Royal Society of Western Australia* 75: in press.

Venning, J. (1988) *Growing Trees for Farms, Parks, and Roadsides*. Lothian, Melbourne.

Wallace, K.J. and Moore, S.A. (1987) Management of remnant bushland for nature conservation in agricultural areas of southwestern Australia – Operational and planning perspectives. In D.A. Saunders, G.W. Arnold, A.A. Burbidge, and A.J.M. Hopkins (eds.) *Nature Conservation: The Role of Remnants of Native Vegetation*. Surrey Beatty and Sons, Chipping Norton, NSW. pp. 259–268.

Weste, G. and Law, C. (1976) The invasion of native forest by *Phytophthora cinnamomi*. III. Threat to the National Park, Wilson's Promontory, Victoria. *Australian Journal of Botany* 21:31–51.

Wills, R. (1992) The ecological impact of *Phytophthora cinnamomi* in the Stirling Range National Park, Western Australia. *Australian Journal of Ecology* 17: in press.

Section IV

Conclusions

New approaches to farming involving revegetation, farming to soil type and breaking away from the traditional rectangular grid system offer hope that sustainable agriculture and nature conservation are achievable.
(Photo: R. Moore)

Section IV

Conclusion

Chapter 11

Conclusions. Can We Reintegrate Fragmented Landscapes?

R.J. Hobbs and D.A. Saunders

11.1 The Western Australian Wheatbelt as a Model Fragmented System

The aim of this book was to draw together available information and viewpoints on the past, present, and future of one particular area, the central wheatbelt of Western Australia, with a view to providing information relevant to other fragmented landscapes world-wide. In order to assess the generality of the discussions in this book, we first have to establish whether the central wheatbelt area is a useful model for other areas.

Clearly, any region will have particular features which make it unique. The outstanding feature of the Western Australian wheatbelt is probably its ancient weathered landscape and lack of strong relief (McArthur, Chapter 2). The long history of erosion, lack of glaciation, leaching of nutrients, and accumulation of salt have produced the current landscape and underlie the problems now faced by agriculture. At the same time, the long, uninterrupted history of the landscape and its isolation have allowed the evolution of a unique and diverse biota adapted to the local conditions.

The ancient soils of southwestern Australia are nutrient poor and differ markedly from those of northern Europe and North America where soils were exposed only following glacial retreat a few thousand years ago (e.g., Pennington 1969). Further, the landscape of the Northern Hemisphere has evolved with the presence of technological humans over thousands of years. Some civilisations were able to persist for long periods without degrading their soils, whereas others, such as in Mesopotamia, were able to persist only until their agricultural practices led to irreversible land degradation (Hillel 1991). Detrimental effects on the biota have occurred in all areas with old, long-developed agricultural landscapes; for instance, most large carnivores have disappeared. The long history of human agriculture and influence on the landscape and biota in Europe and the Middle East contrasts with the very short period in which European-style agriculture has been practiced in the Western Australian wheatbelt (Hobbs and Hopkins 1990). Here, agriculture developed on ancient soils and led to massive land degradation in under 100 years. The rapidity with which the fragmentation process occurred also differs markedly from other areas in the Northern Hemisphere (Curtis 1956; Webb 1980; Wilcove et al. 1986) but has parallels in the tropics (e.g., Malingreau and Tucker 1988). Vegetation types also differ greatly between regions, and one can expect differences in response to fragmentation between the relatively species-poor temperate forests of the Northern Hemisphere, the species-rich, semi-arid vegetation of southwestern Australia, and the rich tropical rainforests of Amazonia. However, although direct comparisons between regions may be difficult, the effects of the fragmentation process should be similar in all cases but affected to varying degrees by different modifying influences (Saunders et al. 1991).

11.2 Ecological, Economic, and Social Constraints

The history of agricultural development in the wheatbelt is one of pioneering and conquering the land, and the prosperity of Western Australia and Australia as a whole largely stems from this agricultural development. Since the settlement by Europeans, Australia has been, and will probably remain, a primary producer and an exporter of that production. We must, therefore, view the development of agriculture in that light. The activities of the early settlers and the later arrivals in the wheatbelt were based on the premise that the land had to be developed for Australia to prosper. Most of the farmers who cleared the land did so in ignorance of the possible consequences and in ignorance of the natural heritage they were destroying. Indeed, overclearing was officially sanctioned and encouraged by the authorities of the day, despite the availability of evidence which pointed to the likely effects.

It seems likely that, as B.Y. Main (Chapter 3) suggested, it was only when people were born into the landscape that any sort of empathy with it developed. The farming community nowadays is a far cry from the red-neck, uncaring image which is frequently portrayed. Most farmers are now aware of conservation values or at least are receptive to information and new ideas. Most can also afford to do something towards soil and nature conservation. In this respect, the situation in Western Australia may be more hopeful than in other parts of the world where land degradation is primarily driven by poverty and rapidly rising human populations making increasing demands on the land. Having said that, however, the economic situation in the wheatbelt is not good at present, and the history of the area has been one of repeated boom-bust cycles. Such cycles are counterproductive both socially and agriculturally. If we are to reintegrate the fragmented wheatbelt landscape, a significant aim must, therefore, be the development of agricultural methods which break the boom-bust cycle by providing a production system which is more buffered against seasonal and trade fluctuations (Lefroy et al., Chapter 8).

Lefroy et al. (Chapter 8) indicated that, for agriculture to be sustainable, there must be an ecological filter on economic activities. They also recognised, however, that the economic filter is equally important. Ecologically sound practices are imperative, but, unless they are also economically and socially sound, they will never be accepted and the spiral of degradation will continue. However, in balancing ecological and economic aspects, it is important to recognise that the ecological constraints are, relatively, much more fixed than the economic and social constraints. Most of the degradation problems encountered today result from a failure to work within the ecological constraints of the region.

Society's values have been, and generally speaking still are, predominantly economic, with short-term economic expediency overriding practical ecological considerations. The anthropocentric nature of current economic theory accounts for its failure to deal with land degradation. Such parameters as soil fertility and biodiversity are not included in economic balance sheets. There are increasing attempts at putting economic values on ecological parameters; for example, how much are clean air, drinkable water, unpolluted beaches, or rare species worth? The correct valuation of natural resources is a difficult process, and it is perhaps somewhat misguided in that it misses the point that there are other values apart from economic. The value of natural resources as they provide essential life support services and as they affect the quality of life are much more realistic parameters to assess than a deduced economic value which is often calculated using doubtful assumptions. Some method of realistically valuing the integrity of essential ecological processes is required. While dialogue between resource managers and economists is essential, it is also essential that important noneconomic values are not subsumed into the current economic dogma, to be added up in a meaningless environmental accounting system.

11.3 Integrating Solutions

The extent of the problems created by rapid clearance presents a daunting task for those seeking solutions (Hobbs et al., Chapter 4; Nulsen, Chapter 5; McFarlane et al., Chapter 6). From the nature conservation point of view, the major feature is the extent and rapidity of the changes. The disappearance of species from entire areas can be mapped over the space of a few years. The impact of clearing and the introduction of exotic species on the native biota have been immense, and the remaining patches of native vegetation are strongly influenced by the surrounding agricultural matrix (Hobbs et al., Chapter 4). There is, nevertheless, an increasing desire by the farming community to retain remnant areas and maintain them in as a good condition as possible. Without this local support, nature conservation efforts would certainly fail.

The short timescale within which fragmentation has occurred has two positive consequences. Firstly, it is possible that the rapidity of the degradation of the landscape can be followed by an equally rapid recovery, if the correct treatments are applied. Secondly, some of the patches of native vegetation left following clearing are still in relatively good condition and, despite species losses already incurred, many of the less resilient species may still be present. This also leads to the hope that, if the restoration of the landscape can be achieved, many of the species still present will be retained. The possibility of rapid recovery and retention of the biota both depend on how reversible the processes of degradation are. In that respect we appear to be at a nexus at the moment; if we act now, we stand a good chance of being successful, but, if we leave it for too much longer, restoration may become increasingly more difficult, if not impossible.

In the agricultural parts of the landscape, the extent of changes in soil structure and composition is alarming (Nulsen, Chapter 5), and solutions are not easily found for some of the most pressing problems. As with hydrological problems (McFarlane et al., Chapter 6), solutions depend on the development of alternatives to current practices. Finding solutions depends on having a basic understanding of how the system works, and this, in turn, requires that we have baselines against which to measure the changes which have taken place (e.g., the studies of hydrology in naturally vegetated catchments, referred to by McFarlane et al., Chapter 6).

It seems clear that there are strong economic, as well as ecological, incentives to farming to soil type; i.e., the planning and management of farms on an ecosystem basis, rather than on the north-south/east-west grid imposed by the early surveys of the wheatbelt region. As pointed out by McFarlane et al. (Chapter 6), matching crops, pastures, and woody perennials to soil types will result in good growth which will maximise returns and use water which would otherwise run off or become recharge. Revegetation using trees and shrubs has significant benefits in terms of hydrology and erosion control and may also be intrinsically beneficial to production in some instances (e.g., for fire-wood, fenceposts, etc.). Using an integrative approach, such revegetation can also be designed so that it is useful for nature conservation, i.e., by using native trees and shrubs, by routing windbreaks so that they link up isolated remnants, and by making strips wide enough so that they have the potential to act as corridors.

When a reintegrated landscape is achieved, it will be an interacting mosaic of natural, seminatural, and agricultural elements (A.R. Main, Chapter 7). Such a system should maximise biodiversity and production. It should be emphasised that conservation and production goals are equally important in the reintegration process, since the loss of species is as much a form of land degradation as erosion or salinity. In a sense, the diversity of the natural system is the landscape's insurance policy against change; the redundancy built into natural communities proffers a degree of resilience which would not otherwise be present. This is an important point to be considered in relation to restoration projects.

How do we judge the success or failure of our attempts to produce a reintegrated landscape? Clearly, the criterion will be whether the processes of degradation currently observed are being slowed down or reversed. Lefroy et al. (Chapter 8) provided a preliminary attempt to produce a set of parameters (based on the integrity of ecological processes) which can be monitored to determine whether current and future practices are sustainable in ecological terms. Any sensible engineer has monitoring devices built into systems which allow the system to be serviced before damage occurs; this is equally important in the context of the sustainable use of our landscapes. We believe that the development of such methods of assessing ecological impacts is essential if the term "sustainability" is going to mean anything in practical terms. Only when there are yardsticks available can sensible decisions be made on which practices will be most beneficial in ecological terms.

11.4 Problems of Scale and Multiple Ownership

The problem of scale is central to the discussion of integrated landscape management (Lefroy et al., Chapter 8). Clearly, the landscape is the scale on which we are focussing here but nowhere in this book has this scale been defined. The problems of definitions have been discussed elsewhere (Forman and Godron 1986; Risser 1987), but it is useful here to consider exactly what we mean when we talk about the "landscape scale." This is the scale of tens to hundreds of square kilometres which contain the mosiac of different patch types, or segments, present in an area. Thus, a typical Western Australian wheatbelt landscape currently contains farm paddocks and remnants of native vegetation, both in blocks and as corridors along roadverges and paddock boundaries (Fig. 11.1). The main reason why this scale is considered important is because it deals with the interactions between the different patch or segment types. No one segment of the landscape is independent of the segments surrounding it. Classical land management ignores this interconnected-

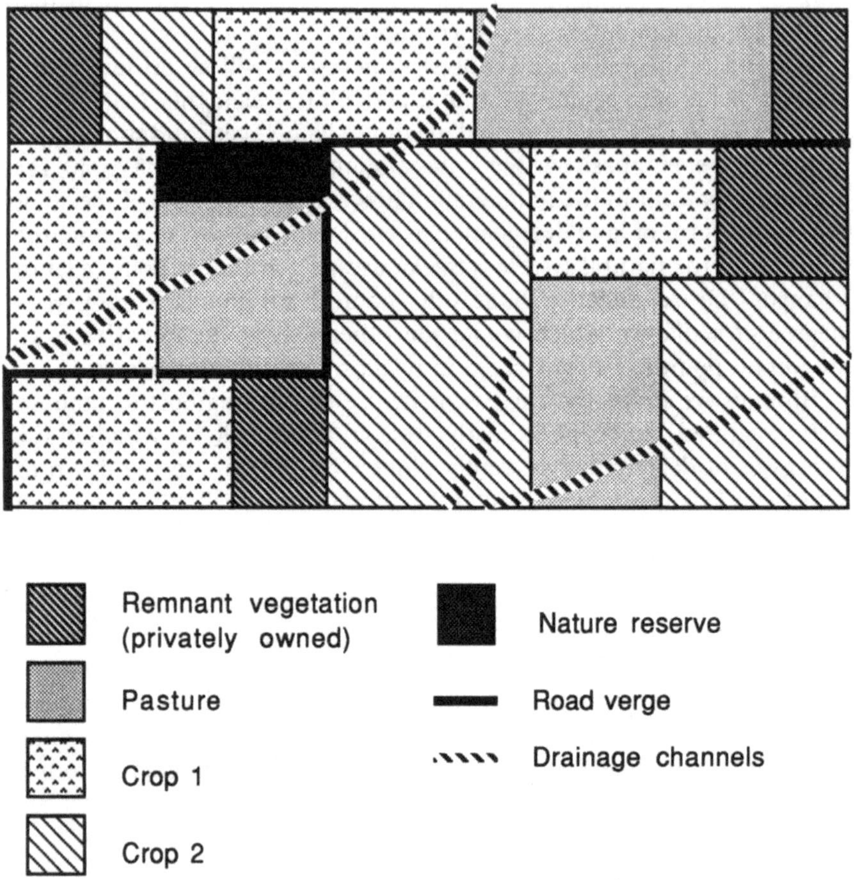

FIGURE 11.1. Hypothetical example of the Western Australian wheatbelt landscape, illustrating the current subdivision based on north-south/east-west axes and including agricultural paddocks, remnant vegetation present as nature reserves and on private land, and roadverges (Compare with Fig. 4.2 and Fig. 8.3).

ness and deals only with individual landscape segments in isolation; thus, the farmer manages his/her individual paddocks, the conservation manager manages his/her nature reserves, and the shire engineer manages the roadverges (Hobbs et al. 1992). Each manager or landowner is primarily concerned with only a small portion of the landscape and, hence, has different perceptions of it. Only when a broader landscape perspective is taken can the implications of actions on one landscape segment for processes in other segments be seen.

The fragmented landscape is thus subject to fragmented management (Hobbs et al. 1992). Throughout this book, it has been indicated that the problems to be tackled are landscape-scale problems, but current management is at the sublandscape or individual segment scale. This mismatch between the scale of problem and scale of treatment is a major obstacle to be overcome if we are to achieve integrated landscape management. It has several components. Firstly, individual motivation is highest when a person clearly "owns" a problem and can tackle it on his/her own land. More widespread problems, or those on neighbours' properties, are "someone else's problem." Secondly, the resources available to individual managers and owners are limited, and hence it may be impossible to consider anything outside the immediate sphere of relevance. Thirdly, solutions to larger-scale problems invariably involve conflicts of interests and resolution of these conflicts involves compromise. Finally, in the case of management and advisory agencies involved in managing adjacent lands with different uses, there may be problems of institutional rivalry for funds, conflicts of aims, and empire building by executives, all of which can reduce the possibility of cooperation (e.g., Grumbine 1990, 1991).

How, then, can these problems be overcome? Clearly, to be successful, integrated landscape management must have the cooperation of the various managers and landowners in the region. A first step in this direction is the development of a strong communications network which promotes a recognition of the importance of landscape-level problems. Farmers need to convince themselves of the importance of remnant vegetation in landscape function, and conservation managers have to recognise the influence of the surrounding landscape on their nature reserves and the relevance of remnant vegetation on private land and of roadverges to the overall conservation network. With a broader understanding of the overall problems and perspectives of other users of the landscape comes a greater possibility of integration.

Cooperation between landowners to overcome large-scale problems is now becoming a reality in Western Australia, and in other parts of Australia, through the development of the Land Conservation Districts and Land Care Groups (Robertson 1988). Farmers are forming groups which aim to deal with exactly the type of landscape-scale problems we have discussed. Although the scheme is supported by the government, it is primarily driven locally and the local people make the decisions, with expert advice from government-appointed advisers. The scheme is, in effect, enlarging the scale of problem which can be "owned" by a group (i.e., internalising cf. externalising it). Working as a group instead of as individuals, Land Care Groups have greater resources at their disposal and hence can tackle larger projects than would have been possible for individuals, e.g., many Landcare Groups now own their own tree planters and are working cooperatively to revegetate areas which cross property boundaries.

Cooperation of this type goes some way to ensuring that landscape-level problems are tackled. However, it should still be noted that the

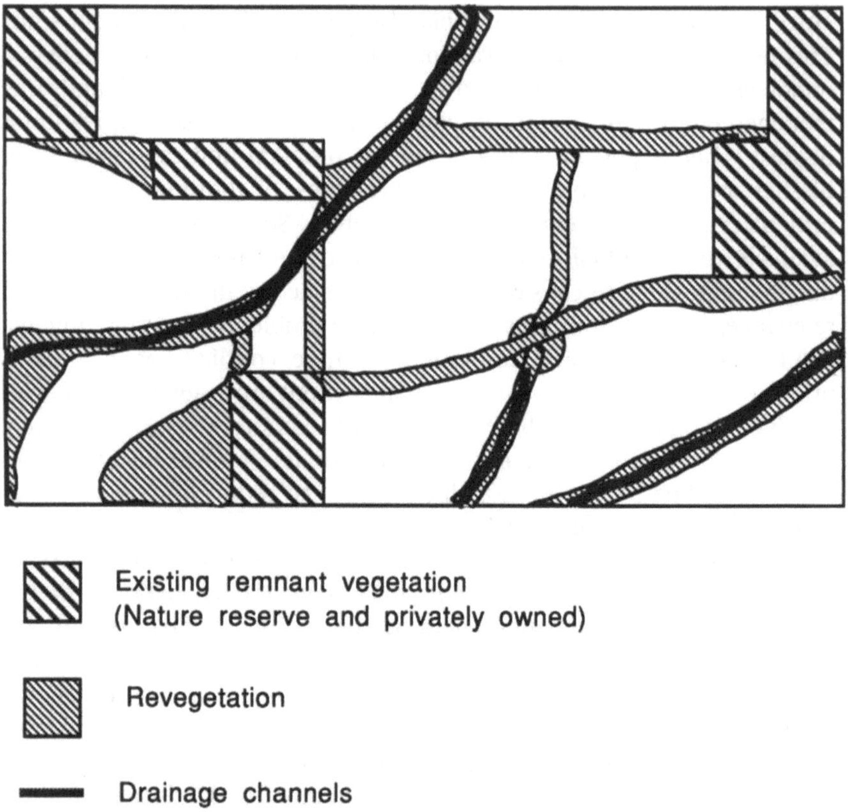

Figure 11.2. Potential future landscape, with farming to soil type and landscape-based revegetation, in which the traditional rectangular agricultural grid is broken down and existing remnant vegetation is linked by revegetated areas (Compare with Fig. 8.3).

basic management units remain the same; i.e., individual farms, remnants, etc. Regional and catchment planning will work only if management actions are planned at the individual farm level. Thus, Kubicki et al. (Chapter 9) applied the principles developed by A. R. Main, (Chapter 7) and Lefroy et al. (Chapter 8) to an individual farm; planning at higher levels again removes the ownership of the problem and nobody is responsible for its implementation.

Clearly, however, it is essential that regional and landscape aspects be built into individual farm plans if larger-scale problems are to be solved. There is scope for the development of landscape planning techniques which give guidance based on geographic information systems and can aid in conflict resolution (e.g., Ive and Cocks 1988; Ruzicka and Miklos 1990). This is where the Landcare Group and its advisors can act to best effect, by providing a broader perspective in which the management of individual farms can be placed. In the same way, nature conservation interests can be catered for; for instance, revegetation planning can incorporate nature conservation features. The principles developed by Lefroy et al. (Chapter 8) could change the landscape depicted in Figure 11.1 into one illustrated in Figure 11.2, in which farming and conservation are more closely linked. When it is recognised that the agricultural productivity and nature conservation value of the landscape are closely interlinked, there will be fewer conflicts between the two. Similarly, the scale of conservation problems is such that the involvement of the local community in nature conservation issues is likely to result in more effective conservation management (Hobbs et al. 1992). Production and conservation must be made compatible because "ecological apartheid" (Bradby 1991) is a recipe for failure. Studies elsewhere have also illustrated the benefits of involving the local people in conservation management (Western 1982; Lewis and Kaweche 1985; Lewis et al. 1990).

11.5 Conclusion

The answer to the question in the title of this chapter is simply that we have to reintegrate fragmented landscapes if we want to retain viable production and conservation systems. The process of reintegration will not be easy, but there is room for optimism, at least in the example of the Western Australian wheatbelt discussed here. In scientific terms, an injection of optimism into ecological thinking would be beneficial to all concerned. Ecology is not only the science of ecological problems and calamities, it is also the science of their solutions. Too often the emphasis is on the problems and little thought is given to possible solutions. No one can ignore the broader global issues of population growth and global change which underlie many of the more localised problems being tackled, but global problems also have to be tackled piecemeal at local scales. Researchers have a role to play in developing solutions to the problems currently besetting many parts of the world. It is not sufficient to stand by as impartial observers and record for posterity the objects of study as they degrade or disappear.

This implies that scientists working in the many fields that impinge on environmental management cannot continue to work in their own little area without putting it into a broader perspective and without making sure that the work they do reaches the people most likely to use the information in a practical way. The reintegration of fragmented landscapes requires the development of new ways of training experts across numerous fields of expertise and a network of communication between researchers, managers, and the public as a whole. Environmental scientists are no longer simply scientists; they are also communicators and ambassadors for the environment. H.G. Wells is reputed to have said "The history of humanity is a race between learning and disaster." We as a group have the opportunity to contribute to that race and to do something positive; the challenge is there.

References

Bradby, K. (1991) A data bank is never enough: The local approach to landcare. In D.A. Saunders and R.J. Hobbs (eds.) *Nature Conservation 2: The Role of Corridors*. Surrey Beatty and Sons, Chipping Norton, NSW. pp. 377–385.

Curtis, J.T. (1956) The modification of midlatitude grasslands and forests by man. In E.L. Thomas (ed.) *Man's Role in Changing the Face of the Earth*. University of Chicago Press, Chicago. pp. 721–736.

Forman, R.T.T. and Godron, M. (1986) *Landscape Ecology*. John Wiley and Son, Chichester.

Grumbine, E.R. (1990) Viable populations, reserve size, and federal lands management: A critique. *Conservation Biology* 4:127–134.

Grumbine, E.R. (1991) Cooperation or conflict? Interagency relationships and the future of biodiversity for US parks and forests. *Environmental Management* 15:27–37.

Hillel, D. (1991) *Out of the Earth: Civilization and the Life of the Soil*. Free Press, New York.

Hobbs, R.J. and Hopkins, A.J.M. (1990) From frontier to fragments: European impact on Australia's vegetation. *Proceedings of the Ecological Society of Australia* 16:93–114.

Hobbs, R.J., Saunders, D.A., and Arnold, G.W. (1992) Integrated landscape management: A Western Australian perspective. *Biological Conservation* in press.

Ive, J.R. and Cocks, K.D. (1988) LUPIS: A decision-support system for land planners and managers. In P.W. Newton, M.A.P. Taylor, and R. Sharpe (eds.) *Desktop Planning: Microcomputer Applications for Infrastructure and Services Planning and Management*. Hargreen, Melbourne. pp. 129–139.

Lewis, D.M. and Kaweche, G. (1985) The Luangwa Valley of Zambia: Preserving its future by integrated managment. *Ambio* 14:362–365.

Lewis, D., Kaweche, G.B., and Mywnya, A. (1990) Wildlife conservation outside protected areas – Lessons from an experiment in Zambia. *Conservation Biology* 4:171–180.

Malingreau, J.P. and Tucker, C.J. (1988) Large-scale deforestation in the southeastern Amazon of Brazil. *Ambio* 17:49–55.

Pennington, W. (1969) *The History of British Vegetation*. English Universities Press, London.

Risser, P.G. (1987) Landscape ecology: State of the art. In M.G. Turner (ed.) *Landscape Heterogeneity and Disturbance*. Springer-Verlag, New York. pp. 3–14.

Robertson, G. (1988) Soil Conservation Districts – Land users solving mutual problems. In P. Newman, S. Neville, and L. Duxbury (eds.) *Case Studies in Environmental Hope*. Environmental Protection Authority, Perth. pp. 89–95.

Ruzicka, M. and Miklos, L. (1990) Basic premises and methods in landscape ecological planning and optimization. In I.S. Zonneveld and R.T.T. Forman (eds.) *Changing Landscapes: An Ecological Perspective*. Springer-Verlag, New York. pp. 233–260.

Saunders, D.A., Hobbs, R.J., and Margules, C.R. (1991) Biological consequences of ecosystem fragmentation: A review. *Conservation Biology* 5:18–32.
Webb, N.R. (1980) The Dorset heathlands: Present status and conservation. *Bulletin d'Ecologie* 11:41–46.
Western, D. (1982) Amboseli National Park: Enlisting landowners to conserve migratory wildlife. *Ambio* 11:302–308.
Wilcove, D.S., McLellan, C.H., and Dobson, A.P. (1986) Habitat fragmentation in the temperate zone. In M.E. Soulé (ed.) *Conservation Biology: The Science of Scarcity and Diversity*. Sinauer, Sunderland, Mass. pp. 237–256.

Index

MIX
Papier aus verantwortungsvollen Quellen
Paper from responsible sources
FSC® C105338

If you have any concerns about our products,
you can contact us on
ProductSafety@springernature.com

In case Publisher is established outside the EU,
the EU authorized representative is:
Springer Nature Customer Service Center GmbH
Europaplatz 3, 69115 Heidelberg, Germany

Printed by Libri Plureos GmbH
in Hamburg, Germany